Solid Waste Management
Volume 2
Biological/Biochemical Approaches

Editors

Surajbhan Sevda
Department of Biotechnology
National Institute of Technology Warangal
Telangana, India

and

Garima Chauhan
Chemical and Materials Engineering
University of Alberta
Edmonton, AB, Canada

CRC Press
Taylor & Francis Group
Boca Raton London New York

CRC Press is an imprint of the
Taylor & Francis Group, an **informa** business

A SCIENCE PUBLISHERS BOOK

Cover credit: The image used on the cover is drawn by the editors.

First edition published 2024
by CRC Press
2385 NW Executive Center Drive, Suite 320, Boca Raton FL 33431

and by CRC Press
4 Park Square, Milton Park, Abingdon, Oxon, OX14 4RN

CRC Press is an imprint of Taylor & Francis Group, LLC

Library of Congress Cataloging-in-Publication Data (applied for)

ISBN: 978-1-032-13568-7 (hbk)
ISBN: 978-1-032-13572-4 (pbk)
ISBN: 978-1-003-22991-9 (ebk)

DOI: 10.1201/9781003229919

Typeset in Times New Roman
by Radiant Productions

Foreword

Management of solid waste(s) in an environmentally acceptable way is a thorny issue due to a variety of reasons. Waste materials in the solid form tend to be bulky and difficult to handle and transport. Its management could happen through physical, thermal, chemical or biological processing stages, with the exact sequence of and their operational optima being decided by the waste composition. Many of the processes that were being followed in earlier years have now become unacceptable due to factors ranging from shortage of land area needed for the process to severe environmental impact such as air pollution or proliferation of toxic compounds in surface water bodies. In addition, we have factors such as the NIMBY (Not In My Back Yard) syndrome to be considered while planning and designing facilities for solid waste management. This book focuses on some of the processes used for managing solid wastes through the biological route. This will enhance the capability of policy makers and engineers and make them aware of the potential application of these processes to manage solid waste(s).

Prof. T.R. Sreekrishnan
Department of Biochemical Engineering & Biotechnology
Indian Institute of Technology Delhi, New Delhi, India -110016

Preface

In recent years, energy crisis and increased waste production are undoubtedly major issues of concern. As we mentioned in the previous volume of this book, defining a waste is crucial in terms of identifying the most adequate approach to recycle or recover resources from the waste. Various chemical, biochemical, and biological approaches are being investigated widely by researchers for waste valorization. Volume 1 of this book covers various chemical approaches used in this direction. Volume 2 of this book brings together the leading researchers working on solid waste management using biological and biochemical approaches.

Chapter 1 discusses the potential reasons, including population growth and lack of adequate treatment methods, responsible for the exponential increase in the waste generation globally. Various biological treatment methods, such as anaerobic digestion, bio-compositing, and mechanical-biological treatment, are discussed. In addition, mechanism for the biological treatment of solid waste is illustrated.

Chapter 2 describes environmental and economic advantages associated with anaerobic digestion of solid organic waste. The process reduces local waste generation through effective recycling along with safe and clean renewable energy production in the form of biogas. In addition, various key factors, including importance of waste characteristics, different anaerobic reactor configurations, factors affecting anaerobic digestion, pre-treatment choices, and co-digestion options, of the digestion process of organic wastes are discussed.

An effective management of municipal solid waste (MSW) has become the necessity across the globe. The MSW accumulates for several years in landfills resulting in huge mountains of waste; therefore, an understanding of various phases of degradation of organic matter inside the landfill layers may assist in identifying ways to solve the issue of waste-accumulation. Chapter 3 discusses kinetics of the organic matter decomposition using various mathematical correlations and model equations, such as first order kinetic model, Monod Kinetics, Michaelis-Menten model, Gompertz modified equation, Logistic equation, and Stover kinetics. These kinetic equations are helpful in determining the kinetic parameters, such as the hydrolysis rate, rate of substrate utilization, maximum methane generation potential, growth rate of the microorganisms, and lag phase of the process.

Solid-state anaerobic digestion (SS-AD), a common approach for biogas production, is explained in Chapter 4. The microorganisms consume the organic substrates present in the waste during the SS-AD process and convert it to biogas. Various operational parameters, such as nutrients, feedstocks, pH, temperature, C/N ratio, inoculums, inhibitors, mixing, and feedstock-inoculum ratio, are discussed that define stability and efficiency of the process. Furthermore, strategies for improving biogas yields are discussed in detail.

Chapter 5 discusses thermal, chemical, biochemical, or biological approaches for solid waste treatment and effect of process parameters on the process efficiency. In addition, strategies to covert waste into value-added products are discussed.

Chapter 6 describes several biological methods to treat the waste generated by leather industry. A leather manufacturing unit using raw hide or skin as a raw material includes various steps including pre-tanning, tanning, post tanning and finishing. These steps utilize various chemicals, such as dehairing agents, tanning agents, dyes, retanning and fat liquoring agents, during the processing and generate a significant amount of waste. The composting of tannery sludge along with different organic waste, such as green or limed fleshing's, green biomass, tertiary coir pith, and paddy straw improve the overall process. Vermicomposting has been shown to stabilize the tannery sludge and yield a nutrient-rich fertilizer

Chapter 7 provides a brief overview of biorefining methods that aim to utilize the generated wastes for the energy production, fuels, and seaweed feeds, with the possible aim to reduce waste pollution to the environment. Many biorefining treatments systems and energy-assisted extraction techniques are suggested for their potential applications in agricultural, municipal, food, and biochemical waste management.

Chapter 8 focuses on sustainable methods for the treatment of MSW and its valorization. A total of eight methods are discussed as alternatives to the traditional waste management methods. These methods are classified into three categories: Biological, Chemical and Bio-Chemical processes.

Chapter 9 focuses on recycling of the urban waste considering the recent environmental awareness, policies, and regulations. The rate of urbanization is faster in developing countries such as India where numerous kinds of industrial, agricultural, and other development activities are very essential. *Distributed modular gasification* technology enables production of hydrogen gas (a clean energy resource) from non-recyclable plastic waste.

Chapter 10 provides the latest insights and a concise summary of work done to valorize agro-food industrial wastes using solid-state fermentation (SSF) for lignocellulolytic enzyme production. SSF favours the production of highly thermostable lignocellulolytic enzymes with improved applicability in commercial settings. The applications and bioreactors used for SSF are discussed.

Chapter 11 focuses on composting and vermicomposting as economical methods to produce humus or compost by utilizing biodegradable wastes comprising kitchen waste, plant waste, weeds, and leaves. Composting and vermicomposting is a sustainable and environmentally sound practices to provide nutrients to plants and help rejuvenate the land. Vermicomposting is an inseparable part of the waste management system. Earthworms can be used to treat a variety of waste and turn them into premium organic fertilizers.

Chapter 12 discusses the potential of agro-residue waste as an easily available and inexpensive substrate to produce valuable products, such as bioethanol, biogas and biohydrogen, using chemical (acid/alkali pretreatment) and biological methods in synergy for the generation of high-yield processes.

Chapter 13 describes the applicability of mechanical biological pre-treatment for municipal solid waste. The primary objectives of mechanical biological treatment (MBT), a combination of mechanical and biological treatment, are to separate waste into several streams and recover materials and energy from the stream. MBT proves to be one-step solution for diversion of potential waste streams from mixed municipal solid waste that end up in ever-increasing landfills.

Chapter 14 integrates a site-specific data into life-cycle assessment (LCA) methods, with a focus on MSW treatment technologies. The authors analyze the performance of the existing and proposed solid waste management system through LCA using site-specific data collected in the Guwahati city, India. The inventory is developed using literature and field sampling to estimate emissions.

Chapter 15 focuses on the biotechnology approaches for treating the plastic waste. Plastics are broken down by bacteria and enzymes, which turns them into biodegradable materials. Compared to conventional techniques used for plastic waste management, the technology described in this chapter offers several advantages, including affordability, environmentally friendly nature, and capacity to manage plastic pollution at the molecular scale. The handling of plastic garbage might be improved by this technique in the future. Moreover, plastic trash may be transformed into biodegradable goods like biofuels and bioplastics using biotechnology methods.

Contents

1

THE USE OF BIOLOGICAL METHODS TO PROCESS SOLID WASTE

Shivani Maddirala,[1] *Vaishnavi Sharma,*[2] *Sudipa Bhadra,*[1] *Garima Chauhan*[2] **and** *Surajbhan Sevda*[1,*]

1. INTRODUCTION

In today's world, energy crisis and increased waste production are undoubtedly major issues of concern. Energy crisis is mainly due to using up of non-renewable sources of energy like coal, petrol, diesel and increased requirement due to increased population, urbanization and hence increased usage. Usage of these non-renewable sources for energy is causing great amount of pollution. Biological solid waste contains a lot of organic matter and nutrients which makes it a potential fertilizer, but it also contains heavy metals, pathogens and other hazardous compounds. Nowadays, waste has increased a lot since the pandemic and also due to increased population. Industrial waste, municipal waste, medical waste and agricultural waste are a matter of concern in this chapter. This increased waste and improper waste management is leading to many problems like greenhouse gas emission, water pollution, soil erosion, loss of fertility in soil and various diseases in humans and animals. It is important to preserve the environment and natural resources (Tuan et al. 2022).

A common method used for waste management is land-filling because of its simple processing, large capacity and low investment (Huang et al. 2022). But landfills occupy plenty of natural resources and have a plenty of negative effects on the ecosystem and public health. Leachate from landfill sites contains various hazardous substances like heavy metals, pathogens, hazardous chemicals and other contaminants. Emission of greenhouse gases from landfills aggravates greenhouse effect. To handle the leachate and gas emissions, an additional setup has to be made, this way the gas released can be collected and used for energy generation (Huang et al. 2022). In this way, usage of the produced waste to produce energy in various forms (Waste to Energy technology) can be used to make use of the waste generated. For this, the waste has to be divided into different categories and various

---
[1] Environmental Bioprocess Laboratory, Department of Biotechnology, National Institute of Technology Warangal, Warangal, Telangana, India-506004.
[2] Department of Civil Engineering, National Institute of Technology Warangal, Warangal-506004, India.
[3] Department of Chemical Engineering, University of Alberta, Edmonton, Alberta, Canada, T6G 2R3.
* Corresponding author: sevdasuraj@gmail.com, sevdasuraj@nitw.ac.in

methods have to be used to convert it into energy (Ding et al. 2021). Some of them are anaerobic digestion, composting and mechanical biological treatment. Waste management technique used depends on location, nutrient composition, physical and chemical composition. Anaerobic digestion and composting are two major biological treatments that could transform organic waste to bioenergy and compost (Uddin et al. 2021). In comparison to land-filling and incineration, anaerobic digestion and composting present more environmental and economic benefits and can generate highly valued products like methane. It is important to assess economic benefits and environmental impacts of waste treatment before choosing. This chapter deals with types of solid wastes and biological methods to treat them to produce energy in the form of steam, heat or electricity.

2. CHARACTERISTICS OF SOLID WASTES

Generally, solid waste is solid or semi-solid waste arising from industries, construction, production and other human activities. It is important to study the characteristics of solid waste to be able to use appropriate methods for treatment and production of required materials from solid waste.

2.1 Physical characteristics

Information and data about the physical properties of solid waste comes under physical characteristics. Material composition, particle shape, particle size, density and moisture content are the physical characteristics of solid waste. These are important for selection and operation of equipment and facilities and in analysis and design of disposal facilities. The waste is examined for its usefulness in producing energy. One of these tests is the degradability assessment (Bayard et al. 2018). In this, the degradability of the solid refuge is assessed by measuring volatile solids (VS), biochemical methane potential (BMP), and cellulose (C), hemicellulose (H), and lignin (L) contents (Panigrahi and Dubey 2019).

2.2 Chemical characteristics

It is important to study the chemical compounds and their characteristics for an understanding of the behaviour of waste throughout the waste management system (Zhao et al. 2022). To us the solid waste as fuels or for the production of fuels, it is important to know its chemical composition. Chemical characteristics include lipid content, carbohydrate content, protein content, heating value and fibre content.

2.3 Biological characteristics

Biological characteristics must be studied to know about the biodegradability of waste, survival of microorganisms and nutrient content. For example, for the solid waste to be used as fertilizer or manure in agriculture, it is important to know about its nutrient content and its ability to grow microorganisms.

3. TYPES OF SOLID WASTE

Solid waste is categorized into various types based on the origin, characteristics, effects and assumed usage.

3.1 Municipal solid waste

Municipal Solid Waste is usually referred to as wastes of daily life, generated from households, schools, offices, shops and hotels (Sipra et al. 2018). Major components of this waste would be food waste, plastic, metal, glass, cloth and ceramics (Tuan et al. 2022). With increased population and urbanization, MSW production has increased. Rate of generation varies from season to season, city to city, living habits, diet, lifestyle, tourism and level of development. Most of the municipal solid waste is burnt or land-filled. A major fraction of this type of waste is spoilable, which can act as a contributor to various environmental problems. Production of greenhouse gases like methane leads to climate change, pollution of water bodies, soil and air which leads to resource depletion. MSW disposal is typically associated with land-filling, composting, open dumping and thermal treatment (Vyas et al. 2022).

3.2 Industrial solid waste

Industrial waste usually refers to waste generated during industrial activities such as factories, mills, and mines. Typically this includes paper, packaging materials, waste from food processing, oils, solvents, resins, paints, sludge, glass, ceramics, stones, metals, cement, plastics, rubber, leather, wood, cloth, straw, abrasives, product residues, kiln dust, slags, and ashes (United Nations Economic and Social Commission for Asia and the Pacific 2002). This is generally divided into general industrial solid waste and hazardous industrial solid waste. General industrial solid waste is that part of the industrial waste that does not pose a threat to the environment or human health, for example plastic, paper, stones, glass, metal, wood and other organic wastes, whereas hazardous industrial waste is that part of industrial solid waste that is harmful to human health and environment, for example paints, solvents, flammable and toxic materials (Vyas et al. 2022).

3.3 Agricultural waste and residues

Agricultural waste is the waste produced as a result of various agricultural operations. Nowadays, there is increased agricultural production like food for human beings, for animals and for industries. Almost all agricultural practices generate waste in large quantities, which when untreated may pose a threat to the human health through environmental pollution. This includes waste from farms, manure, harvest waste, animal waste, waste from slaughter houses and poultry houses, fertilizer run off, pesticides entering into air, water and soil from field (United Nations Economic and Social Commission for Asia and the Pacific 2002). In many cases, agricultural waste is burnt or dumped in opening places resulting in emission of greenhouse gases, soil contamination, smoke, ash all of which can degrade the environment. Some of these wastes can be made useful for ethanol production, biogas generation, as animal feed, organic fertilizer, compost and others (Vyas et al. 2022).

3.4 Hazardous waste

Hazardous waste is the waste that has the ability to pose potential threat to the public health and environment. These are toxic substances produced from industries, health care facilities, agricultural practices (pesticides, herbicides, insecticides and chemical fertilizers), nuclear establishments, hospitals and manufacturing facilities. They can be explosive, inflammable, corrosive and react when exposed to other substances. High volumes of industrial hazardous waste is produced from petrochemical industries, petroleum industries, wood treatment plants, paper industry, leather industry, textile industry, metal plants, energy generating plants, automobile industries, hospitals, pesticide users and dry cleaners. Like other types of wastes, hazardous waste produced is increasing as a result of urbanization and development as many new chemicals and toxic substances are being introduced and used in large quantities. Open dumping of such waste causes soil pollution, contamination of surface and ground water supplies while burning causes air pollution and mixing of ash in water resources causes water pollution. Decomposition of organic fraction of the hazardous waste produces methane, a greenhouse gas contributing to greenhouse effect. Bio-medical waste is a potential carrier of pathogens which could spread due to open dumping. Disposal of oils, chemicals, used batteries, paints, toxic effluents and biocides can cause adverse effects on human health and environment (Vyas et al. 2022).

4. SOLID WASTE TREATMENT BY BIOLOGICAL METHODS

4.1 Anaerobic digestion of organic waste

The organic fraction of solid waste is a valuable resource that can be used for the production of useful products. This can be done using microorganisms. Anaerobic digestion is the process of biological conversion of organic waste to biogas and other organic compounds that are energy rich, in the absence of oxygen (Khalid et al. 2011). It is considered as the best option for biological production of methane. Anaerobic digestion is applicable to organic fractions of a wide range of wastes like agricultural, municipal and industrial. It is a highly complicated biological interaction between biotic and abiotic components and involves a series of metabolic reactions. They are hydrolysis, acidogenesis, acetogenesis and methanogenesis (Khalid et al. 2011).

A variety of microorganisms catalyze the anaerobic digestion process. They convert macromolecules into low molecular weight compounds. Sludge is commonly used as inoculums for waste treatment. Naturally selected or artificially mixed strains of microorganism are also used. Cell aggregates in the form of flocs, biofilms, granules and mats, with dimensions in the range of 0.1 to 100 mm may also be used for treatment (Jeong et al. 2010). It is very unusual for a biological process to rely on a single strain, hence a microbial consortium is used for anaerobic digestion process (Fantozzi and Buratti 2009). According to Ike et al. (2010), a group of microorganisms such as actinomyces, *Thermomonospora*, *Ralstonia* and *Shewanella* are involved in the degradation of food waste into volatile fatty acids, but *Methanosarcina* and *Methanobrevibacter/Methanobacterium* mainly contribute in methane production (Ogbonna et al. 2015). Similarly, Charles et al. (2009) reported the presence *of Methanosarcina thermophila*, *Methanoculleus thermophilus*, and *Methanobacterium formicicum* during anaerobic digestion (Charles et al. 2009). Using denaturing gradient gel electrophoresis and DNA sequencing techniques, Trzcinski and Stuckey (2019) found hydrogenotrophic species (mainly, *Methanobrevibacter* sp., *M. formicicum* and *Methanosarcina* sp.) active in methane synthesis (Trzcinski and Stuckey 2009). Classification of anaerobic digestion of organic waste (Uddin et al. 2021):

a) Based on the overall solid content, it is categorized as a dry or wet process.

b) According to the feeding mode, it is categorized as a batch or continuous process.

c) According to the operating temperature, it is categorized as mesophilic or thermophilic.

d) According to the number of the stages, it is categorized as single-stage or multi-stage.

e) According to the digester, it is classified as a fixed dome, floating dome, balloon, and garage type AD process.

Advantage of using AD process for different solid waste treatment are as follows (Park et al. 2005):

a) Much time required for biostabilization.

b) Sensitive to high levels of ammonia, which results in the degradation of protein components.

c) Long retention time.

d) Low removal efficiency of organic compound.

Disadvantage of using AD process for different solid waste treatment are as follows:

a) Much time required for biostabilization.

b) Sensitive to high levels of ammonia, which results in the degradation of protein components.

c) Long retention time.

d) Low removal efficiency of organic compound.

e) Needs preventative treatment to enhance methanol yield and post-treatment to eliminate hazardous substances.

Table 1 shows the major advantage and disadvantage of the anaerobic digestion process for solid waste treatment.

Environmental benefits of anaerobic digestion of organic fraction of solid waste: reduction in air and water pollution, green energy production, organic waste treatment, replacement of inorganic fertilizer, reduction in greenhouse gas emission.

4.1.1 Hydrolysis

It is the first step of the digestion process. Hydrolysis is the process of conversion of complex substrates to soluble molecules. This step converts carbohydrates into sugars, lipids into long chain fatty acids and proteins into amino acids (Mir et al. 2016). Extracellular enzymes called hydrolases are used in this process. These include esterases, peptidases and glycosidases (Khalid et al. 2011). Hydrolysis of complex substances is the rate limiting step in anaerobic digestion as it is very slow, hence it is important to improve hydrolysis to optimize the digestion process (Marin et al. 2010, Liew et al. 2011).

4.1.2 Acidogenesis

Here the soluble molecules from the hydrolysis stage are transformed into volatile fatty acids, lactate, alcohol and carbon dioxide by fermentative bacteria (Silva et al. 2013). Acetic acid, propionic acid and ethanol are the main products of this step (Zhou et al. 2018). Inorganic compounds such as CO_2, H_2, H_2S and NH_3 are also produced (Zhou et al. 2018).

4.1.3 Acetogenesis

Acetogenic bacteria oxidises the volatile fatty acids and alcohols into acetic acid and hydrogen (Mir et al. 2016). Bacteria that form the acetate by using butyrates and propionates, respectively, are known as *Syntrophobacter wolinii* and *Smithella propionica. Pelotomaculum schinkii* and *Clostridium aceticum* is another microorganism that develops H_2 and CO_2 acetate (Capson-Tojo et al. 2018).

4.1.4 Methanogenesis

Methane and carbon dioxide are derived from the products of acetogenesis by the methanogenic bacteria. Mainly acetrophic methanogens and hydrogenotrophic methanogens are used to accomplish formation of biogas from organic materials. Approximately 70% methane is derived from acetate (Acetrotrophic methanogens degrade acetate to methane and carbon dioxide) and 30% from hydrogenotrophic methanogens (use CO_2 and H_2 to produce CH_4) (Zhang et al. 2015, Mir et al. 2016).

Table 1. Shows advantage and disadvantage for solid waste treatment using anaerobic digestion process.

		Pros	Cons
Solid Content	Dry	Low volatile acid formation	Volatile fatty acids are aggregated
		Higher coefficient of methane generation	Long operating time to get methane and organic matter degradation
		Decreased growth of microorganisms	Enhancement of specific growth rates of microorganisms
	Wet	Low sludge production	Less technical diffusion
		High organic loading rate	Stability issues
		Used at landfill sites	
Feeding Mode	Batch	Low technology needed	Clogging
		Simple	Small OLR
		Robust	Need for bulking agent
	Continuous	Low operating cost	High initial investment
		Less land area	Technical difficulty associated with pump in loading
		Uninterrupted digestion	Requires high internal fluidity for smooth feedstock intake
Temperature	Mesophilic	More stable and easier to maintain	Low biogas production
		Low operating cost	Longer retention time
		Operates with robust microorganisms	
	Thermophilic	High pathogen destruction	Difficult to maintain the system
		Increased OLR	More energy is needed for heating
		Increased methane generation	Responsive to toxins
Stage	Single stage	Robust system	Expensive equipment
		Long OLR	Waste under 20% TS cannot be handled alone
		Less expensive treatment	Less dilution possibility with fresh water
	Multi stage	Design flexibility	Larger investment
		More reliable for cellulose poor kitchen waste	Complex system
		Only reliable for C/N < 20	Lower biogas yield if solids are not digested
Digester type	Garage	Simple design	Inoculation is needed for every new batch, thus reducing capacity for fresh feedstock
		Only little amount of additional water is needed	Gas tightness of opening difficulty
		Easy treatment of digestate	
	Floating drum	Simple and easy operation	Corrosion of steel parts
		Constant gas pressure	High material cost for steel drum
		Relatively easy construction	Shorter life span compared to fixed-dome digester
	Fixed dome	Long lifespan	Skilled technician is required for construction
		No moving parts	Difficult to construct in bedrock
		Low manufacturing costs	Fluctuating gas pressure depending on the volume of stored gas
	Balloon	Easy to construct	Susceptible to mechanical damage
		Low construction cost	Relatively short lifespan
		Easy transportation	Material is usually not available locally

4.1.5 Optimizing hydrolysis

The complex lignocelluloses materials of the organic fraction of solid waste hinders its biodegradation. It is seen that conversion of complex organic matter to simpler compounds is the rate limiting step in the degradation of waste with higher solid content. To solve this problem, it is important to use various pre-treatment methods (physical, chemical and enzymatic) to increase substrate solubility and rate of biodegradation of organic matter. Improvement of the hydrolysis step will facilitate biogas recovery, reduce greenhouse gases emissions, and facilitate low cost environmental management. Different chemical, mechanical, thermal, biological and combined pre-treatments are used to optimize hydrolysis.

Mechanical pre-treatment mainly focuses on reduction of particle size and crystallinity of lignocellulosic material (Richard et al. 2019). It increases biological process kinetics by changing chemical composition, polymerisation, releasing dissolved organic matter, increase of surface area to volume ratio and substrate pore volume (Hansen 2007, Jain et al. 2015). Some of the mechanical treatment methods are sonication, bead-milling and high-pressure homogenizer (Jain et al. 2015).

Thermal pre-treatment is used to enable the conversion of lignocellulosic biomass before anaerobic digestion. High temperatures melt lignin and free it from shielding cellulose and hemi-cellulose framework. It improves digestion process by improving dewaterability, deflocculating molecules, solubilising refractory molecules, disinfecting by sterilization and reducing pollution (Liu and Bond 2012, Jin et al. 2016, Liu et al. 2012). However, thermal pre-treatment can also cause inactivation of methanogenic bacteria. From various studies, it was seen that thermal pre-treatment increased biogas generation and methane yield.

Chemicals such as acids, alkalis and ozone are used for pre-treatment of lignocellulosic substances. In alkali pre-treatment, acetyl groups are substituted with uronic acid; this increases cellulose and hemicelluloses accessibility to hydrolytic enzymes. Internal surface area is increased in alkaline pretreatment, lignin is disrupted and the bond between lignin and carbohydrates is broken (Bazargan et al. 2015). Liew et al. (2011) saw a 20% increase of methane yields after pre-treatment of fallen leaves with 3.5% NaOH compared to unpretreated ones (Liew et al. 2011). Acid pre-treatment is used to disrupt covalent hydrogen bonds and Van der Waals forces, leading to solubilisation and hydrolysis of hemicelluloses and reduction of cellulose (Sarto et al. 2019). It is highly suitable for lignocellulose rich substrates (Mancini et al. 2018). One of the drawbacks of chemical pre-treatment is formation of inhibitory products such as phenolic compounds, furans and carboxylic acids that inhibit the growth of methanogenic bacteria (Behera et al. 2014). Ozone is a strong oxidizing agent. Ozone treatment enhances hydrolysis by solubilizing lignin. Ozonation can be performed at normal temperature and pressure, and does not produce any inhibitory or toxic by-products. Despite these advantages, ozonation has a disadvantage of high operational cost (Cesaro and Belgiorno 2013).

Biological pre-treatment is an environmental friendly pre-treatment and it can be applied to a vast range of feedstocks. Enzymatic pre-treatment, precomposting, mature compost addition, brown, white and soft rot pre-treatment are a few biological pre-treatment methods. A particular enzyme is required for each basic component of the substate; therefore, it is necessary to use an optimum mixture of enzymes for enzymatic pre-treatment. b-glucanase, hemicellulase, xylanase, arabinase, glucoamylase are commonly used enzymes for the improvement of hydrolysis (Cesaro and Belgiorno 2014). Microbial pretreatments are also used to treat lignocellulosic substances. In various studies, increase in biogas production when compared to untreated biomass is observed. Less corrosiveness and production of less harmful products due to absence of chemicals are a few advantages of this pre-treatment method (Wagner and Illmer 2018). But these processes are very slow and require several enzymes because of the heterogeneous composition of the organic fraction of waste. Fungal pre-treatment provides better nutrients and increases organic acid availability to anaerobes, hence improving anaerobic digestion process. Each particular type of fungi has affinity towards a particular substance. Brown and soft rot fungi degrade cellulose more effectively than lignin, whereas white rot degrades lignin effectively (Fdez et al. 2011b, c, a).

Each individual pre-treatment has affinity to a particular substrate, hence a particular pre-treatment cannot be used for a heterogeneous organic fraction of solid waste. In this case, combined pre-treatment technique comes in handy. This technique has advantages like high methane yield, better biomass utilization and low pre-treatment severity; however, there may be an increase in process cost (Fdez et al. 2011b). For example, addition of acid or alkali while carrying out thermal pre-treatment enhances pre-treatment effectiveness. Thermochemical pre-treatment can enhance solubilisation of organic carbon, energy recovery and reduce hydraulic retention time (HRT) (Fdez et al. 2011c).

4.1.6 Optimizing acidogenesis and acetogenesis

Absence of trace elements is considered as a contributing factor for process instabilities and failure. Hence, addition of trace elements improves process stability of the system, improves enzyme activity and growth of methanogens. Trace elements can be added directly as external trace elements or through co-digestion with substrate rich in trace elements.

Results of the study by Banks et al. indicated that supplementation of Se and Co improved process stability and prevented process failures (Banks et al. 2012, Zhang et al. 2015). On addition of Fe, Co and Ni in the anaerobic digestion of food waste, Zhang et al. (2015), observed decrease in volatile fatty acids (VFA) inhibition, increase in methane production and process stability of the digesters (Zhang et al. 2015). These results indicate that shortage of trace elements in substrates is one of the contributors of VFA accumulation and inhibition.

Addition of granular activated carbon (GAC) is used to improve acidogenesis and acetogenesis phases of anaerobic digestion process. This is due to the pores on their surface. GAC is used for immobilization of syntrophic microorganisms, promotion of direct interspecies electron transfer and adsorption of inhibitors (Capson-Tojo et al. 2018).

Capson-Tojo et al. (2018) and Dang et al. (2017) in their study observed enhanced VFA consumption, increased production of methane, promoted growth of methanogens and reduced lag phase, when GAC is added to anaerobic digestion of food waste and municipal solid waste, respectively (Dang et al. 2017, Capson-Tojo et al. 2018). Xu et al. 2018 studied effect of GAC on methanogenic degradation of VFAs (Xu et al. 2018). The results indicated that addition of GAC accelerated degradation of propionate and butyrate in conditions of high organic load, which subsequently increased production of methane.

4.1.7 Optimizing methanogenesis

Ammonia is an essential nutrient for the growth of microorganisms, but at higher concentrations it is toxic to microorganisms and it inhibits methanogenesis. In anaerobic digestion process, ammonia is produced through biological degradation of nitrogenous matter like proteins and amino acids. The toxicity of ammonia is due to unionized ammonia (Rajagopal et al. 2013). It has the capability to penetrate the cell membrane of the microorganisms leading to disturbance of potassium and proton balances. Optimization of process parameters is important to control the concentration of unionized ammonia. At optimal concentration, ammonia forms $NH_4(HCO_3)$ by combining with CO_2 and H_2O; this decreases buffering capacity and maintains stability of the digestion process. Trace elements' optimization, acclimatization and blending of feedstock are a few strategies that can be used to counteract ammonia inhibition (Rajagopal et al. 2013, Zhang et al. 2013, Sun et al. 2016).

4.2 Factors affecting anaerobic digestion

Various physical and chemical factors affect the mechanism including pH, temperature, C:N ratio, VFA, HRT, organic loading rate (OLR), seeding and stirring. Any changes in these parameters lead to disturbance in the environment of microbes and movement inside the digester. For the digester to run smoothly and to maximize biogas production, it is important to optimize and control the affecting parameters.

4.2.1 Temperature

Temperature has significant effect on microbial community, stability, process kinetics, yield and thermodynamic equilibrium of biochemical reactions. Lower temperatures in the process decrease microbial growth, substrate utilization rates and biogas production (Kim et al. 2006). They may also lead to leakage of intracellular substances or complete lysis of the cell (Kashyap et al. 2003). On the contrary, at high process temperatures, biogas yield is lower due to the production of volatile gases which suppresses methanogenic activity, for example, ammonia (Fezzani and Cheikh 2010).

Generally anaerobic digestion at mesophilic temperatures is more stable, has better effluent quality and requires a smaller energy expense. In a research by Castillo et al. (2006), the best operational temperature was found to be 35°C with an 18 day digestion period while a little fluctuation from 35°C to 30°C caused a reduction in biogas production rate (Castillo et al. 2006). A temperature range of 35°C to 37°C is considered suitable for methane production. Briski et al. (2007) reported that for biodegradation, the temperature must be below 65°C because above 65°C denaturation of enzymes occurs (Briški et al. 2007). However, thermophilic conditions have certain advantages like faster degradation of organic waste, higher biomass production, higher gas production, higher pathogen reduction, low HRT and less effluent viscosity.

Optimal temperatures for various methanogenic bacteria as shown by Ward et al. (2008): 37–45°C for *mesophilic Methanobacterium*, 37–40°C for *Methanobrevibacter*, 35–40°C for *Methanolobus*, *Methanococcus*, *Methanoculleus*, *Methanospirillum* and *Methanolobus*, 30–40°C for *Methanoplanus* and *Methanocorpusculum* and 50–55°C for thermophilic *Methanohalobium* and *Methanosarcina* (Ward et al. 2008).

4.2.2 pH

Researchers have reported a range of pH values that are suitable for anaerobic digestion. Optimal pH for methanogenesis was found to be 7.0. Agdag and Sponza (2007) reported that pH of 7.0–7.2 is suitable for last 50 days of anaerobic digestion for industrial sludge (Ağdağ and Sponza 2007). Ward et al. (2008) suggested 6.8–7.2 as the ideal pH range for anaerobic digestion (Ward et al. 2008). Lee et al. (2009) observed that methanogenesis occurs efficiently at pH 6.5–8.2 whereas acidogenesis and hydrolysis occur at pH 6.5 and 5.5, respectively (Lee et al. 2009). Park et al. (2008) showed that 6–7 is the appropriate pH for thermophilic acidogens (Park et al. 2008). Dong et al. (2009) observed maximum hydrogen production when initial pH of the biosystem is maintained at 9 (Dong et al. 2009). However, Kapdan and Kargi (2006) observed similar results at pH 5–6 (Kapdan and Kargi 2006). Liu et al. (2008) showed 6.5–7.5 as the most favourable range to obtain maximum biogas yield in anaerobic digestion (Conant et al. 2008).

4.2.3 Moisture

Usually, high moisture contents facilitate the anaerobic digestion. High water contents affect the process performance by readily dissolving degradable organic matter. Water promotes diffusion of nutrients and substrates towards bacterial sites. But it is difficult to maintain the same availability of water throughout the digestion cycle. As the process of anaerobic digestion proceeds, water added at a higher level is dropped to a certain lower level. Bouallagui et al. (2003) reported that the highest methane production rates occur at 60–80% humidity (Bouallagui et al. 2003).

4.2.4 Substrate

The type, availability and complexity of substrate strongly affect the rate of anaerobic digestion. Different type of sources support different groups of microorganisms. Before starting the process, it is important to assess the source of its carbohydrate, lipid protein and fibre contents. The substrate should be characterized for the amount of methane that can be produced under anaerobic conditions. Fernandez et al. (2008) reported that the initial concentration and total solid content of substrate can significantly affect the process performance and the amount of methane produced during the process (Fernández et al. 2008).

Carbohydrates are considered as the most important organic component of the municipal solid waste for biogas production. However, starch can act as an effective low cost substrate in comparison to glucose and sucrose.

4.2.5 Nitrogen

As we already know, nitrogen is an essential component for protein synthesis and is a primary nutrient required by the microorganisms in anaerobic digestion. In organic wastes, nitrogen is usually present in the form of protein, which is converted to ammonium by anaerobic digestion. Ammonium stabilizes the pH in the bioreactor (Parida et al. 2022). Microorganisms assimilate ammonium for the production of new cell mass.

Ammonia in high concentrations may lead to inhibition of the biological process and inhibit methanogenesis at higher concentrations, approximately greater than 100 mM (Fricke et al. 2007). The amount of ammonia may also affect hydrogen production and removal of volatile solids.

4.2.6 C/N ratio

Unbalanced nutrients is an important factor limiting anaerobic digestion of organic wastes. Carbon is the energy source for anaerobes and nitrogen has a major importance in increasing microbial population. C/N ratio indicates total ammonia nitrogen, nutrient level of feedstock and accumulation of volatile fatty acids inside the digester. For the improvement of C/N ratios and nutrition, co-digestion of organic wastes is used (Khalid et al. 2011). Co-digestion of fish waste, abattoir wastewater and waste activated sludge with fruit and vegetable waste facilitates balancing of the C/N ratio. Their greatest disadvantage is in the buffering of the organic loading rate and ammonia production which reduces the limitations of fruit and vegetable waste digestion. The C/N ratio of 20–30 may provide sufficient nitrogen for the process (Ge et al. 2016). Bouallagui et al. (2009) suggested that a C/N ratio between 22 and 25 seemed to be best for anaerobic digestion of fruit and vegetable waste (Bouallagui et al. 2009). Guermoud et al. (2009) and Lee et al. (2009) reported that the optimal C/N ratio for anaerobic degradation of organic waste was 20–35 (Guermoud et al. 2009, Lee et al. 2009).

4.2.7 Organic loading rate

As OLR increases, biological yield typically increases. OLR is influenced by various factors such as solids content, active microbial concentration and hydraulic retention time. Very low OLRs may contribute to deprivation and affect anaerobic digestion adversely. Conversely, too high OLRs may generate insufficient products that promote microbial growth, while large loads contribute to accumulation of volatile fatty acids in the fermenter that prevents microbial growth (Uddin et al. 2021).

4.2.8 Hydraulic retention time

HRT influences the association between substrates and microbes, thereby making treatment more effective. Shorter HRT is beneficial because it is concerned with reduction in capital costs and enhancement in process efficiency. Shi et al. (2017) reported that a minimum of 10 days HRT is required to prevent washing off of bacteria. Longer HRTs are more advantageous in order to produce biogas and methane (Shi et al. 2017).

4.2.9 Volatile fatty acids

These are crucial properties affecting anaerobic digestion process. Approximately 90% VFA are dissolved at pH 6.0, while approximately 99.9% are dissolved at pH 8.0 (Forgács 2012). A rise in VFAs in anaerobic digestion prevents methanogenesis.

4.2.10 Stirring

Stirring is an important factor that influences anaerobic digestion and it can be performed both electrically and mechanically. Stirring ensures uniformity of microbes and equal distribution of heat and hence improves interaction between bacteria and food. It also removes gas bubbles, preventing formation of layer and settlement (Tian et al. 2015).

4.3 Composting

It is a biological decomposition of organic waste, majorly in aerobic conditions, carried out by various groups of different kinds of microorganisms and nematodes (Pergola et al. 2018). Though composting involves hydrolysis of organic matter into humus and is thought to be an oxygen-demanding process, anaerobic organisms such as *clostridium* have been implicated in the process. The organic matter/waste is decomposed into a stable humus-like product with reduced toxicity and pathogens, which is called the compost. Composting provides sanitized and stabilized products that are a potential source of organic fertilizers or in soil amendments. It is a reliable waste treatment option that reduces organic substances into smaller volumes in a continuous manner. Factors influencing composting are temperature, pH, particle size, moisture content, aeration and electrical conductivity (Farrell and Jones 2009).

In composting, organic matter is converted into useful products by using the degradative attributes of microorganisms. It is widely used for transforming organic waste to organic manure and thus recycles mineral nutrients (nitrogen, phosphorus and potassium) which can be used for agricultural purposes. The microbial consortia in the compost has the ability to degrade various antimicrobial compounds, release heat, compete with pathogens in the soil and affect their viability, hence they suppress the development of plant diseases. Because of the important roles played by microorganisms in the composting process, various formulations of inoculants are commercially available for the agronomic enhancement of end product, such as "Microbial activator Super LDD 1" and "Effective Microorganism" (Lim et al. 2016).

According to the different physiochemical conditions of different composting stages, participating microbial consortia varies (Onwosi et al. 2017). At the initial stage, mesophiles are predominant. This stage involves decomposition of readily degradable soluble compounds and is a moderate-temperature stage. The metabolic activities and growth of mesophiles leads to a rapid rise in temperature. The rapid rise in temperature causes thermophilic bacteria to succeed mesophiles. This is a high-temperature stage where polysaccharides, proteins and fats are decomposed. Also, the high temperature leads to death of soil-borne pathogens and weed seeds, causing sanitization. The last stage involves cool down and maturation, where mesophiles are dominant. In this curing stage, the compost is stabilized for plant use. An indicator of the stage and compost quality is the presence of certain organisms.

Along with the various benefits of composting, there are a few drawbacks. Composition of substrate and operating conditions have a great effect on composting. Depending on these factors, methane is evolved by obligate anaerobic microorganisms. Use of this compost for agriculture may lead to introduction of pathogens into soil. Composting facilities may become a significant source of aerosolized pathogens that cause different infections. Contaminated compost or irrigation water can be a source for transmission of various microorganisms that can cause outbreaks (Farrell and Jones 2009).

4.3.1 Factors affecting composting

Temperature, aeration, pH, moisture content, C/N ratio and particle size influence the effectiveness of composting.

4.3.1.1 Aeration/turning frequency

Most common procedure for aeration is turning of compost material. This makes the material to be readily available for utilization by microbes and thus results in emission of gases (Tian et al. 2015). Improved aeration at early stages of degradation causes shortening of time taken for waste stabilization, but excess aeration could lead to loss of important components of composting (Awasthi et al. 2014). Hence, in order to retain relevant nutrients and to achieve other targets such as higher rate of hygienization, it is important to optimize the turning regime (Willts 2014). There is a connection between turning frequency and some physicochemical variables, which can be used as indicators of maturity of compost. Turning frequency affects pH, C/N ratio, total nitrogen, total carbon, moisture content, temperature, dry matter and various other variables (Getahun et al. 2012, Onwosi et al. 2017).

4.3.1.2 Temperature

Composting can be monitored by using various parameters, one of them is temperature. Temperature is a function of the process. A temperature gradient occurs in the process because of non-linear mass and energy balances. During composting, temperature could rise as a result of accelerated biodegradation of organic matter by microorganisms, making it an exothermic process which depends on initial temperature and substrate biodegradability (Raut et al. 2008). As the temperature rises, efficiency of composting process decreases. Temperature can be used to determine the relative dominance of some microbial population over another. Temperatures between 52°C and 60°C favours elimination of pathogens, parasites, ensures sanitary conditions and maintains the greatest thermophilic activity in composting system (Vuorinen and Saharinen 1997). Temperatures should not become too elevated as high temperature and excess ammonia can destroy microbial population essential in degradation, inhibit growth and activity of nitrifying bacteria and cease the process (Huang et al. 2004). Excess heat during composting could also be a fire hazard (Onwosi et al. 2017).

4.3.1.3 C/N ratio

Microbes obtain nutrients and energy for metabolism by breaking down organic compounds. C, N, P and K are the major nutrients needed by the microbial population in composting. Of these, the most crucial ones are C and N. C is energy source and N is used for building the structure of the cell. If N is limiting, growth of microbes decreases and decomposition of available C slows down. Contrarily, if excess N is present, then it is volatilized as ammonia gas. From this, it can be understood that C/N ratio is an indicator of degree of decomposition of organic matter. This ratio decreases during composting as the rate of mineralization of organic N is lower than that of organic C (Yang et al. 2015). Various studies have shown that adjustment of raw materials to achieve C/N ratio of 25–30:1 is great for active composting, although initial C/N ratio of 20–40:1 has given good yield. It implies that microbes use up C 30–35 times faster than the rate of conversion of N (Huang et al. 2004.

At lower C/N ratio, N will be liberated as ammonia as N will be in excess, which results in undesirable odour. Compost with lower C/N liberated large amount of basic salt, making the soil unfavourable for plant growth (Awasthi et al. 2014), whereas at higher ratio, there is not enough N for optimal growth of microorganisms, because of which the decomposition of organic matter happens at a slow rate and the compost remains cool comparatively. Generally, bulking agents such as sawdust, rice husk, peanut shells, and wood chip are added to organic material to adjust the C/N ratio of feedstock and to enhance porosity. This also supports odour control as excess moisture is absorbed due to co-composting of manure with bulking agents (Zhang and Sun 2016a, b, Onwosi et al. 2017).

4.3.1.4 Moisture content

Moisture content influences free air space, oxygen uptake rate, microbial activity and temperature of the process (Petric et al. 2012). Various studies have been conducted to understand the optimal moisture content during composting. In those studies, it is pointed out that the optimum moisture content required for composting is approximately 40–70% of the compost weight ([CSL STYLE ERROR: reference with no printed form], Bernal et al. 2009). Optimum moisture content depends on the physicochemical and biological properties of feedstock for composting. The oxygen uptake rate and rate of diffusion of gas reduces as moisture content increases. Because of the restricted activity, the process might become anaerobic. Sudharsan and Kalamdhad (2015) observed that temperature and moisture content are inversely related; as temperature increases, moisture content decreases and low temperature was observed when moisture content was high ([CSL STYLE ERROR: reference with no printed form]). During composting, moisture content is important for distribution of nutrients that are water soluble for the metabolic activity of microbes. Low moisture content might cause early dehydration, whereas increasing moisture content during composting may form water logs that could lead to anaerobic conditions that halts active composting activities. Loss of moisture during composting process is a strong indicator of decomposition rate of organic matter (Onwosi et al. 2017).

Petric and Selimbasic (2008) observed optimal initial moisture content in composting poultry manure and wheat straw to be around 70% (Petric and Selimbašić 2008). For pig slurry, Ros et al. (2006) used about 60–70% moisture content (Ros et al. 2006). However, at moisture content greater than 60%, movement of oxygen is delayed as pore spaces are closed and the composting process moves towards anaerobiosis (Ros et al. 2006).

4.3.1.5 Electrical conductivity

Electrical Conductivity represents the ability of a material to conduct electricity. It is numerical expression of conduction of electric current and is a measure of how easily a material allows electric current to flow through it. Electrical Conductivity indicates salinity/total salt content of the compost. This reflects the quality of compost to be used as a fertilizer (Onwosi et al. 2017).

Electrical conductivity might increase during the process due to the formation of mineral salts and phosphates through transformation of organic matter. Electrical conductivity might decrease due to ammonia volatilization and precipitation of mineral salts. For the manure compost to support plant growth, it must have very low electrical conductivity. High electrical conductivity will have a negative effect on germination of seed or plant growth. If compound with high electrical conductivity has to be used, then, before using it must be mixed well with soil or other materials with low electrical conductivity. Chan et al. (2016) explained that zeolite can be used to reduce salinity during composting, resulting in decrease of the electrical conductivity as it has the ability to accommodate and allow free exchange and absorption of ions on its surface (Chan et al. 2016).

4.3.1.6 Aeration

Aeration rate affects the microbial activity during composting and quality of the compost (Gao et al. 2010). The effect of aeration rate on distribution of temperature, organic matter decomposition rate and other parameters depends largely on time, location within composting mass and ambient condition (Bari and Koenig 2012).

Basically, composting is an aerobic process, where oxygen is consumed and gaseous carbon dioxide and water are released. Aeration provides oxygen needed for oxidation of organic matter, evaporates excess moisture and is the major factor influencing the stability of compost (Petric and Selimbašić 2008). As discussed earlier, too little aeration leads to anaerobic conditions and excessive aeration results in cooling, hence preventing thermophilic conditions needed for decomposition of organic matter (Gao et al. 2010). The relationship between aeration and pH can be described as follows: high oxygen concentration leads to low concentrations of organic acids and rapid decomposition of acids in the compost, leading to high pH levels (Sundberg and Jönsson 2008, Onwosi et al. 2017).

4.3.1.7 pH

Like temperature, pH follows a pattern during composting. In early stages of composting, decline in pH levels can be observed, whereas in later stages, elevation in pH levels is observed (Turan 2008). The rise in pH observed after the initial decline is due to the decomposition of organic acids. Increase in pH causes an increase in NH_3/NH_4^+ ratio resulting in elevated volatilization rates ([CSL STYLE ERROR: reference with no printed form]). Aerobic composting process has high pH, possibly due to higher potassium release. Decrease in pH can be caused by microbial decomposition of organic matter and production of acids, volatilization of NH_4^+-N and release of H^+ as a result of nitrification, mineralization of nitrogen and phosphorous compounds. Low pH could be challenging as it negatively affects the transition from mesophilic to thermophilic phase. Degradation of proteins leads to release and accumulation of ammonia, resulting in increase in pH. Alkalization of organic matter interrupts the survival of microorganisms that are sensitive to pH, hence contributing to sanitation.

To raise the pH, addition of alkaline such as $CaCO_3$ is a well-known method. An alternative to this is to cool the compost to < 40C until utilization of organic acids, which leads to increase in pH. Zeolite can also be used to increase the pH. The reason for using such methods for pH control over alkaline amendments is because alkaline amendments have disadvantages such as increased ammonia emissions (Sundberg et al. 2004). It has to be noted that the influence of pH on composting sometimes is not independent of temperature. Different microbial groups exist at different combinations of pH and temperature (Onwosi et al. 2017).

4.3.1.8 Particle size

Particle size distribution of compost determines water-holding capacity, gas and water exchange. It is important for maintenance of adequate porosity for proper aeration. Particle size of compost should not be too large as they will decompose slowly (Zhang and Sun 2014). It should not be too small as they reduce porosity of compost substrate by forming a compact mass. Sieving is the principle method for determining the particle size distribution (Ge et al. 2015, Onwosi et al. 2017).

4.4 Mechanical-Biological Treatment (MBT)

As the name suggests, mechanical-biological treatment is a combination of two treatments, mechanical and biological, which are performed one after the other. Mechanical-Biological Treatment is the integration of various mechanical processes used in waste management facilities, composting or anaerobic digestion (Fei et al. 2018). It is the mechanical separation of different types of waste with stabilization of organic matter through biological stabilization using processes such as anaerobic digestion or composting. A large percentage of recyclables can be recovered from mixed waste streams by MBT. Hence, it compliments other waste management technologies as a part of the integrated waste management systems. Main purpose of the biological treatment is to produce material with low environmental impact fit for disposal or land application (Abdeljaber et al. 2022).

Other objectives: volume of the waste to be dumped is reduced, hence reducing and preserving the required landfill volume; biological activity of the organic matter of household waste is reduced so that as little landfill gas as possible is emitted from the dumpsite; harmful substances that might enter into the groundwater are minimized. The aim of MBT plants is to separate the organic fraction that is biodegradable and recover the recyclables from the mixed waste (Abdeljaber et al. 2022). Generally, MBT plants are designed to process mixed household wastes as well as industrial and commercial wastes. They contain techniques and processing operations based on the market requirement of end products, thus MBT systems vary greatly in their functionality and complexity (Farrell and Jones 2009).

The products of MBT are: biogas (due to biological treatment of the organic fraction), stabilized organic waste, refuse derived fuel, recyclable materials such as metals, plastic, paper, and glass, inert wastes that are disposed to sanitary landfills. Effective utilization and disposal is important for the MBT outputs, else there is a risk of further environmental pollution. Outputs of MBT after separation needs to be treated with more energy efficient follow-up technologies to achieve improved performance.

5. CONCLUSION

This chapter provides an overview of solid wastes, types and methods to treat them. Solid wastes may contain potential pathogenic microorganisms, rodents, insects, organic pollutants and harmful physical objects. It is important to treat waste to avoid environmental contamination, retrieve recyclables and make use of the organic and biodegradable fraction of waste to produce useful products such as biofuel and biogas. This way pollution and greenhouse gas emissions can be reduced. For this process, the methods discussed in this chapter, which are anaerobic digestion, composting and mechanical-biological treatment can be used individually or in combination of two or more depending on the quality and composition of waste. It also depends on the objectives of treatment. Here we have discussed the biochemical methods for waste treatment: (1) Anaerobic digestion, (2) Composting (3) Mechanical-Biological treatment, but there are other thermochemical methods such as incineration, gasification and pyrolysis which are commonly used. Each method has its own set of advantages and disadvantages as discussed earlier, hence it is important to combine treatment methods accordingly. Quite a few researches are performed on solid waste management, which can be studied for further understanding of solid waste management.

REFERENCES

Abdeljaber, A., Zannerni, R., Masoud, W. et al. 2022. Eco-efficiency analysis of integrated waste management strategies based on gasification and mechanical biological treatment. Sustainability 14: 3899. https://doi.org/10.3390/su14073899.

Ağdağ, O.N. and Sponza, D.T. 2007. Co-digestion of mixed industrial sludge with municipal solid wastes in anaerobic simulated landfilling bioreactors. J. Hazard. Mater. 140: 75–85. https://doi.org/10.1016/j.jhazmat.2006.06.059.

Ariunbaatar, J., Panico, A., Yeh, D.H. et al. 2015. Enhanced mesophilic anaerobic digestion of food waste by thermal pretreatment: Substrate versus digestate heating. Waste Manag. 46: 176–181. https://doi.org/10.1016/j.wasman.2015.07.045.

Awasthi, M.K., Pandey, A.K., Khan, J. et al. 2014. Evaluation of thermophilic fungal consortium for organic municipal solid waste composting. Bioresour. Technol. 168: 214–221. https://doi.org/10.1016/j.biortech.2014.01.048.

Banks, C.J., Zhang, Y., Jiang, Y. et al. 2012. Trace element requirements for stable food waste digestion at elevated ammonia concentrations. Bioresour. Technol. 104: 127–135. https://doi.org/10.1016/j.biortech.2011.10.068.

Bari, Q.H. and Koenig, A. 2012. Application of a simplified mathematical model to estimate the effect of forced aeration on composting in a closed system. Waste Manag. 32: 2037–2045. https://doi.org/10.1016/j.wasman.2012.01.014.

Bayard, R., Benbelkacem, H., Gourdon, R. et al. 2018. Characterization of selected municipal solid waste components to estimate their biodegradability. J. Environ. Manage. 216: 4–12. https://doi.org/10.1016/j.jenvman.2017.04.087.

Bazargan, A., Bazargan, M. and Mckay, G. 2015. Optimization of rice husk pretreatment for energy production. Renew. Energy 77: 512–520. https://doi.org/10.1016/j.renene.2014.11.072.

Behera, S., Arora, R., Nandhagopal, N. et al. 2014. Importance of chemical pretreatment for bioconversion of lignocellulosic biomass. Renew. Sustain. Energy Rev. 36: 91–106. https://doi.org/10.1016/j.rser.2014.04.047.

Bernal, M.P., Alburquerque, J.A. and Moral, R. 2009. Composting of animal manures and chemical criteria for compost maturity assessment. A review. Bioresour. Technol. 100: 5444–5453. https://doi.org/10.1016/j.biortech.2008.11.027.

Bouallagui, H., Ben Cheikh, R., Marouani, L. et al. 2003. Mesophilic biogas production from fruit and vegetable waste in a tubular digester. Bioresour. Technol. 86: 85–89. https://doi.org/10.1016/S0960-8524(02)00097-4.

Bouallagui, H., Lahdheb, H., Ben Romdan, E. et al. 2009. Improvement of fruit and vegetable waste anaerobic digestion performance and stability with co-substrates addition. J. Environ. Manage. 90: 1844–1849. https://doi.org/10.1016/j.jenvman.2008.12.002.

Briški, F., Vuković, M., Papa, K. et al. 2007. Modelling of composting of food waste in a column reactor. Chem. Pap. 61: 24–29. https://doi.org/10.2478/s11696-006-0090-0.

Capson-Tojo, G., Moscoviz, R., Ruiz, D. et al. 2018. Addition of granular activated carbon and trace elements to favor volatile fatty acid consumption during anaerobic digestion of food waste. Bioresour. Technol. 260: 157–168. https://doi.org/10.1016/j.biortech.2018.03.097.

Castillo, M.E.F., Cristancho, D.E. and Victor Arellano, A. 2006. Study of the operational conditions for anaerobic digestion of urban solid wastes. Waste Manag. 26: 546–556. https://doi.org/10.1016/j.wasman.2005.06.003.

Cesaro, A. and Belgiorno, V. 2013. Ultrasonics Sonochemistry Sonolysis and ozonation as pretreatment for anaerobic digestion of solid organic waste. Ultrason - Sonochemistry 20: 931–936. https://doi.org/10.1016/j.ultsonch.2012.10.017.

Cesaro, A. and Belgiorno, V. 2014. Pretreatment methods to improve anaerobic biodegradability of organic municipal solid waste fractions. Chem. Eng. J. 240: 24–37. https://doi.org/10.1016/j.cej.2013.11.055.

Chan, M.T., Selvam, A. and Wong, J.W.C. 2016. Reducing nitrogen loss and salinity during "struvite" food waste composting by zeolite amendment. Bioresour. Technol. 200: 838–844. https://doi.org/10.1016/j.biortech.2015.10.093.

Charles, W., Walker, L. and Cord-Ruwisch, R. 2009. Effect of pre-aeration and inoculum on the start-up of batch thermophilic anaerobic digestion of municipal solid waste. Bioresour. Technol. 100: 2329–2335. https://doi.org/10.1016/j.biortech.2008.11.051.

Conant, T., Karim, A. and Datye, A. 2008. Coating of steam reforming catalysts in non-porous multi-channeled microreactors. Bioresour. Technol. 99: 882–888. https://doi.org/10.1016/j.biortech.2007.01.013.

Dang, Y., Sun, D., Woodard, T.L. et al. 2017. Stimulation of the anaerobic digestion of the dry organic fraction of municipal solid waste (OFMSW) with carbon-based conductive materials. Bioresour. Technol. 238: 30–38. https://doi.org/10.1016/j.biortech.2017.04.021.

Ding, Y., Zhao, J., Liu, J.W. et al. 2021. A review of China's municipal solid waste (MSW) and comparison with international regions: Management and technologies in treatment and resource utilization. J. Clean. Prod. 293. https://doi.org/10.1016/j.jclepro.2021.126144.

Dong, L., Zhenhong, Y., Yongming, S. et al. 2009. Hydrogen production characteristics of the organic fraction of municipal solid wastes by anaerobic mixed culture fermentation. Int. J. Hydrogen Energy 34: 812–820. https://doi.org/10.1016/j.ijhydene.2008.11.031.

Fantozzi, F. and Buratti, C. 2009. Biogas production from different substrates in an experimental Continuously Stirred Tank Reactor anaerobic digester. Bioresour. Technol. 100: 5783–5789. https://doi.org/10.1016/j.biortech.2009.06.013.

Farrell, M. and Jones, D.L. 2009. Critical evaluation of municipal solid waste composting and potential compost markets. Bioresour. Technol. 100: 4301–4310. https://doi.org/10.1016/j.biortech.2009.04.029.

Fdez, L.A., Álvarez-gallego, C., Márquez, D.S. et al. 2011a. Biological pretreatment applied to industrial organic fraction of municipal solid wastes (OFMSW): Effect on anaerobic digestion. 172: 321–325. https://doi.org/10.1016/j.cej.2011.06.010.

Fdez, L.A., Álvarez-gallego, C., Sales, D. et al. 2011b. Determination of critical and optimum conditions for biomethanization of OFMSW in a semi-continuous stirred tank reactor. 171: 418–424. https://doi.org/10.1016/j.cej.2011.03.096.

Fdez, L.A., Álvarez-gallego, C., Sales, D. et al. 2011c. The use of thermochemical and biological pretreatments to enhance organic matter hydrolysis and solubilization from organic fraction of municipal solid waste (OFMSW). 168: 249–254. https://doi.org/10.1016/j.cej.2010.12.074.

Fei, F., Wen, Z., Huang, S. et al. 2018. Mechanical biological treatment of municipal solid waste: Energy efficiency, environmental impact and economic feasibility analysis. J. Clean. Prod. 178: 731–739. https://doi.org/10.1016/j.jclepro.2018.01.060.

Fernández, J., Pérez, M. and Romero, L.I. 2008. Effect of substrate concentration on dry mesophilic anaerobic digestion of organic fraction of municipal solid waste (OFMSW). Bioresour. Technol. 99: 6075–6080. https://doi.org/10.1016/j.biortech.2007.12.048.

Fezzani, B. and Cheikh R. Ben. 2010. Two-phase anaerobic co-digestion of olive mill wastes in semi-continuous digesters at mesophilic temperature. Bioresour. Technol. 101: 1628–1634. https://doi.org/10.1016/j.biortech.2009.09.067.

Forgács, G. 2012. Biogas Production from Citrus Wastes and Chicken Feather: Pretreatment and Co-digestion.

Fricke, K., Santen, H., Wallmann, R. et al. 2007. Operating problems in anaerobic digestion plants resulting from nitrogen in MSW. Waste Manag. 27: 30–43. https://doi.org/10.1016/j.wasman.2006.03.003.

Gao, M., Li, B., Yu, A. et al. 2010. The effect of aeration rate on forced-aeration composting of chicken manure and sawdust. Bioresour. Technol. 101: 1899–1903. https://doi.org/10.1016/j.biortech.2009.10.027.

Ge, J., Huang, G., Huang, J. et al. 2015. Mechanism and kinetics of organic matter degradation based on particle structure variation during pig manure aerobic composting. J. Hazard. Mater. 292: 19–26. https://doi.org/10.1016/j.jhazmat.2015.03.010.

Ge, X., Xu, F. and Li, Y. 2016. Solid-state anaerobic digestion of lignocellulosic biomass: Recent progress and perspectives. Bioresour. Technol. 205: 239–249. https://doi.org/10.1016/j.biortech.2016.01.050.

Getahun, T., Nigusie, A., Entele, T. et al. 2012. Effect of turning frequencies on composting biodegradable municipal solid waste quality. Resour. Conserv. Recycl. 65: 79–84. https://doi.org/10.1016/j.resconrec.2012.05.007.

Guermoud, N., Ouadjnia, F., Abdelmalek, F. et al. 2009. Municipal solid waste in Mostaganem city (Western Algeria). Waste Manag. 29: 896–902. https://doi.org/10.1016/j.wasman.2008.03.027.

Hansen, T.L. 2007. Effects of pre-treatment technologies on quantity and quality of source-sorted municipal organic waste for biogas recovery. 27: 398–405. https://doi.org/10.1016/j.wasman.2006.02.014.

Huang, D., Du, Y., Xu, Q. et al. 2022. Quantification and control of gaseous emissions from solid waste landfill surfaces. J. Environ. Manage. 302: 114001. https://doi.org/10.1016/j.jenvman.2021.114001.

Huang, G.F., Wong, J.W.C., Wu, Q.T. et al. 2004. Effect of C/N on composting of pig manure with sawdust. Waste Manag. 24: 805–813. https://doi.org/10.1016/j.wasman.2004.03.011.

Jain, S., Jain, S., Tim, I. et al. 2015. A comprehensive review on operating parameters and different pretreatment methodologies for anaerobic digestion of municipal solid waste. 52: 142–154. https://doi.org/10.1016/j.rser.2015.07.091.

Jeong, E., Kim, H.W., Nam, J.Y. et al. 2010. Enhancement of bioenergy production and effluent quality by integrating optimized acidification with submerged anaerobic membrane bioreactor. Bioresour. Technol. 101: S7–S12. https://doi.org/10.1016/j.biortech.2009.04.064.

Jin, Y., Li, Y. and Li, J. 2016. Influence of thermal pretreatment on physical and chemical properties of kitchen waste and the efficiency of anaerobic digestion. J. Environ. Manage. 180: 291–300. https://doi.org/10.1016/j.jenvman.2016.05.047.

Kapdan, I.K. and Kargi, F. 2006. Bio-hydrogen production from waste materials. 38: 569–582. https://doi.org/10.1016/j.enzmictec.2005.09.015.

Kashyap, D.R., Dadhich, K.S. and Sharma, S.K. 2003. Biomethanation under psychrophilic conditions: A review. Bioresour. Technol. 87: 147–153. https://doi.org/10.1016/S0960-8524(02)00205-5.

Khalid, A., Arshad, M., Anjum, M. et al. 2011. The anaerobic digestion of solid organic waste. Waste Manag. 31: 1737–1744. https://doi.org/10.1016/j.wasman.2011.03.021.

Kim, J.K., Oh, B.R., Chun, Y.N. et al. 2006. Effects of temperature and hydraulic retention time on anaerobic digestion of food waste. J. Biosci. Bioeng. 102: 328–332. https://doi.org/10.1263/jbb.102.328.

Lee, D.H., Behera, S.K., Kim, J.W. et al. 2009. Methane production potential of leachate generated from Korean food waste recycling facilities: A lab-scale study. Waste Manag. 29: 876–882. https://doi.org/10.1016/j.wasman.2008.06.033.

Liew, L.N., Shi, J. and Li, Y. 2011. Enhancing the solid-state anaerobic digestion of fallen leaves through simultaneous alkaline treatment. Bioresour. Technol. 102: 8828–8834. https://doi.org/10.1016/j.biortech.2011.07.005.

Lim, S.L., Lee, L.H. and Wu, T.Y. 2016. Sustainability of using composting and vermicomposting technologies for organic solid waste biotransformation: Recent overview, greenhouse gases emissions and economic analysis. J. Clean Prod. 111: 262–278. https://doi.org/10.1016/j.jclepro.2015.08.083.

Liu, X., Wang, W., Gao, X. et al. 2012. Effect of thermal pretreatment on the physical and chemical properties of municipal biomass waste. Waste Manag. 32: 249–255. https://doi.org/10.1016/j.wasman.2011.09.027.

Liu, Y. and Bond, D.R. 2012. Long-distance electron transfer by *G. sulfurreducens* biofilms results in accumulation of reduced c-type cytochromes. 1047–1053. https://doi.org/10.1002/cssc.201100734.

Mancini, G., Papirio, S., Lens, P.N.L. et al. 2018. Increased biogas production from wheat straw by chemical pretreatments. Renew. Energy 119: 608–614. https://doi.org/10.1016/j.renene.2017.12.045.

Marin, J., Kennedy, K.J. and Eskicioglu, C. 2010. Effect of microwave irradiation on anaerobic degradability of model kitchen waste. Waste Manag. 30: 1772–1779. https://doi.org/10.1016/j.wasman.2010.01.033.

Mir, M.A., Hussain, A. and Verma, C. 2016. Design considerations and operational performance of anaerobic digester: A review. Cogent. Eng. 3. https://doi.org/10.1080/23311916.2016.1181696.

Ogbonna, C., Berebon, D. and Onwuegbu, E. 2015. Relationship between Temperature, Ph and population of selected microbial indicators during anaerobic digestion of Guinea Grass (Panicum Maximum). Am. J. Microbiol. Res. 3: 14–24. https://doi.org/10.12691/ajmr-3-1-3.

Onwosi, C.O., Igbokwe, V.C., Odimba, J.N. et al. 2017. Composting technology in waste stabilization: On the methods, challenges and future prospects. J. Environ. Manage. 190: 140–157. https://doi.org/10.1016/j.jenvman.2016.12.051.

Panigrahi, S. and Dubey, B.K. 2019. A critical review on operating parameters and strategies to improve the biogas yield from anaerobic digestion of organic fraction of municipal solid waste. Renew. Energy 143: 779–797. https://doi.org/10.1016/j.renene.2019.05.040.

Parida, V.K., Sikarwar, D., Majumder, A. et al. 2022. An assessment of hospital wastewater and biomedical waste generation, existing legislations, risk assessment, treatment processes, and scenario during COVID-19. J. Environ. Manage. 308: 114609. https://doi.org/10.1016/j.jenvman.2022.114609.

Park, C., Lee, C., Kim, S. et al. 2005. Upgrading of anaerobic digestion by incorporating two different hydrolysis processes. J. Biosci. Bioeng. 100: 164–167. https://doi.org/10.1263/jbb.100.164.

Park, Y.J., Tsuno, H., Hidaka, T. et al. 2008. Evaluation of operational parameters in thermophilic acid fermentation of kitchen waste. J. Mater. Cycles Waste Manag. 10: 46–52. https://doi.org/10.1007/s10163-007-0184-y.

Pergola, M., Persiani, A., Palese, A.M. et al. 2018. Composting: The way for a sustainable agriculture. Appl. Soil Ecol. 123: 744–750. https://doi.org/10.1016/j.apsoil.2017.10.016.

Petric, I. and Selimbašić, V. 2008. Development and validation of mathematical model for aerobic composting process. Chem. Eng. J. 139: 304–317. https://doi.org/10.1016/j.cej.2007.08.017.

Petric, I., Helić, A. and Avdić, E.A. 2012. Evolution of process parameters and determination of kinetics for co-composting of organic fraction of municipal solid waste with poultry manure. Bioresour. Technol. 117: 107–116. https://doi.org/10.1016/j.biortech.2012.04.046.

Rajagopal, R., Massé, D.I. and Singh, G. 2013. Bioresource technology A critical review on inhibition of anaerobic digestion process by excess ammonia. Bioresour. Technol. 143: 632–641. https://doi.org/10.1016/j.biortech.2013.06.030.

Raut, M.P., Prince William, S.P.M., Bhattacharyya, J.K. et al. 2008. Microbial dynamics and enzyme activities during rapid composting of municipal solid waste—A compost maturity analysis perspective. Bioresour. Technol. 99: 6512–6519. https://doi.org/10.1016/j.biortech.2007.11.030.

Richard, E.N., Hilonga, A., Machunda, R.L. et al. 2019. A review on strategies to optimize metabolic stages of anaerobic digestion of municipal solid wastes towards enhanced resources recovery. Sustain. Environ. Res. 1: 1–13. https://doi.org/10.1186/s42834-019-0037-0.

Ros, M., Klammer, S., Knapp, B. et al. 2006. Long-term effects of compost amendment of soil on functional and structural diversity and microbial activity. Soil Use Manag. 22: 209–218. https://doi.org/10.1111/j.1475-2743.2006.00027.x.

Sarto, S., Hildayati, R. and Syaichurrozi, I. 2019. Effect of chemical pretreatment using sulfuric acid on biogas production from water hyacinth and kinetics. Renew. Energy 132: 335–350. https://doi.org/10.1016/j.renene.2018.07.121.

Shi, X.S., Dong, J.J., Yu, J.H. et al. 2017. Effect of hydraulic retention time on anaerobic digestion of wheat straw in the semicontinuous continuous stirred-tank reactors. Biomed. Res. Int. 2017. https://doi.org/10.1155/2017/2457805.

Silva, F.C., Serafim, L.S., Nadais, H. et al. 2013. Acidogenic fermentation towards valorisation of organic waste streams into volatile fatty acids. Chem. Biochem. Eng. Q 27: 467–476.

Sipra, A.T., Gao, N. and Sarwar, H. 2018. Municipal solid waste (MSW) pyrolysis for bio-fuel production: A review of effects of MSW components and catalysts. Fuel Process Technol. 175: 131–147. https://doi.org/10.1016/j.fuproc.2018.02.012.

Sun, C., Cao, W., Banks, C.J. et al. 2016. Bioresource technology biogas production from undiluted chicken manure and maize silage: A study of ammonia inhibition in high solids anaerobic digestion. Bioresour. Technol. 218: 1215–1223. https://doi.org/10.1016/j.biortech.2016.07.082.

Sundberg, C. and Jönsson, H. 2008. Higher pH and faster decomposition in biowaste composting by increased aeration. Waste Manag. 28: 518–526. https://doi.org/10.1016/j.wasman.2007.01.011.

Sundberg, C., Smårs, S. and Jönsson, H. 2004. Low pH as an inhibiting factor in the transition from mesophilic to thermophilic phase in composting. Bioresour. Technol. 95: 145–150. https://doi.org/10.1016/j.biortech.2004.01.016.

Tian, L., Zou, D., Yuan, H. et al. 2015. Identifying proper agitation interval to prevent floating layers formation of corn stover and improve biogas production in anaerobic digestion. Bioresour. Technol. 186: 1–7. https://doi.org/10.1016/j.biortech.2015.03.018.

Trzcinski, A.P. and Stuckey, D.C. 2009. Continuous treatment of the organic fraction of municipal solid waste in an anaerobic two-stage membrane process with liquid recycle. Water Res. 43: 2449–2462. https://doi.org/10.1016/j.watres.2009.03.030.

Tuan, A., Sabev, P., Ni, S. et al. 2022. Perspective review on Municipal Solid Waste-to-energy route: Characteristics, management strategy, and role in circular economy. 359. https://doi.org/10.1016/j.jclepro.2022.131897.

Turan, N.G. 2008. The effects of natural zeolite on salinity level of poultry litter compost. Bioresour. Technol. 99: 2097–2101. https://doi.org/10.1016/j.biortech.2007.11.061.

Uddin, M.N., Siddiki, S.Y.A., Mofijur, M. et al. 2021. Prospects of bioenergy production from organic waste using anaerobic digestion technology: A mini review. Front. Energy Res. 9. https://doi.org/10.3389/fenrg.2021.627093.

United Nations Economic and Social Commission for Asia and the Pacific (2002) Chapter 8 Types of wastes. United Nations ESCAP Libr 170–194.

Vuorinen, A.H. and Saharinen, M.H. 1997. Evolution of microbiological and chemical parameters during manure and straw co-composting in a drum composting system. Agric. Ecosyst. Environ. 66: 19–29. https://doi.org/10.1016/S0167-8809(97)00069-8.

Vyas, S., Prajapati, P., Shah, A.V. et al. 2022. Opportunities and knowledge gaps in biochemical interventions for mining of resources from solid waste: A special focus on anaerobic digestion. Fuel 311: 122625. https://doi.org/10.1016/j.fuel.2021.122625.

Wagner, A.O. and Illmer, P. 2018. Biological Pretreatment Strategies for Second-Generation Lignocellulosic Resources to Enhance Biogas Production. https://doi.org/10.3390/en11071797.

Wang, W., Hou, H., Hu, S. et al. 2010. Bioresource Technology Performance and stability improvements in anaerobic digestion of thermally hydrolyzed municipal biowaste by a biofilm system. Bioresour. Technol. 101: 1715–1721. https://doi.org/10.1016/j.biortech.2009.10.010.

Ward, A.J., Hobbs, P.J., Holliman, P.J. et al. 2008. Optimisation of the anaerobic digestion of agricultural resources. Bioresour. Technol. 99: 7928–7940. https://doi.org/10.1016/j.biortech.2008.02.044.

Willts, H. 2014. International Journal of Waste. Longdom.org 4: 1–5.

Xu, S., Han, R., Zhang, Y. et al. 2018. Differentiated stimulating effects of activated carbon on methanogenic degradation of acetate, propionate and butyrate. Waste Manag. 76: 394–403. https://doi.org/10.1016/j.wasman.2018.03.037.

Yang, F., Li, G., Shi, H. et al. 2015. Effects of phosphogypsum and superphosphate on compost maturity and gaseous emissions during kitchen waste composting. Waste Manag. 36: 70–76. https://doi.org/10.1016/j.wasman.2014.11.012.

Zhang, L. and Sun, X. 2014. Changes in physical, chemical, and microbiological properties during the two-stage co-composting of green waste with spent mushroom compost and biochar. Bioresour. Technol. 171: 274–284. https://doi.org/10.1016/j.biortech.2014.08.079.

Zhang, L. and Sun, X. 2016a. Improving green waste composting by addition of sugarcane bagasse and exhausted grape marc. Bioresour. Technol. 218: 335–343. https://doi.org/10.1016/j.biortech.2016.06.097.

Zhang, L. and Sun, X. 2016b. Influence of bulking agents on physical, chemical, and microbiological properties during the two-stage composting of green waste. Waste Manag. 48: 115–126. https://doi.org/10.1016/j.wasman.2015.11.032.

Zhang, T., Nie, H., Bain, T.S. et al. 2013. Environmental science improved cathode materials for microbial electrosynthesis. 217–224. https://doi.org/10.1039/c2ee23350a.

Zhang, W., Wu, S., Guo, J. et al. 2015. Performance and kinetic evaluation of semi-continuously fed anaerobic digesters treating food waste: Role of trace elements. Bioresour. Technol. 178: 297–305. https://doi.org/10.1016/J.BIORTECH.2014.08.046.

Zhao, H.X., Zhou, F.S., Evelina, L.M.A. et al. 2022. A review on the industrial solid waste application in pelletizing additives: Composition, mechanism and process characteristics. J. Hazard. Mater. 423: 127056. https://doi.org/10.1016/j.jhazmat.2021.127056.

Zhou, M., Yan, B., Wong, J.W.C. et al. 2018. Enhanced volatile fatty acids production from anaerobic fermentation of food waste: A mini-review focusing on acidogenic metabolic pathways. Bioresour. Technol. 248: 68–78. https://doi.org/10.1016/j.biortech.2017.06.121.

2

ANAEROBIC DIGESTION OF SOLID ORGANIC WASTE TO GENERATE BIOGAS

Bella K., Mothe Sagarika, Kommuri Ravi Chandra Reddy and P. Venkateswara Rao*

1. INTRODUCTION

Waste is defined as the material which can't be used for intended purpose. Solid waste generation is a never ending process. Population is directly proportional to the solid waste generation; in other words, solid waste generation is more in metro cites than rural areas. A population of around 1.38 billion in India generates approximately 1,47,613 metric tonnes of solid waste per day (Singh 2020). Solid waste generation from 46 metro cities is given in Table 1 for over a decade, from 1999–2000 to 2015–2016. It clearly shows that population is proportional to solid waste generation, like Mumbai city with population of 1,24,42,373 generated 11,000 tonnes per day whereas Ranchi with a population of 10,73,440 generated 150 tonnes per day. Cities like Delhi, Bangalore, Chennai, Hyderabad generated more solid waste (> 3500 TPD) whereas cities like Raipur, Gwalior, Dhanbad generated less amount of solid waste (< 250 TPD). It is clearly evident that waste generation depends upon the population and peoples' living standards. Solid waste generation rate increased per person over the years. In the year 2015, one person generated around 450 g/capita/day. The municipal solid waste of India comprises of 40–60% compostable portion, 30–50% inert and 10–30% recyclable (Sharma and Jain 2019). Figure 1 clearly displays that over a decade of municipal solid waste composition data from 1996–2011, more than half of the waste composition is biodegradable waste. Paper and rubber waste composition increased immensely over a decade. Paper and rubber waste increased from 3.63% and 0.6% to 13.8% and 7.89%, respectively. Composition of solid waste varies from place to place and from region to region significantly. It mainly depends upon the culture, habits and waste management rules at the waste generation place. Composition of solid waste is crucial

Department of Civil Engineering, National Institute of Technology Warangal.
* Corresponding author: pvenku@.nitw.ac.in

Table 1. Solid waste generation in 46 metro cities.

Rank	City	Waste generation (TPD[1])				
		1999–2000	2004–05	2010–11	2015–16	(2015–16) Kg/Capita/day
1	Mumbai	5355	5320	6500	11,000	0.88
2	Delhi	400	5922	6800	8700	0.79
3	Bangalore	200	1669	3700	3700	0.44
4	Chennai	3124	3036	4500	5000	0.71
5	Hyderabad	1566	2187	4200	4000	0.59
6	Ahmedabad	1683	1302	2300	2500	0.45
7	Kolkata	3692	2653	3670	4000	0.89
8	Surat	900	1000	1200	1680	0.38
9	Pune	700	1175	1300	1600	0.51
10	Jaipur	580	904	310	1000	0.33
11	Luck now	1010	475	1200	1200	0.43
12	Kanpur	1200	1100	1600	1500	0.54
13	Nagpur	443	504	650	1000	0.42
14	Visakhapatnam	300	584	334	350	0.17
15	Indore	350	557	720	850	0.43
16	Thane	--	-	-	700	0.38
17	Bhopal	546	574	350	700	0.39
18	Pimpri-Chinchwad	-	-	-	700	0.40
19	Patna	330	511	220	450	0.27
20	Vadodara	400	357	600	700	0.42
21	Ghaziabad	-	-	-	-	
22	Ludhiana	400	735	850	850	0.53
23	Coimbatore	350	530	700	850	0.53
24	Agra	-	654	520	790	0.50
25	Madurai	370	275	450	450	0.29
26	Nashik	-	200	350	500	0.34
27	Vijayawada	-	374	600	550	0.37
28	Faridabad	-	448	700	400	0.28
29	Meerut	-	490	520	500	0.38
30	Rajkot	-	207	230	450	0.35
31	Kalian-Dombivali	-	-	510	650	0.52
32	Vasai-Virar	-	-	-	600	0.49
33	Varanasi	412	425	450	500	0.42
34	Srinagar	-	428	550	550	0.46
35	Aurangabad	-	-	-	-	
36	Dhanbad	-	77	150	180	0.15
37	Amritsar	-	438	550	600	0.53

Table 1 contd. ...

---

[1] TPD-Tonnes per day.

...Table 1 contd.

Rank	City	Waste generation (TPD[1])				
		1999–2000	2004–05	2010–11	2015–16	(2015–16) Kg/Capita/day
38	Navi Mumbai	-	-	-	675	0.60
39	Allahabad	-	509	350	450	0.40
40	Ranchi	-	208	140	150	0.14
41	Howrah	-	-	-	740	0.69
42	Jabalpur	-	216	400	550	0.52
43	Gwalior	-	-	285	300	0.28
44	Jodhpur	-	-	-	-	
45	Raipur	-	184	224	230	0.23
46	Kota	-		-	-	

Source: Data collected from CPCB (2020).

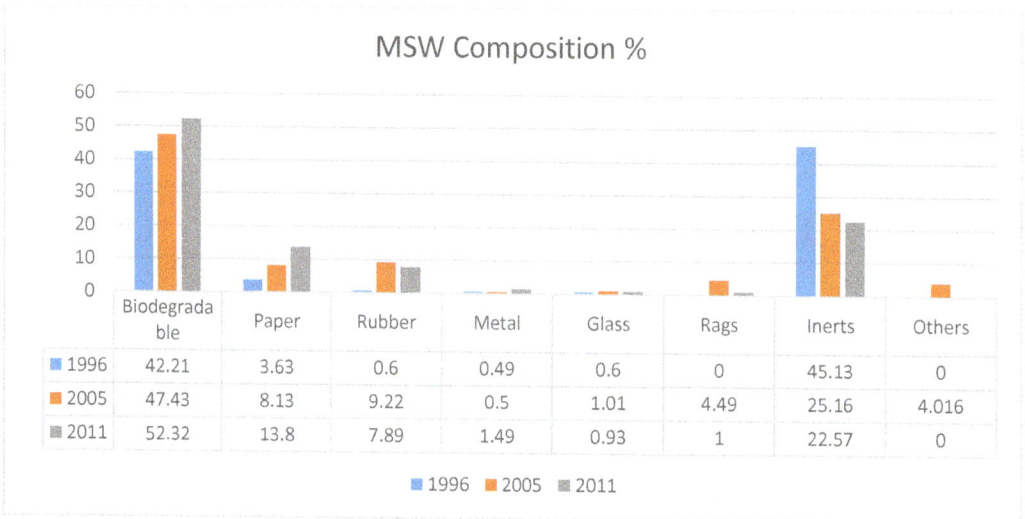

Figure 1. Change in Municipal Solid Waste composition over years 1996–2011.
Sources: Data from (Joshi and Ahmed 2016, Planning Commission 2014, Zhu et al. 2008).

to know because management of solid waste depends upon the composition. The 18 nations in European Union have decided to stop landfilling of recyclable materials by 2025, and many other countries imposed taxes on landfilling to make it a less attractive alternative for waste disposal (World Energy Council 2016). Most of the wastes dumped into landfill sites contain high amount of organic matter which will lead to source of environmental pollution by causing leachate flow. Moreover, landfilling is the least preferred option for waste disposal as it requires large space and is not environmental friendly (Sondh et al. 2022). The growing concern about the disposal options for the large organic wastes generated has led to the utilisation of such wastes for energy recovery options. Anaerobic digestion (AD) is considered as a viable energy recovery method which can generate biogas which is a renewable energy source. The process of AD can be explained by a series of steps: hydrolysis, acidogenesis, acetogenesis and methanogenesis. AD provides useful products like biofuels, and fuel for vehicles, and the digestate rich in nutrients that can be used as a soil conditioner. Compared to other biological processes, AD is economically cheaper and requires less maintenance cost. The fuels generated - methane and hydrogen - are cleaner than that generated from fossil fuels. The AD of solid organic waste is not as widely applied as aerobic treatment processes because of the longer time required for achieving bio-stabilisation and increased sensitivity of the process. The recent advancements in several bio-reactors' designs have increased the applicability of AD. Still there are some areas which necessitates attention like various operational parameters, and suitability of different wastes in AD. This paper is intended to review various factors associated with AD like reactor configurations, process parameters, pre-treatment options and co-digestion.

2. PHYSICAL CHARACTERISTICS OF SOLID WASTE

The physico-chemical characteristics of a particular solid waste depends on its source of generation and type of waste. Physical characteristics include density, moisture content, particle size, colour, voids, field capacity, and size distribution. The knowledge about the characteristics of these wastes helps in the selection of proper treatment option, its design and operation. Density, size distribution and moisture content are the important physical characteristics among them all.

2.1 Density

Density is defined as the total mass per unit volume (kg/m^3). It is a very important parameter to know because the selection of the collection vehicle and area of the land required for the landfill site is based on the density of the waste. The volume of wastes can be reduced by 75% by using a typical compact machine, resulting in a density increase from $100 \ kg/m^3$ to $400 \ kg/m^3$. The density of solid wastes varies greatly depending on geographic location, season, and storage time, choosing average values should be done with caution. Table 2 given below shows wide variations in density values of different cities in India. The typical value of municipal solid garbage delivered in compaction vehicles was determined to be around $300 \ kg/m^3$ (Hartwell et al. 2021).

Table 2. Density values of Municipal Solid Waste in major cities (Bhide and Sundaresan 1983).

S. no.	City	Density (kg/m^3)
1.	Hyderabad	369
2.	Bangalore	390
3.	Delhi	422
4.	Jaipur	537
5.	Baroda	457

The design part of a sanitary landfill is equally important as the collection, transportation and storage of garbage. The garbage transported to the site has to be compacted well to its optimum density for the efficient landfill operation (Ramachandra et al. 2018).

2.2 Moisture content

The moisture content can be calculated from the ratio of weight of water to the total wet weight of the solid waste. Following formula can be used to determine the moisture content of the solid waste.

$$\text{Moisture content (\%)} = \left[\frac{Wet \ weight - Dry \ weight}{Wet \ weight} \right] * 100 \qquad \text{Eq. 1}$$

where Dry weight is the weight of sample after oven drying at 105°C.

Organic wastes like fruits and vegetable wastes possess higher content of moisture ranging from 88–94% (Jain et al. 2019). Moisture content is an important characteristic of organic wastes as it influences the microbial activity in anaerobic digestion process. High moisture content is normally favourable for growth of anaerobes.

2.3 Field capacity

The total quantity of water that can be held in a waste sample when subjected to gravitational force is known as field capacity of Municipal Solid Waste (MSW). It is mainly used to know the leachate generation from the landfill. It's a crucial safeguard because water that exceeds the capacity of the field would generate leachate, which may be a serious issue in landfills.

2.4 Compressibility of MSW

The degree of physical change occurring to a waste pile (in filter cake form) when subjected to pressure can be measured using the following formula:

$$\Delta H_T = \Delta H_i + \Delta H_c + \Delta h_\alpha \qquad \text{Eq. 2}$$

where, ΔH_T = total settlement
ΔH_i = immediate settlement
ΔH_c = consolidation settlement
ΔH_α = secondary compression or creep

$$C' \, \alpha = \frac{\Delta H}{\left(H_0 * Log\left(\dfrac{t_2}{t_1}\right) \right)} = \frac{C\alpha}{1 + e_0} \qquad \text{Eq. 3}$$

where, C_α = Secondary compression index
C'_α = Modified Secondary Compression index;
t_1, t_2 = Starting and ending time of secondary settlement, respectively

3. CHEMICAL CHARACTERISTICS OF SOLID WASTE

The ability to determine the efficacy of any treatment technique requires a thorough understanding of the chemical properties of waste (Ramachandra et al. 2018). The chemical characteristics include carbohydrates, proteins, fats, lipids, synthetic organic material, natural fibres and non-combustible materials. All these compounds are found in food waste, yard waste and kitchen waste. Most of the above-mentioned materials are biodegradable. Compounds like natural fibres are rich in lignin, cellulose and these compounds are resistant to biodegradation. Synthetic organic material (plastic) is very hard to degrade and it will take millions of years to degrade it fully.

Study on Solid Waste: Analysis of solid waste is essential for knowing the individual concentration present in the waste and it is useful for directing the solid waste for the appropriate treatment process accordingly.

3.1 Ultimate analysis

This is a waste analysis that determines the proportions of carbon, hydrogen, oxygen, nitrogen, and sulphur for calculating the mass balance for a thermal or chemical process (Zaher et al. 2009). The presence of metals such as cadmium, chromium, mercury, nickel, lead, tin, and zinc makes it vital to evaluate ash fraction because of its potentially severe environmental impacts. Other non-hazardous metals like iron, potassium, and magnesium are also present (Zaher et al. 2009).

3.2 Proximate analysis

This is critical when assessing the combustion qualities of wastes or waste-derived fuels. The interest fractions are as follows.

- Moisture content: adds weight to the garbage without increasing its heating value, and evaporation of water, which reduces the amount of heat emitted by the fuel.
- Ash: doesn't produce heat at combustion, but adds weight.
- Volatile matter: portion which gets converted to gases or vapour during or after combustion.
- Fixed carbon: refers to the carbon that remains as charcoal on the surface grates. To ensure complete combustion, a waste or fuel with a high proportion of fixed carbon requires a longer retention period on the furnace grates than a waste or fuel with a low proportion of fixed carbon.

A proximate analysis results of typical municipal solid waste collected from various studies are shown in the table below. Some of the other characteristics of organic wastes reported in studies are presented in Table 3. Table 4 shows the characteristics of different organic wastes reported in literatures.

Table 3. Proximate analysis of solid organic waste.

Moisture (wt %)	Volatile matter (wt %)	Fixed carbon (wt %)	Ash (wt %)	Reference
59.67	30.02	7.02	3.29	(Salwa Khamis et al. 2019)
6.33	2.67	88.7	2.33	(Ibikunle et al. 2018)
NA	71.01	13.41	15.58	(Johari et al. 2012)
55.01	31.36	4.37	9.26	(Kathirvale et al. 2003)
NA	25	5	70	(Parikh et al. 2005)

Table 4. Summary of data collected on characteristics of organic wastes.

Organic waste	Density (kg/m³)	Moisture%	TS%	VS	pH	C:N	Reference
Food waste	510		84.7	72.4	5.9	37	(Forster-Carneiro et al. 2008a)
Food waste	-		10.3	9.2	3.77	-	(Nagao et al. 2012)
MS-OFMSW	295		17.2	7.31	7.9	11.9	(Pérez and Romero 2008)
Food waste	500		89.8	69.8	7.6	35.4	(Forster-Carneiro et al. 2008b)
OFMSW	-	80.06	19.94	19.19	5.9		(Pavi et al. 2017)
Fruits and vegetable waste	-	80.46	19.54	18.80	4.66		(Pavi et al. 2017)
OFUSW	864		16.8	15.8		32.7	(Edgar et al. 2006)
Food waste	-	78.8	21.2	19.67	4.7	13.4	(Dai et al. 2013)
WS-OFMSW	933		18.4	11.3	5.3	11.4	(Dong et al. 2010)
OFMSW	-	70.3	29.7	22.3	-	-	(Campuzano and González-Martínez 2015)
Food waste			29.4	28	4.1	14.2	(Browne and Murphy 2013)
OFMSW			18.7	16.9	6.2		(Sajeena Beevi et al. 2015)
MS-OFMSW		70.8	29.2	24.9	5.26		(Lópeza et al. 2010)

4. DIFFERENT CONFIGURATIONS OF ANAEROBIC DIGESTERS

4.1 Covered lagoon

A covered lagoon digester is digester with long retention time and high dilution factor. The components of covered-lagoon digester are: (i) Solids separator, (ii) Lagoons, (iii) Floating lagoon cover and (iv) Biogas utilization system. Figure 2 illustrates a covered lagoon.

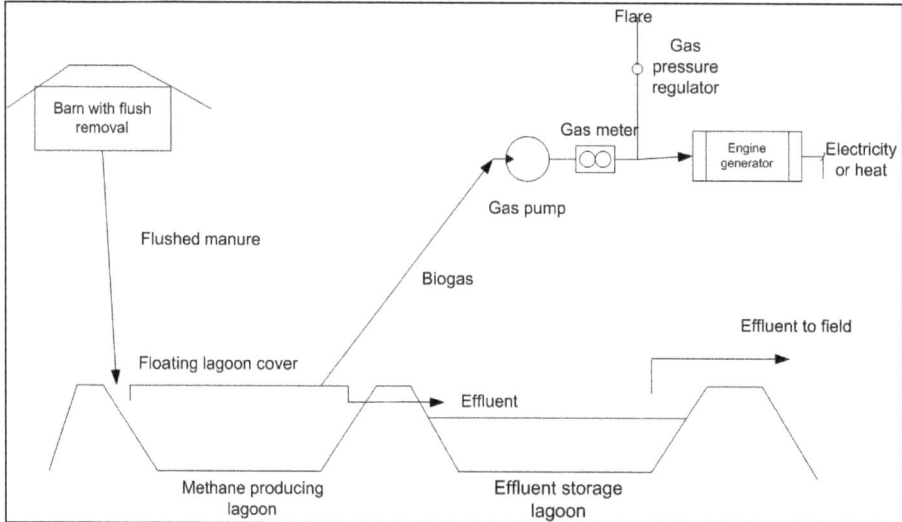

Figure 2. Covered lagoon with Methane utilization (Source: Tauseef et al. 2013).

4.1.1 Design variables

a) Soil and foundation: The lagoons are desirable to place in soils with low to moderate permeability or in soils that can be sealed by sedimentation and biological action. Over cracked or cavernous rock avoid gravelly soils and shallow soils.

b) Depth: The primary lagoon should be dug as deep as possible due to soil and geological factors. The primary lagoon's depth is critical for effective operation, but the secondary lagoon's depth is less critical. Deep lagoons help to preserve bacterial growth-friendly temperatures. Increased depth allows for a smaller surface area, which reduces the amount of rain and the size of the cover, lowering floating cover expenses. The major lagoon should have a minimum liquid depth of 12 feet.

c) Loading rate: The higher one between the minimal hydraulic retention time (HRT) value and volatile solids loading rate is used to size the primary anaerobic lagoon. The Volatile solids loading rate (VSLR) is a design number that is used to decide the size of the lagoon which provides enough time for bacteria to digest manure. It is dependent mostly on climate. The principal lagoon inlet and outlet should be placed so that the distance between them across the lagoon is as short as possible.

d) Cover materials: Agricultural and industrial lagoons have been covered with a variety of materials. The dimensions of floating covers are usually unrestricted. Gas can be stored under a floating cover. Cover materials should be UV resistant, hydrophobic, tear and puncture resistant, bacteria-free, and with a bulk density close to that of water. When selecting a cover material, consider the material's availability, serviceability, and cost. Thin materials are less expensive in general; however, they may not have the same proven or assured life as thicker materials. Although fabric reinforced materials are stronger than unreinforced materials, the thickness, serviceability, cost, and estimated life of the materials may compensate for the lack of reinforcement.

4.2 Plug flow digester

It is a narrow type, long, insulated, heated tank which is made of reinforced concrete, steel, or fibreglass that captures biogas through a gas tight cover ('Logan and Viswanathan 2019). They are run as continuous digesters with high hydraulic retention time, same as solids retention time and operated at temperatures 35°C–55°C (Fardin et al. 2018). No external mixing provisions are provided in plug flow digesters and are considered as low-rate systems. A total solids content of 10–15% should be maintained in this type of digesters. The digester produces biogas, which is used to heat it to the necessary temperature (Kalia 1988).

4.2.1 Components of plug flow digester

The components include collection/mix tank, plug-flow digester, biogas utilization system, methane recovery system, inflatable cover, floating cover.

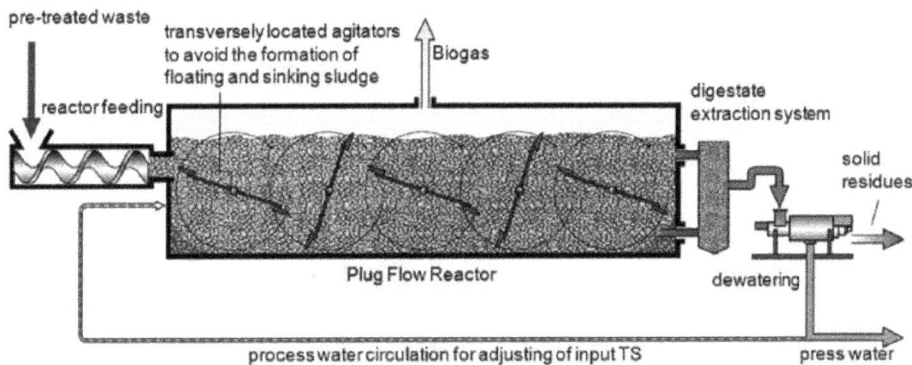

Figure 3. Plug flow digester (Source: Logan and Visvanathan 2019).

4.2.2 Plug flow digester design variables

Dimensions: A plug-flow digester's depth can range from 8 to 16 feet, depending on soil conditions and tank volume requirements. In most cases, the width-to-depth ratio is larger than 1 but less than 2.5. The ratio of length to width should be between 3.5–5.

Hydraulic retention time: An HRT will work with a plug-flow digester for 12 to 80 days. However, a HRT of 15 to 20 days is most typically utilised to produce 70 percent to 80 percent of the final methane production economically.

Operating temperature: Most digesters operate at mesophilic temperature between 35–37°C.

4.3 Fixed film digester

A fixed film digester is a column filled with media like wood chips or small plastic rings. Microorganisms that produce methane thrive on the media. Liquid manure passes through the media. Attached growth digesters or anaerobic filters are other names for these digesters. Biofilm refers

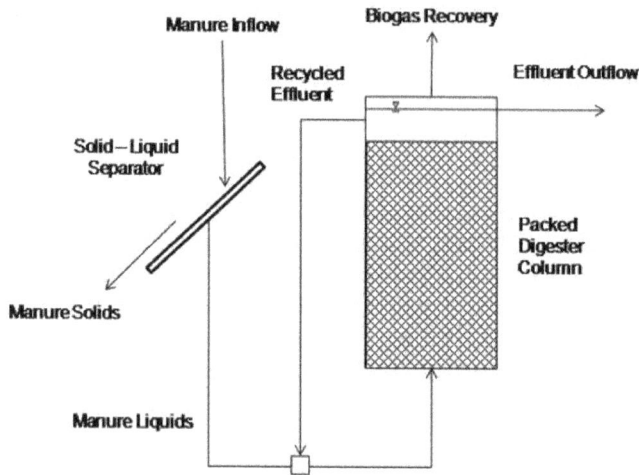

Figure 4. Fixed film digester (Source:Chen and Neibling 2014).

to the slimy development that coats the media. Fixed film digesters can have retention durations of under five days, resulting in digesters that are quite tiny. To maintain a continuous upward flow, wastewater is typically recycled. The fact that manure solids might fill the media in fixed film digesters is a disadvantage. To remove particles from manure before feeding it to the digester, a solid separator is required. The effectiveness of the system is determined by the performance of the solid separator; as a result, the influent manure concentration should be adjusted to maximise separator performance (typically 1% to 5% total solids). When manure solids are removed, some potential biogas is lost (Sivaprakasam and Balaji 2020). The schematic diagram of fixed film digester is shown in Fig. 4.

5. PRE-TREATMENT METHODS FOR SOLID WASTE

The pre-treatment methods are used mainly to alter the composition and structure of the substrate by easing the hydrolysis stage. The main aim of pre-treatment technology on organic waste is to change or alter the structural and compositional changes to hydrolysis (Kaur and Phutela 2016). Pre-treatment helps in the easier degradation of complex substrates which increases its better utilization in all the stages. The type of pre-treatment method to be used depends on the complexity of the feedstock. Substrates containing cellulose, hemicellulose, protein, lipids, and fats are difficult to undergo hydrolysis without pre-treatment. During pre-treatment, waste material gets hydrolysed into simpler monomers like monosaccharides, fatty acids, and alcohols (Kasinath et al. 2021). Pre-treatment methods are classified into physical, chemical, thermal, and biological methods primarily. The addition of metal supplements like Fe^{2+}, Zn^{2+}, Ni^{2+}, Mg^{2+}, is also categorized under pre-treatment methods. The primary classification of pre-treatment methods is illustrated in Fig. 5.

Physical pretreatment methods primarily encompass processes such as milling, grinding, and chipping. Chemical pretreatment involves the use of substances such as acids (e.g., sulfuric acid, acetic acid, hydrochloric acid, nitric acid, and ionic acids), alkalis (e.g., sodium hydroxide, potassium hydroxide, calcium hydroxide, magnesium hydroxide, sodium bicarbonate, and ammonium carbonate), oxidants (e.g., hydrogen peroxide and ozone), and various additives. Additionally, there is biological pretreatment, which involves microorganisms, fungi, enzymatic treatments, and their combinations, known as combined pretreatment. The production of methane from complex organic compounds is difficult as it is recalcitrant to enzymatic or microbial degradation, because of its composition and structure (Hendriks and Zeeman 2009). The above pre-treatment processes result in physical and/or

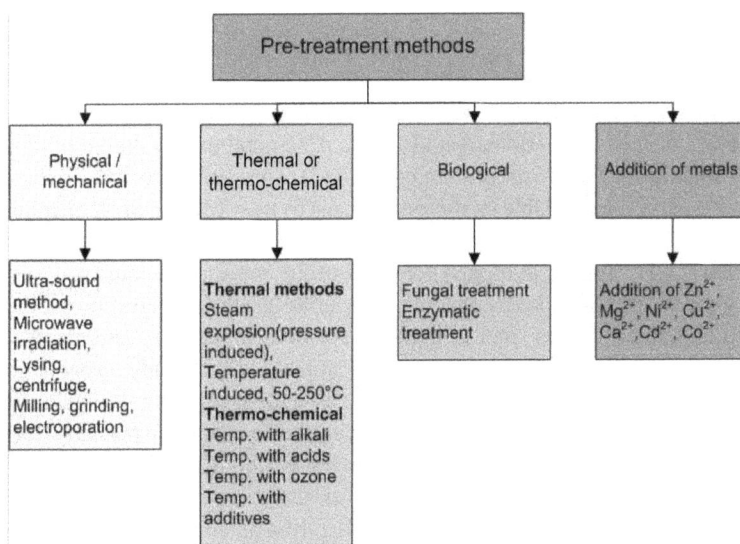

Figure 5. Classification of pre-treatment methods, modified from ref (Carrere et al. 2016).

chemical changes in the biomass (Mosier et al. 2005) as the biomass is recalcitrant to digestion. The organic fraction of municipal solid waste (OFMSW) inhibits the hydrolysis process in anaerobic digestion because of the recalcitrant matter in biomass as it is mainly from cardboard and paper waste (Mahmoodi et al. 2018). Pre-treatment process enhances the cellulose accessibility in hydrolysis step (which is the first step in anaerobic digestion) of the biomass structure to the enzymes (Song et al. 2013). Physical and chemical pre-treatment of OFMSW improved the biogas production, volatile solids reduction and raised biomass solubilisation (Zamri et al. 2021). The pre-treatment process is selected based on the substrate characteristics, pre-treatment mechanism, and end requirements (Kumar and Samadder 2020) and suggested to be careful in producing the intermediate hindering compounds due to excessive hydrolysis (Kumar and Samadder 2020).

5.1 Physical pre-treatment

The physical pre-treatment increases the pore size and available surface area of the biomass, and change and/or alter the degree of polymerization and crystallinity in biomass (Harmsen et al. 2010). In physical pre-treatment, grinding, milling, shredding, rotary drum comes under mechanical treatment which reduces the substrate size and removes the inert particles that comes in between the process. Milling has reduced the size of food waste to finer particles which improved the biogas production, but much finer size may cause high amounts of volatile fatty acids (VFA) (Izumi et al. 2010). Rotary drum reactor pre-treatment has produced moisture feedstock and works on low capital and low energy demand with minimal retention time. Thermal pre-treatment is preferred for improving the substrate solubilisation, removing pathogens and reducing digested viscosity (Amiri et al. 2017). Fernández-Rodríguez et al. (2016) has worked on heating shell reactor (thermal pre-treatment) of OFMSW which improved the hydrolysis and reduced recalcitrant matter like lignin; however, long exposure to heat or higher temperatures (> 150°) has generated complex compounds that are difficult to degrade. Microwave pre-treatment on OFMSW has improved substrate solubilisation and soluble chemical oxygen demand (sCOD) but the process is not economical as the methane generated was not enough for energy demand (Shahriari et al. 2012). But physical pre-treatment is comparatively ineffective in improving substrate digestibility (Chang et al. 1998). Even after improving results in biogas with physical pre-treatment, maintenance and high energy demands are the disadvantages for opting to biomass.

5.2 Chemical pre-treatment

Chemical pre-treatment alters lignin and hemicellulose structure in substrate, which in turn reduces crystallinity and degree of polymerization of the cellulosic part of the substrate which improves the biodegradability of feedstock (Behera et al. 2014). Chemical pre-treatment has gained attention because of its clean and simple process (Song et al. 2013). The main objective of chemical acid pre-treatment is making cellulose available to microorganisms by hemicellulose solubilization by breaking ether bonds in lignin. Alkali pre-treatment (NaOH, KOH, $Ca(OH)_2$, $Mg(OH)_2$), acidic pre-treatment (H_2SO_4, HCL, HNO_3, CH_3COOH), hydrogen peroxide, ozone, ionic pre-treatment and additives (heavy metals) are the general chemical pre-treatments used in the anaerobic digestion process. Hydrolysis with strong acid and alkali is also used to improve the biodegradability and stabilize the microbe accessibility in the digestion process (Modenbach and Nokes 2012). A study on ozonation of OFMSW has improved biodegradability of substrate by getting rid of toxic compounds which inhibit anaerobic system; however, it needs high maintenance cost and substrate should be partially degraded (Cesaro et al. 2019). Alkali pretreatment has been effective in increasing the biomass's surface area, thereby facilitating better access for microorganisms. However, it may be more suitable for substrates with a high lignin content (Manser et al. 2015). The solubilization of substrate in alkali depends upon the chemical used and its concentration, which was suggested as $Ca(OH)_2 < Mg(OH)_2 < KOH < NaOH$ (Li et al. 2008, Lin et al. 2009, Neumann et al. 2016, Saratale and Oh 2015). Alkaline pre-treatment raised the methane productivity by 21.4% and is more productive than addition of heavy metals like Co, Ni and Fe (Mancini et al. 2018), whereas Romero-Güiza et al. (2016) has stated that addition of cobalt, nickel and iron has improved methane generation along with system stability, but excessive addition may lead to system instability.

5.3 Biological pre-treatment

Biological pre-treatment is eco-friendly technique that works with aerobic, anaerobic, and enzymatic methods to digest and improve hydrolysis in anaerobic digestion which gained attention for its advantages than physical and chemical pre-treatments like less energy demands, no generation of toxic or inhibitory compounds, surface and reaction specificity and majorly significant production of desired products (Yuan et al. 2014). Biological pre-treatment is mainly related to the reaction of fungi such as white, brown, and soft rot fungi with biomass (Cianchetta et al. 2014). Microorganism fungi pre-treatment on OFMSW has enhanced the biodegradation of substrate with no production of toxic or inhibitory compounds in digestion; however, opting for the biological pre-treatment for full scale anaerobic digestion is expensive and due to slow reaction of enzymes and microorganisms, the digestion period may take longer time than usual (Carrere et al. 2016). Micro aeration on food waste has resulted in improvement of carbohydrate and protein hydrolysis along with reduction in lipid toxic and inherent compounds that lead system instability (Lim and Wang 2013).

5.4 Combined pre-treatment

The combined pre-treatment is combination of physical, chemical and biological pre-treatments. Microwave and autoclave pre-treatment has improved the biodegradability of high refractory matter along with reduction of contaminants (Zhang et al. 2018). Romero-Güiza et al. (2014) conducted experiments involving physico-chemical pretreatment of OFMSW, which resulted in increased methane production and enhanced system stability. However, it also led to the generation of propionic acid and hydrogen sulfide within the digester. Combined pre-treatment of chemical-biological treatment in full scale has produced high biogas and generated high nutrient rich solid and liquid digestate but a bit complex in controlling (López Torres and Espinosa Lloréns 2008).

Zamri et al. (2021) has suggested combined pretreatment for better biomethane production and high biodegradability of substrate. However, selecting the combination of pretreatments, substrate moisture content and incubating time of treatment alter the results and stability of anaerobic system.

The selection of the pre-treatment method for a specific substrate depends on the type and composition of the feedstock. Apart from these, it should be economically feasible and environmentally friendly. The details on various pre-treatment methods used in the AD of different organic wastes are given in Table 5. The thermal pre-treatment method requires heat and electricity, but the heat required for heating can be availed from the combustion of biogas. Compared to the thermal method, mechanical pre-treatment consumes higher energy even though it is less sensitive towards substrate specificities. Chemical treatment methods cause ultimate chemical contamination of substrate and risk of recalcitrant compounds. Among all the methods, biological methods are the least energy-consuming and can be synthesized in the laboratory only. But biological methods offer only a moderate increase in methane yield. Hence, the appropriate method has to be chosen by considering feedstock characteristics, energy requirement, cost constraints, and operational expertise.

Table 5. Effect of various pretreatment methods in AD of various solid organic wastes.

Feedstock	Pre-treatment method	Anaerobic digestion conditions	% increase in biogas yield	Reference
Waste activated sludge	Thermal pre-treatment: 170°C, HRT-40 minutes, pressure-7.6 bar	CSTR, T-35°C, HRT-10 d, HRT of control reactor-20 d	82% at HRT 10 d and 17% at HRT 20 d	(Souza et al. 2013)
Mixed sludge	Thermal pre-treatment: 70°C, 9–48 hour	CSTR, T-55°C, HRT-10 d,	20%	(Ferrer et al. 2009)
Mixed sludge	Physical pre-treatment-Sonication, 20 kHz, 13.7 W/cm², Sludge flow: 8.33 m³/s HRT: 1.5s	Egg shape digester, T-29 to 33°C, HRT-22.5 d	45%	(Xie et al. 2007)
Thickened mixed sludge	Physical: Focused pulsed technique: Opencel 20–30 kV, Treatment of 85% sludge flow	CSTR, Mesophilic temperature, HRT-30 to 35 d	40%	(Rittmann et al. 2008)
Sewage sludge	Physical: Lysing centrifuge, 39 m³/h 3, 140 rpm	CSTR, Mesophilic temperature, HRT-40d	26%	(Zábranská et al. 2006)
Ensiled sorghum forage	Thermo-chemical: Alkaline: 40°C, 24 h, 10 g NaOH/100 g TS	CSTR, 35°C, HRT-21 d	25%	(Sambusiti et al. 2013)
Maize and sorghum silage	Biological: Enzymatic method	HRT: 63 d, OLR: 5–5.8 kgVS/ m³ d	NA	(Schimpf et al. 2013)

6. THEORETICAL METHANE PRODUCTION OF BIOMASS

Estimation of methane, carbon dioxide and miscellaneous gases from anaerobic digestion of organic waste based on the elemental analysis, the organic fraction of substrate can be represented with formulation of $C_cH_hO_oN_nS_s$. The theoretical methane and carbon dioxide yield of organic waste can be calculated from the following equation (Chen et al. 2015).

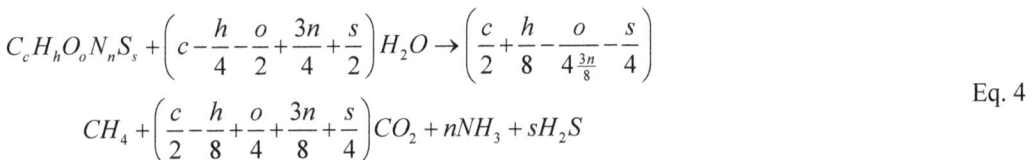

$$C_cH_hO_oN_nS_s + \left(c - \frac{h}{4} - \frac{o}{2} + \frac{3n}{4} + \frac{s}{2}\right)H_2O \rightarrow \left(\frac{c}{2} + \frac{h}{8} - \frac{o}{4\frac{3n}{8}} - \frac{s}{4}\right)$$

$$CH_4 + \left(\frac{c}{2} - \frac{h}{8} + \frac{o}{4} + \frac{3n}{8} + \frac{s}{4}\right)CO_2 + nNH_3 + sH_2S$$

Eq. 4

The above equation gives the theoretical methane, carbon dioxide and trace gases from anaerobic digestion of organic waste and this is to compare the experimental results from anaerobic digestion of sole organic waste, and with several co-digestion and pre-treatment techniques. However, co-digestion and pre-treatment effect on the above equation has not been found in literature and can be explored further.

7. METHANE PRODUCTION AND BIODEGRADABILITY

In AD process, among the four steps - hydrolysis, acidogenesis, acetogenesis and methanogenesis - hydrolysis step is the limiting step. OFMSW has similar biochemical properties (carbohydrates, fats, proteins, minerals) as it obtained from municipal solid waste segregation (or source selection), although varies in characteristics depending on its origin which give it a high biochemical methane potential (BMP) (Cabbai et al. 2013). When considering OFMSW sources, the one comprising food waste from canteens and restaurants exhibits biomethane productivity, while municipal solid waste derived from OFMSW yields the highest biomethane productivity (Fisgativa et al. 2016). The anaerobic co-digestion of OFMSW and sewage sludge, combined with both physical and chemical pre-treatments, resulted in a 65% reduction in volatile solids. This approach yielded substantial methane production attributed to the balanced C/N ratio and favorable nutrient content (Sedighi et al. 2022). Co-digestion of OFMSW with food waste, newsprint paper and branches has improved the methane yield by 22% due to synergism between substrate (Zhou et al. 2021). Allegue et al. (2020) has demonstrated the thermal hydrolysis for solid fraction of OFMSW and enhanced biogas productivity by 62% because of the biodegradability of substrate. In a real-time anaerobic plant with a capacity of 45 m³, OFMSW is digested alongside landfill leachate, resulting in a notable 56.4% increase in biogas production through the reduction of volatile solids (Jayanth et al. 2020). Not only the parameters of anaerobic digestion inoculum source, but adaptation strategy is also significant in anaerobic digestion process for OFMSW (Rocamora et al. 2020). In a two-stage anaerobic digestion process of OFMSW with the addition of hydrolytic enzymes, there was an improvement of 19% in methane production compared to the control. The system also achieved 67% of the theoretical methane yield (Nasir et al. 2020), whereas in wheat bran fermentation and enzymatic hydrolysis of OFMSW, there is high methane productivity improvement of around 220.8% and 255.1%, respectively, with change in COD solubilization and reduction in sugars (Mlaik et al. 2019). Dastyar et al. (2021) has studied effect of feedstock to inoculum ratio and percolate recirculation time, at F/I – 1 86% of methane recovery whereas at F//I – 2, only 16% methane recovery because of high VFA accumulation. From the above cited studies, it is clear that AD is one of the cheap and efficient waste management method to treat OFMSW. The major observations are that selecting appropriate co-substrate is important, pre-treatment improves the degradability of substrate but combination treatment, concentration of solution, time of reaction and toxic or inhibitory compound production must be monitored. Even after having various advantages of anaerobic digestion of OFMSW over other disposal techniques, still there is scope for the process improvement and treatment of the biomass generated from the process. Summary of AD studies on OFMSW and other organic wastes is given in Table 6.

8. PARAMETERS AFFECTING BIOGAS YIELD

The successful working of an anaerobic digestion system is governed by various process parameters. A broad list of such parameters influencing AD includes pH, temperature, mixing, organic loading, and retention time. These parameters are linked to the design and type of reactor chosen. Apart from them, inoculum type, co-digestion, and pre-treatment method employed are some of the properties which are directly linked to the feedstock type. A brief description of the effect of each of the aforementioned parameters on the overall AD process is covered in the following sections.

Table 6. Effect of process parameters on the anaerobic digestion processes of OFMSW.

Feedstock	Operational conditions	Additional treatment	Action of digestion	Influencing factor	Reference
OFMSW and sewage sludge	2 mm particle size, 1L volume, at 35°C	Physical pre-treatment (crushing), chemical pre-treatment (alkalinity for pH)	65% VS reduction and high methane yield	Balanced C/N ratio, good nutrient content	(Sedighi et al. 2022)
OFMSW, food waste, newsprint paper, branches	2 × 8 mm particle size, 1 mm particle size, 1 L volume. 37 ± 1°C	Cutting	Methane yield improved by 22%	Synergism between substrates	(Zhou et al. 2021)
OFMSW	100 mL, volume, 37 ± 0.5°C	Thermal hydrolysis	Biogas improved by 62%	Biodegradability by hydrolysis	(Allegue et al. 2020)
OFMSW and landfill leachate	45 m³ volume, 4 mm particle size	Sieving	Biogas productivity improvement by 56.4%	Volatile solid reduction	(T.A.S et al. 2020)
OFMSW	4.5 L volume, at 37°C	Two stage dry anaerobic process	Theoretical methane yield increases by 67%	Hydrolytic enzymes	(Nasir et al. 2020)
OFMSW	1–3 mm particle size, 37°C	Enzymatic pre-treatment	Methane potential increment of 255.1%, 34.6% change in soluble COD and 40.6% of reducing sugars	Reduction in COD and sugars	(Mlaik et al. 2019)
OFMSW	1–3 mm particle size, 37°C	Wheat bran fermentation	Methane potential increment of 220.8%, 37.1 change in soluble COD and 50% of reducing sugars	Reduction in COD and sugars	(Mlaik et al. 2019)
OFMSW	2mm particle size, 14.5 L volume, 37 ± 2°C	Cutting	At F/I -1, 86% of theoretical methane potential recovery	Less VFA accumulation	(Dastyar et al. 2021)

8.1 pH

pH is advised to be monitored throughout the AD process as a minor variation in its value may create substantial change in the performance, reaction kinetics, and biogas production rates. The pH value represents the health and condition of microorganisms in an AD system (Chen and Neibling 2014). The optimum range of pH for hydrolysis to occur is reported in the neutral range. The acidogenesis stage is also feasible in the range of 5.5 to 6.5 (Noxolo et al. 2014). In spite of this, methanogens are very sensitive to pH changes. A pH value below 6.3 and above 7.8 can harmfully affect the growth of methanogens which might lead to the digester failure (Liu et al. 2008). The volatile fatty acid (VFA) accumulation during acidogenesis often leads to a pH drop (< 3) which eventually results in the collapse of the system. Hence, VFA should be effectively metabolized into other products. Higher ammonia production during hydrolysis can lead to a pH increase as it supplies alkalinity to the system. Normally, hydrolysis of protein-rich substrates results in the accumulation of ammonia. Extreme acidic or alkaline conditions are inhibitory to acidogenesis and hydrolysis.

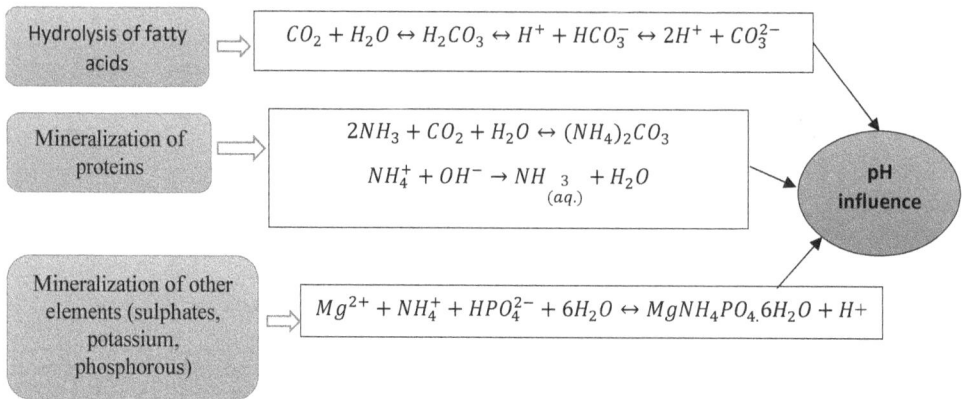

Figure 6. Various pathways showing degradation of substrates and influence on pH.

Many biochemical interactions occurring in the digester lead to intermediate product formation. The variation in digester pH is observed mainly due to the release of such products as ammonium carbonate or the reduction of CO_2 (Georgacakis et al. 1982). Reduction of multivalent anions like SO_4^{-2}, $Fe(OH)_3$) and production of basic cations like Mg^{2+}, Ca^{2+}, K^+ also results in a rise of pH. The influence of various ions formed during mineralization and hydrolysis of substrates on pH is shown in Fig. 6. The fluctuations in pH can be controlled by using an automatic pH controller which helps to balance the pH by adding buffers or neutralizing agents. pH can be balanced by controlling other parameters like organic loading (Yu et al. 2016), co-digestion (Nayono et al. 2010), recirculation of digestate (Wu et al. 2018), and varying HRTs (Demirel and Yenigun 2004).

8.2 *Temperature*

The temperature ranges at which AD occurs are classified as psychrophilic, mesophilic, thermophilic, and extremophilic temperatures ranging from 4–25°C, 30–40°C, and 50–60°C and > 60°C, respectively (Visvanathan 2010). Even though AD is possible at a temperature less than 20°C, it is found that degradation rates are much lower at T < 10°C. Dry or solid-state digestion studies are mostly reported in mesophilic and thermophilic temperature ranges as better performance of digester and higher biogas yield are foreseen. The thermophilic temperature range is more beneficial compared to the mesophilic one due to faster solid degradation rates, better solid-liquid separation, and promoting the growth of microorganisms (Kim et al. 2006). Moreover, higher biogas production at a shorter retention time and higher loading capacity can be achieved under thermophilic conditions. In the study conducted by Kim et al. (Kim et al. 2006), a 50% rise in biogas yield was obtained during wet AD of food waste at 12% TS concentration at temperatures 45–50°C than at 40°C. Another study on dry AD of OFMSW by Fernandez et al. (Fernández-Rodríguez et al. 2013) reported a 27% increase in biogas production within 20 days of operation when the operating temperature was raised from 35°C to 55°C.

Despite the benefits, thermophilic AD has drawbacks like higher energy requirements and the need for scrupulous process control as micro-organisms are highly sensitive to temperature variations (Visvanathan 2010). Kim and Speece (Kim and Speece 2002) studied the presence of possible inhibitory agents by increasing loading from 2 to 10 kg VS/m³/day under thermophilic AD conditions and accumulation of propionate was experienced at 7.5 kg VS/m³/day. Thermophilic AD can be implemented more effectively if a proper process control facility is available along with a better understanding of the nature of the feedstock.

8.3 Mixing

Proper maintenance of all other parameters like pH, temperature, HRT is dependent on mixing. Mixing serves as a critical factor in the overall AD process as proper mixing enables the micro-organisms and substrate to remain in close contact within the digester volume which helps in increasing reaction kinetics and methane yield. In solid-state AD, biogas recirculation and mechanical mixing are the two mechanisms followed. Biogas recirculation can be carried out by pumping gas through the bottom of the digester, while mixing is done either internally (Karim et al. 2005)impeller mixing, and slurry recirculation or externally (Karthikeyan and Visvanathan 2013). Especially for solid substrates, mixing is important as TS contents are high in them. Karim et al. (Karim et al. 2005) reported a 20% increase in methane yield during AD of cow manure (TS > 5%) when mixing was employed. Biogas recirculation helps in COD and VFA reduction as well as increased biogas generation potential and biomass holding capacity (Siddique et al. 2015). Few authors reported the negative effect of mixing stating that vigorous mixing might disturb the syntrophic relation between microbes and shear forces created may break microbial flocs (Singh et al. 2019). Karim et al. (Karim et al. 2005) reported a reduction in methane production when mechanical mixing was opted for in a 4 L digester during the start-up of the digestion. Kaparaju et al. (Kaparaju et al. 2008) found out that minimal mixing of 10 minutes is more promising than continuous mixing during the AD of cattle manure at 8% TS concentration.

8.4 Hydraulic retention time

In the continuous AD of solid organic wastes, HRT stands for the average time substrate stays inside the reactor and it helps to determine the contact period of micro-organisms with the substrate. HRT can affect the rate at which VFAs are converted into other intermediate products or biogas (Dareioti and Kornaros 2014, Ho et al. 2014). The operational HRT should be neither too long nor too short but should be adequate to provide contact time between microbes and substrate. Longer HRTs have drawbacks like larger digester size, high investment requirement, and high operational cost (Speece 1983). On the other hand, shorter HRTs might create VFA accumulation problems leading to slower methanogenic activity and loss of nutrients (Shi et al. 2017). Retention time often represents both HRT and solid retention time (SRT). HRT stands for the retention time of the liquid phase while SRT represents the retention time of microbial culture. In cases when both microbial and feedstock lie in the same phase, HRT equals SRT. HRT and SRT can be found using the following equations (Eqs. 1 and 2).

$$HRT = \frac{V}{Q} \qquad\qquad \text{Eq. 5}$$

$$SRT = \frac{VX}{Q_x X_x} \qquad\qquad \text{Eq. 6}$$

where, V: Reactor volume in m^3, Q: Influent flow rate in m^3/d, X: Mixed liquid suspended solids (mg/l), Q_x: Excess bio-solids removal rate in m^3/d, X_x: Mixed liquid suspended solids in excess bio-solids flow (mg/l).

Dinsdale et al. (Dinsdale et al. 2000) studied the working of a 2-stage anaerobic digester co-digesting secondary sewage sludge with fruit and vegetable waste which yielded biogas of 0.37 m^3/kg $_{added}$ at a loading of 5.7 g VS/d and HRT of 13 days. In another study by Dareioti and Kornaros (2014), the effect of HRT on hydrogen and methane production during AD of cheese whey, olive mill waste, and cattle manure was studied. The highest hydrogen yield was obtained when the acidogenic reactor was operated at an HRT of 0.75d. At higher HRTs, accumulation of VFAs is found which reduced both hydrogen and methane production. The choice of HRT depends on various factors like substrate composition, volume of digester, biochemical interactions inside digester, and other processes. For substrates like sugar and starch, having higher hydrolysis rates require shorter retention time compared to fibre and cellulose type substrates which are harder to degrade. Hence, many studies recommended a HRT ranging from 10 to 25 days, even though longer HRTs like 50 days can be used for operating digesters in colder climate (Buysman 2009).

8.5 C/N ratio

Theoretically, C/N represents the ratio of total organic carbon to the total nitrogen present in the substrate or feedstock. C/N ratio is a critical factor affecting methane production in AD as it can influence the interactions between acetogens and methanogens in the reaction media. At a higher C/N ratio, methanogens consume nitrogen at higher rates, resulting in low methane production, whereas at a low C/N ratio, the higher ammonia concentration inhibits the methanogenic activity (Li et al. 2019). Hence, maintaining an optimum C/N ratio is necessary for the successful operation of an anaerobic digester. This is done by adding more substrates which help in balancing the C/N ratio within a specific range. The optimum range of the C/N ratio is reported between 20 to 30 (Bouallagui et al. 2009).

8.6 Co-digestion

Co-digestion indicates one or more substrate utilization in AD. Co-digestion helps in increasing the biogas yield in multiple ways like maintaining optimum C/N ratio, supplying additional micro and macro-nutrients, and balancing the pH. The co-substrates added should be checked for (i) nutrient value (ii) moisture content and pH (iii) presence of toxic substances, and (iv) individual methane productivity (Alhraishawi and Alani 2018). Co-digestion also helps in reducing the lag time faced during the start-up period. Some examples of co-digestion studies are presented in Table 7. In many studies, cow manure is used as co-substrate as it can supply additional buffering capacity along with nutrients. Manure is found well suiting in co-digestion with organic wastes like food wastes (El-Mashad and Zhang 2010), seaweed (Sarker et al. 2014), agricultural residues (Huttunen and Rintala 2007), straw wastes (Wang et al. 2014), fruit and vegetable wastes (Callaghan et al. 2002), and molasses (Sarker and Møller 2013).

Table 7. Biomethane generation potential of various organic wastes in anaerobic co-digestion.

Feedstock	Mix ratio	Operational conditions	Methane yield	Reference
Sewage sludge (SS) and OFMSW	SS:OFMSW-1:1	Inoculum to substrate ratio - 2, 4, 6. TS-3%, 5%, 7% T-35°C	448 L/KgVS	Sedighi et al. 2022
Coffee pulp (CP) and cattle manure (CM), food waste (FW) and sewage sludge (SS)	CP:CM-1:0, 4:1, 2:1, 4:3, and 0:1. FW:SS-1:0, 4:1, 2:1, 4:3, and 0:1	Inoculum to substrate ratio - 1. V-180 mL T-37°C	448.8 mL/gVS (@CP:CM -1:0) 795.04 mL/gVS (@FW:SS-1:0)	Karki et al. 2022
Olive-mill waste (OW) and cattle manure (CM)	OW:CM-60:40	Inoculum to substrate ratio - 2. T-37°C C/N ratio-20–30	380mL/gVS	Rubio el al. 2022
Cheese whey (CW) and poultry waste (PW)	CW:PW-25:75, 75:25, 50:50	Inoculum to substrate ratio - 1. V-2.25 L T-37°C OLR-1 gCOD/L/d	395 mL/gCOD @ CW:PW-25:75	Abdallah et al. 2022
Sewage sludge (SS) and orange peel	Orange peel added in amounts 1.5 and 3 g	T-55°C V-1.8L	458.6 mLCH$_4$/g	Szaja et al. 2022

Generally, the mechanism lying behind digestion is biodegradation of the substrate by several groups of micro-organisms involved in hydrolysis, acidogenesis, acetogenesis, methanogenesis, and subsequent production of biogas. The products of each step often result in creating inhibition of methanogenesis. For example, the accumulation of VFA in acidogenesis is a serious problem encountered in systems having low buffering capacity. Co-digestion helps in supplying adequate buffering by the addition of suitable wastes as co-substrates. Also, the digester can be designed to operate at optimum C/N ratio by using co-digestion. In addition, the availability of macro-nutrients can be ensured by adjusting the fractions of carbon, sulphur, phosphorous, and nitrogen.

9. INFLUENCES OF SUBSTRATE COMPOSITION AND INOCULUM SOURCE ON THE MICROBIAL COMMUNITY

The whole process of anaerobic digestion is run by some group of microbes, namely, hydrolysers, acidogens, acetogens, and methanogens. They are classified under domain bacteria and archaea. All these grow and work in syntrophy through several intra-dependent pathways (Carballa et al. 2015). The conversion pathways for product formation in each phase of AD and examples of associated microbes are given in Table 8. The performance of an anaerobic digester lies in the status of this

Table 8. Details on product formation in each phase and bacteria groups.

Phase	Bacterial group	Conversion pathway	Type of bacteria
Hydrolysis	Hydrolyser	Carbohydrate $\longrightarrow$ Simple sugars	Clostridium, Proteus vulgaris, Peptococcus, Bacteriodes, Bacillus, Vibrio
		Proteins $\longrightarrow$ Amino acids	Clostridium, Acetovibrio celluliticus, Staphylococcus, Bacteroides
		Lipids $\longrightarrow$ VFA or alcohols	Clostridium, Micrococcus, Staphylococcus,
Acidogenesis	Acidogens	Aminoacids $\longrightarrow$ Fatty acids, acetate, NH_3	Lactobacillus, Escherichia, Staphylococcus, Pseudomonas, Bacillus, Desulfovibrio, Selenomonas, Sarcina, Veillonella.
		Simple sugars $\longrightarrow$ Alcohols, acids, CO_2, H_2S	Clostridium, Eubacterium limosum, Streptococcus
Acetogenesis	Acetogens	Higher fatty acids $\longrightarrow$ H_2 or acetate	Clostridium, Syntrophomonas wolfeii, Syntrophomonas wolinii.
Methanogenesis	Methanogens	H_2 and CO_2 $\longrightarrow$ Methane	Methanobacterium, Methanoplanus
		Acetate $\longrightarrow$ Methane and CO_2	Methanosaeta, Methanosarcina

complex symbiotic relationship. The growth and functioning of microbes are dependent on various factors like temperature, pH, loading rate, and mixing. Among these, substrate composition and inoculum supplied are important ones that can't be generalized based on a few studies alone. The following section briefly describes the characteristics of various substrates and inocula and their influence on the microbial community.

9.1 Substrate characteristics

In recent years, many studies have been reported in utilisation of organic wastes for production of biogas through AD. These organic wastes include food wastes, agricultural wastes, sewage sludge, industrial wastes, OFMSW, and aquatic biomass and various energy crops. Various feedstock from different sectors are illustrated in Fig. 7. Even though many studies were reported in this domain,

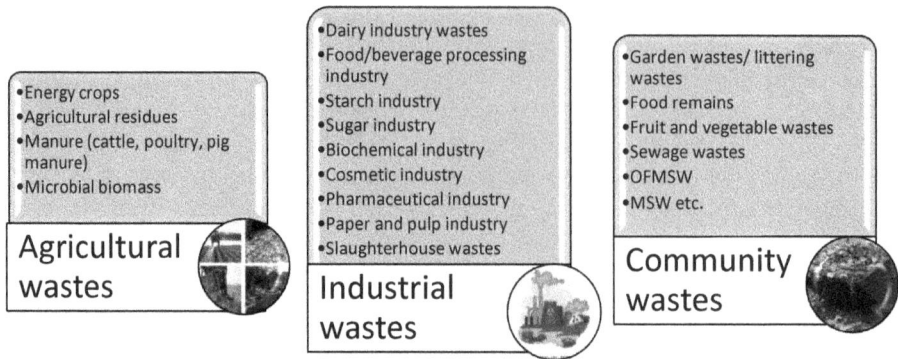

Figure 7. Feedstock from various fields used in AD (Steffen 1998).

there is little knowledge on the effect of various substrates on the microbial community of AD. The biogas production potential of a solid organic waste depends on its rate of biodegradation. Main components of a substrate are lipids (oils, fats, glycerol, proteins, and carbohydrates), cellulose, lignin, and hemicellulose. All these components are not readily biodegradable. For example, fats and proteins dissociate in days, cellulose takes weeks, while VFA and alcohols break down in few hours (Steffen et al. 1998). The biodegradable portion usually varies from 70–90% of the dry matter content. The characteristics of substrates are discussed below by categorising them into lipid rich substrates, ligno-cellulose rich and carbon rich substrates.

9.1.1 Lipid-rich substrates

Lipid-rich substrates are mainly generated from food industries, dairy industries, vegetable oil or fat refineries, slaughterhouses wastes, olive oil mills, food remains, and hotel oil traps. Lipids are first hydrolysed into free long-chain fatty acids and glycerol. This step is catalysed by acidogens through extracellular lipase secretion. Glycerol is converted to acetate by acidogens. Long-chain fatty acid (LCFA) conversion occurs through β-oxidation process which is slower. LCFA gets converted into acetate and hydrogen through syntrophic acidogenesis. If odd number of carbon atoms are present in LCFA, other products like propionate, butanol are produced (Weng and Jeris 1976). The function of hydrogenotrophic methanogens is to utilise hydrogen produced for methane formation. Due to the mismatch found in the production and consumption of LCFA in this process, growth and function of methanogens are often affected.

The methane yield from lipids is more than that of carbohydrates. The methane generation potential of lipids is around 70%, while that of carbohydrates and proteins is 50% and 65%, respectively (Leung and Wang 2016). Studies reported that AD of substrates rich in fats and oils helped in increasing methane yield by promoting archaeal and bacterial growth (Yang et al. 2016). The waste produced by food processing industries, slaughterhouses, and restaurants often contains elevated levels of oil and fats. When utilized in AD, these wastes have an impact on the production of ammonia and sulfides. High ammonia levels create imbalance in the AD by altering pH and buffering. In addition, problems like sludge floatation, pipes and pump blockage, foaming, clogging in gas collectors are likely to occur while dealing with lipid rich substrates. Co-digestion of lipid-rich substrates with other organic wastes like paper waste (Xiao et al. 2019), rice husk (Sahoo and Rao 2019), sewage sludge (Latha et al. 2019), and municipal waste (Long et al. 2012) can overcome these issues to an extent and enhance the methane production. The functioning of bacterial community involved in the hydrolysis reaction was found increased when co-digestion is employed. When sewage sludge is co-digested with 10% of fat and oil, the richness in bacterial population is noticed (Simpson index increased from 6.2 to 8.5) (Amha et al. 2017).

Similarly in another study, the growth of bacteria Syntrophomonas was profoundly increased after 3–14% of fat, oil and grease FOG addition (Ziels et al. 2016).

9.1.2 Lignocellulose-rich substrates

Lignocellulosic biomass are also good sources of biogas production. The biogas production potential of sorghum, wheat straw, and grass is very high. The main components of lignocellulosic biomass are cellulose, hemicellulose and lignin. Out of these, lignin is extremely resistant to biodegradation because of its strong bonds. Hence, high lignin content opposes the substrate hydrolysis (Teghammar 2013). Lignocellulosic wastes are mainly seen in garden wastes, paper wastes, bagasse, hay and agricultural residues. Many studies have reported suggesting the pre-treatment requirement for these substrates to maximise the biodegradation rates (Amin et al. 2017, Kucharska et al. 2018, Saritha et al. 2012). The effect of lignocellulosic substrates on microbial community is not well documented. But reports have shown that due to the recalcitrant nature, microbial population is less diverse in case of AD of lignocellulosic substrates. Wilkins et al. (2015) has reported that the microbial community involved in the AD process of cellulose, xylan and xylose was less diverse than that in the AD of food waste.

9.1.3 Carbon-rich substrates

When carbon-rich and nutrient-rich substrates are used for co-digestion, the overall process efficiency of AD is found to be increasing (Nghiem et al. 2017). The higher biodegradability rates of food waste make it a good substrate for AD. The methane potential values of food wastes are reported in ranges $0.3–1.1$ m^3 CH$_4$/kgVS$_{added}$ which is normally higher than that of lignocellulosic wastes and manures (Mao et al. 2015). Food waste affects the microbial community and structure strongly. A C: N ratio ranging between 20 and 30 is found effective to maintain proper nutrient levels in anaerobic digestion. In co-digestion of two or more substrates, addition of carbon-rich substrates has to be done carefully as increased organic loading rate might lead to reduction in microbial diversity. Addition of food waste to sewage sludge in anaerobic co-digestion study by Qi et al. (2018) has shown reduced diversity indices of microbes.

9.2 Effect of inoculum sources

Different sources of inocula and its characteristics influence the start-up of the AD process. Other factors in association with inoculum affecting AD are concentration of nutrients, acclimatisation period, incubation temperature, storage conditions and specific methanogenic activity (Holliger et al. 2016, Raposo et al. 2006). The essential nutrients and other trace elements are traditionally being supplied through co-substrate addition. Addition of inoculum is meant to provide suitable environment and microbial consortia for initialising the start-up phase of AD. Selection of inoculum type is an important criterion, as different inoculum supplies different group of microbes. The rate of biodegradation and initial lag phase time depends on the population and concentration of microorganisms. Majorly, the digestate from biogas plants are found to be used as inoculum in many studies which ensure the presence of rich anaerobes (Elbeshbishy et al. 2012a). Other organic wastes like different manure types (cow manure, pig manure, poultry wastes), sewage sludge, and septic tank sludge are also used as inocula.

In a study by Rajput et al. (2019), 3 inocula types, namely, septage, acclimatised sludge and cow manure are used as inocula for AD of sunflower meal and wheat straw. The activity of inocula is studied by analysing solids removal, biogas yield, and buffering capacity. Cattle manure showed better performance compared to other two. Angelidaki et al. (2009) has suggested that the inoculum selected should be fresh, uniform, sieved and pre-incubated to have a good population of hydrolytic and methanogenic microbes. They also stressed to conduct check tests with substrates like cellulose and acetic acid. The effect of inoculum on substrate digestion was studied by Elbeshbishy et al. (2012b) for primary sludge and food waste and Chamy and Ramos (2011) for turkey manure digestion. The methane yield was found to be positively affected by inoculum sources used in both studies.

10. ENERGY POTENTIAL OBTAINED FROM BIOGAS

Biogas technology has got multiple benefits like greenhouse gas reduction along with immense energy generation potential (Angelidaki et al. 2009, Tricase and Lombardi 2009). Biogas generated through AD process can be used to generate heat, electricity or both. An upgraded form of methane is pure methane or bio-methane or renewable natural gas, which is made by removing hydrogen sulphide, carbon dioxide, water and other elements from biogas. This upgraded form is just like the natural gas which can be transported through pipeline grid in liquid or compressed form. According to the report by World Biogas Association (WBA) (Jain 2019), the major feedstock all over the world can generate energy of around 10,100 to 14,000 Twh, which can meet 6–9% of world's primary energy consumption and can substitute 23–32% of energy production from coal. If it is used in form of electricity, it has the ability to meet 16–22% of world's electricity consumption. Among the various feedstock, wastes from livestock sector and crop residues have the greatest energy generation potential which is followed by energy crops. The high energy generation from livestock sector can be connected to the well-established dairy sector in countries like India. The sheer volume of manure produced from the dairy farms can be utilised for this. In urban settings, food waste can be employed.

The ministry of renewable energy of India has implemented two central sector schemes, namely, New National Biogas and Organic Manure Programme (NNBOMP) and Biogas Power Generation (Off-grid) and Thermal energy application Programme (BPGTP). Both schemes have a power generation capacity in the range of 3 KW to 250 KW. In BPGTP programme, organic wastes from food manufacturing industries, animal manures, poultry wastes, and food and vegetable wastes are used as feedstock for AD. This scheme has helped in meeting power requirements of individual poultry and dairy plants, and heating requirements of various processes of industries. Other purposes in which the produced energy is used are milk chilling operations, cooking, pumping and lighting. The bio-methane production potential can be employed to increase the market share of biogas. Biomethane is used as an alternative to gas oil which helps in improving the air quality with regards to oxides of nitrogen and particulate matter.

11. FUTURE PERSPECTIVES

The detailed process parameters of AD process, various reactor configurations, pre-treatment methods and co-digestion options are reviewed in the write-up. AD is an efficient technology to treat highly concentrated organic wastes like wastewater generated from dairy industries and oil refineries, which helps in generating energy along with waste reduction. Food industries can utilise the energy developed through AD to meet their local energy demand. Other than this, biogas has immense energy potential as it can substitute energy generated from fossil fuels for heat and power generation, and also as a vehicle fuel. One of the big challenges in utilisation of biogas is the presence of gases like H$_2$S, and CO, which is harmful and can harm the metal parts of heating equipment and power units (Irini Angelidaki et al. 2018). Hence, preventive measures have to be taken to reduce the intrusion of harmful gases in biogas; also, adopting post-treatment methods to eliminate H$_2$S might be feasible. More studies on climate effects on AD of different organic wastes has to be conducted. Established information on various operating parameters of anaerobic reactors, data regarding availability of local feedstock, and universal design of reactors are required for implementing a biogas plant at large scale. The potential production of methane through biogas plants should be aimed at low capital cost.

REFERENCES

Abdallah, M., Greige, S., Beyenal, H. et al. 2022. Investigating microbial dynamics and potential advantages of anaerobic co-digestion of cheese whey and poultry slaughterhouse wastewaters. Scientific Reports 12(1): 1–12. https://doi.org/10.1038/s41598-022-14425-1.

Alhraishawi, A.A. and Alani, W.K. 2018. The co-fermentation of organic substrates: A review performance of biogas production under different salt content. Journal of Physics: Conference Series 1032(1). https://doi.org/10.1088/1742-6596/1032/1/012041.

Allegue, L.D., Puyol, D. and Melero, J.A. 2020. Novel approach for the treatment of the organic fraction of municipal solid waste: Coupling thermal hydrolysis with anaerobic digestion and photo-fermentation. Science of the Total Environment 714: 136845. https://doi.org/10.1016/j.scitotenv.2020.136845.

Amha, Y.M., Sinha, P., Lagman, J. et al. 2017. Elucidating microbial community adaptation to anaerobic co-digestion of fats, oils, and grease and food waste. Water Research 123: 277–289. https://doi.org/10.1016/j.watres.2017.06.065.

Amin, F.R., Khalid, H., Zhang, H. et al. 2017. Pretreatment methods of lignocellulosic biomass for anaerobic digestion. AMB Express 7(1). https://doi.org/10.1186/s13568-017-0375-4.

Amiri, L., Abdoli, M.A., Gitipour, S. et al. 2017. The effects of co-substrate and thermal pretreatment on anaerobic digestion performance. Environmental Technology (United Kingdom) 38(18): 2352–2361. https://doi.org/10.1080/09593330.2016.1260643.

Angelidaki, I., Alves, M., Bolzonella, D. et al. 2009. Defining the biomethane potential (BMP) of solid organic wastes and energy crops: A proposed protocol for batch assays. Water Science and Technology 59(5): 927–934. https://doi.org/10.2166/wst.2009.040.

Angelidaki, Irini, Treu, L., Tsapekos, P. et al. 2018. Biogas upgrading and utilization: Current status and perspectives. Biotechnology Advances 36(2): 452–466. https://doi.org/10.1016/j.biotechadv.2018.01.011.

Anjan, K. Kalia. 1988. Development and evaluation of a fixed dome plug flow anaerobic digester. Biomass 16: 225–235.

Behera, S., Arora, R., Nandhagopal, N. et al. 2014. Importance of chemical pretreatment for bioconversion of lignocellulosic biomass. Renewable and Sustainable Energy Reviews 36: 91–106. https://doi.org/10.1016/j.rser.2014.04.047.

Bhide, A.D. and Sundaresan, B.B. 1983. Solid waste management in developing Countries. Waste Management and Research 3(1): 176–176. doi:10.1177/0734242X8500300119.

Bouallagui, H., Lahdheb, H., Romdan, E. Ben et al. 2009. Improvement of fruit and vegetable waste anaerobic digestion performance and stability with co-substrates addition. Journal of Environmental Management 90(5): 1844–1849. https://doi.org/10.1016/j.jenvman.2008.12.002.

Browne, J.D. and Murphy, J.D. 2013. Assessment of the resource associated with biomethane from food waste. Applied Energy 104: 170–177. https://doi.org/10.1016/j.apenergy.2012.11.017.

Buysman, E. 2009. Anaerobic digestion for developing Countries with cold climates - Utilizing solar heat to address technical challenges and facilitating dissemination through the use of carbon finance. April, 1–165. https://doi.org/10.13140/RG.2.1.4848.8807.

Cabbai, V., Ballico, M., Aneggi, E. et al. 2013. BMP tests of source selected OFMSW to evaluate anaerobic codigestion with sewage sludge. Waste Management 33(7): 1626–1632. https://doi.org/10.1016/j.wasman.2013.03.020.

Callaghan, F.J., Wase, D.A.J., Thayanithy, K. et al. 2002. Continuous co-digestion of cattle slurry with fruit and vegetable wastes and chicken manure. Biomass and Bioenergy 22(1): 71–77. https://doi.org/10.1016/S0961-9534(01)00057-5.

Campuzano, R. and González-Martínez, S. 2015. Extraction of soluble substances from organic solid municipal waste to increase methane production. Bioresource Technology 178: 247–253. https://doi.org/10.1016/j.biortech.2014.08.042.

Carballa, M., Regueiro, L. and Lema, J.M. 2015. Microbial management of anaerobic digestion: Exploiting the microbiome-functionality nexus. Current Opinion in Biotechnology 33: 103–111. https://doi.org/10.1016/j.copbio.2015.01.008.

Carrere, H., Antonopoulou, G., Affes, R. et al. 2016. Review of feedstock pretreatment strategies for improved anaerobic digestion: From lab-scale research to full-scale application. Bioresource Technology 199: 386–397. https://doi.org/10.1016/j.biortech.2015.09.007.

Cesaro, A., Belgiorno, V., Siciliano, A. et al. 2019. The sustainable recovery of the organic fraction of municipal solid waste by integrated ozonation and anaerobic digestion. Resources, Conservation and Recycling 141: 390–397. https://doi.org/10.1016/j.resconrec.2018.10.034.

Chamy, R. and Ramos, C. 2011. Factors in the determination of methanogenic potential of manure. Bioresource Technology 102(17): 7673–7677. https://doi.org/10.1016/j.biortech.2011.05.044.

Chang, V.S., Nagwani, M. and Holtzapple, M.T. 1998. Lime pretreatment of crop residues bagasse and wheat straw. Applied Biochemistry and Biotechnology - Part A Enzyme Engineering and Biotechnology 74(3): 135–159. https://doi.org/10.1007/BF02825962.

Chen, L. and Neibling, H. 2014. Anaerobic digestion basics. University of Idaho Extension, 6.

Cheng-nan Weng and Jeris, J.S. 1976. Biochemical mechanisms in the methane. Water Research 10(1): 9–18.

Cianchetta, S., Di Maggio, B., Burzi, P.L. et al. 2014. Evaluation of selected white-rot fungal isolates for improving the sugar yield from wheat straw. Applied Biochemistry and Biotechnology 173(2): 609–623. https://doi.org/10.1007/s12010-014-0869-3.

CPCB. 2020. Annual Report (2019–20) on Implementation of Solid Waste Management Rules, 2016. https://cpcb.nic.in/uploads/MSW/MSW_AnnualReport_2019-20.pdf.

Dai, X., Duan, N., Dong, B. et al. 2013. High-solids anaerobic co-digestion of sewage sludge and food waste in comparison with mono digestions: Stability and performance. Waste Management 33(2): 308–316. https://doi.org/10.1016/j.wasman.2012.10.018.

Dareioti, M.A. and Kornaros, M. 2014. Effect of hydraulic retention time (HRT) on the anaerobic co-digestion of agro-industrial wastes in a two-stage CSTR system. Bioresource Technology 167: 407–415. https://doi.org/10.1016/j.biortech.2014.06.045.

Dastyar, W., Mohammad Mirsoleimani Azizi, S., Dhadwal, M. et al. 2021. High-solids anaerobic digestion of organic fraction of municipal solid waste: Effects of feedstock to inoculum ratio and percolate recirculation time. Bioresource Technology 337: 125335. https://doi.org/10.1016/j.biortech.2021.125335.

Demirel, B. and Yenigun, O. 2004. Anaerobic acidogenesis of dairy wastewater : the effects of variations in hydraulic retention with no pH control. Journal of Chemical Technology and Biotechnology 79(7): 755–760. https://doi.org/10.1002/jctb.1052.

Dinsdale, R.M., Premier, G.C., Hawkes, F.R. et al. 2000. Two-stage anaerobic co-digestion of waste activated sludge and fruit/vegetable waste using inclined tubular digesters. Bioresource Technology 72(2): 159–168. https://doi.org/10.1016/S0960-8524(99)00105-4.

Domalski, E.S.D., Churney, K.L., Ledford, A.E. et al. 1991. Assessing the credibility of the calorific content of municipal solid waste. Pure and Applied Chemistry 63(10): 1415–1418. https://doi.org/10.1351/pac199163101415.

Dong, L., Zhenhong, Y. and Yongming, S. 2010. Semi-dry mesophilic anaerobic digestion of water sorted organic fraction of municipal solid waste (WS-OFMSW). Bioresource Technology 101(8): 2722–2728. https://doi.org/10.1016/j.biortech.2009.12.007.

Edgar Fernando Castillo M., Diego Edison Cristancho and A. Victor Arellano. 2006. Study of the operational conditions for anaerobic digestion of urban solid wastes. Waste Management 26: 46–556.

El-Mashad, H.M. and Zhang, R. 2010. Biogas production from co-digestion of dairy manure and food waste. Bioresource Technology 101(11): 4021–4028. https://doi.org/10.1016/j.biortech.2010.01.027.

Elbeshbishy, E., Nakhla, G. and Hafez, H. 2012a. Biochemical methane potential (BMP) of food waste and primary sludge: Influence of inoculum pre-incubation and inoculum source. Bioresource Technology 110: 18–25. https://doi.org/10.1016/j.biortech.2012.01.025.

Elbeshbishy, E., Nakhla, G. and Hafez, H. 2012b. Biochemical methane potential (BMP) of food waste and primary sludge: Influence of inoculum pre-incubation and inoculum source. Bioresource Technology 110: 18–25. https://doi.org/10.1016/j.biortech.2012.01.025.

Fardin, J.F., de Barros Jr, O. and Dias, A.P. 2018. Biomass: Some basics and biogas. In Advances in Renewable Energies and Power Technologies (pp. 1–37). Elsevier.

Fernández-Rodríguez, J., Pérez, M. and Romero, L.I. 2013. Comparison of mesophilic and thermophilic dry anaerobic digestion of OFMSW: Kinetic analysis. Chemical Engineering Journal 232: 59–64. https://doi.org/10.1016/j.cej.2013.07.066.

Fernández-Rodríguez, J., Pérez, M. and Romero, L.I. 2016. Semicontinuous Temperature-Phased Anaerobic Digestion (TPAD) of Organic Fraction of Municipal Solid Waste (OFMSW). Comparison with single-stage processes. Chemical Engineering Journal 285: 409–416. https://doi.org/10.1016/j.cej.2015.10.027.

Ferrer, I., Serrano, E., Ponsá, S. et al. 2009. Enhancement of thermophilic anaerobic sludge digestion by 70° pre-treatment: Energy considerations. Journal of Residuals Science and Technology 6(1): 11–18.

Fisgativa, H., Tremier, A. and Dabert, P. 2016. Characterizing the variability of food waste quality: A need for efficient valorisation through anaerobic digestion. Waste Management 50: 264–274. https://doi.org/10.1016/j.wasman.2016.01.041.

Forster-Carneiro, T., Pérez, M. and Romero, L.I. 2008a. Anaerobic digestion of municipal solid wastes: Dry thermophilic performance. Bioresource Technology 99: 8180–8184. https://doi.org/https://doi.org/10.1016/j.biortech.2008.03.021.

Forster-Carneiro, T., Pérez, M. and Romero, L.I. 2008b. Influence of total solid and inoculum contents on performance of anaerobic reactors treating food waste. Bioresource Technology 99(15): 6994–7002. https://doi.org/10.1016/j.biortech.2008.01.018.

Georgacakis, D., Sievers, D.M. and Iannotti, E.L. 1982. Buffer stability in manure digesters. Agricultural Wastes 4(6): 427–441. https://doi.org/10.1016/0141-4607(82)90038-5.

Harmsen, P., Huijgen, W., López, L. et al. 2010. Literature review of physical and chemical pretreatment processes for lignocellulosic biomass. Food and Biobased Research, 1–49.

Hartwell, J., Mousavi, M.S., Eun, J. et al. 2021. Evaluation of depth-dependent properties of municipal solid waste using a large diameter-borehole sampling method. In Journal of the Air and Waste Management Association 71(4): 433–446.

Hendriks, A.T.W.M. and Zeeman, G. 2009. Pretreatments to enhance the digestibility of lignocellulosic biomass. In Bioresource Technology 100(1): 10–18. https://doi.org/10.1016/j.biortech.2008.05.027.

Ho, D., Jensen, P. and Batstone, D. 2014. Effects of temperature and hydraulic retention time on acetotrophic pathways and performance in high-rate sludge digestion. Environmental Science and Technology 48(11): 6468–6476. https://doi.org/10.1021/es500074j.

Holliger, C., Alves, M., Andrade, D. et al. 2016. Towards a standardization of biomethane potential tests. Water Science and Technology 74(11): 2515–2522. https://doi.org/10.2166/wst.2016.336.

Huttunen, S. and Rintala, J.A. 2007. Laboratory investigations on co-digestion of energy crops and crop residues with cow manure for methane production: Effect of crop to manure ratio. Resources, Conservation and Recycling 51(9): 591–609. https://doi.org/10.1016/j.resconrec.2006.11.004.

Ibikunle, R., Titiladunayo, I., Akinnuli, B. et al. 2018. Modelling the energy content of municipal solid waste amd determination of its physicochemical correlation, using multiple regression analysis. International Journal of Mechanical Engineering and Technology (IJMET) 9(11): 220–232.

Izumi, K., Okishio, Y. ki, Nagao, N. et al. 2010. Effects of particle size on anaerobic digestion of food waste. International Biodeterioration and Biodegradation 64(7): 601–608. https://doi.org/10.1016/j.ibiod.2010.06.013.

Jain, M.S., Daga, M. and Kalamdhad, A.S. 2019. Variation in the key indicators during composting of municipal solid organic wastes. Sustainable Environment Research 29(1): 1–8.

Jain, S. 2019. Global Potential of Biogas. World Biogas Association, UK, 56.

Johari, A., Hashim, H., Mat, R. et al. 2012. Generalization, formulation and heat contents of simulated MSW with high moisture content. Journal of Engineering Science and Technology 7(6): 701–710.

Joshi, R. and Ahmed, S. 2016. Status and challenges of municipal solid waste management in India: A review. Cogent Environmental Science 2(1): 1–18. https://doi.org/10.1080/23311843.2016.1139434.

Kaparaju, P., Buendia, I., Ellegaard, L. et al. 2008. Effects of mixing on methane production during thermophilic anaerobic digestion of manure: Lab-scale and pilot-scale studies. Bioresource Technology 99(11): 4919–4928. https://doi.org/10.1016/j.biortech.2007.09.015.

Karim, K., Hoffmann, R., Klasson, T. et al. 2005. Anaerobic digestion of animal waste: Waste strength versus impact of mixing. Bioresource Technology 96(16): 1771–1781. https://doi.org/10.1016/j.biortech.2005.01.020.

Karki, R., Chuenchart, W., Surendra, K.C. et al. 2022. Anaerobic co-digestion of various organic wastes: Kinetic modeling and synergistic impact evaluation. Bioresource Technology 343: 126063.

Karthikeyan, O.P. and Visvanathan, C. 2013. Bio-energy recovery from high-solid organic substrates by dry anaerobic bio-conversion processes: A review. Reviews in Environmental Science and Biotechnology 12(3): 257–284. https://doi.org/10.1007/s11157-012-9304-9.

Kasinath, A., Fudala-ksiazek, S., Szopinska, M. et al. 2021. Biomass in biogas production: Pretreatment and codigestion. Renewable and Sustainable Energy Reviews 150: 111509. https://doi.org/10.1016/j.rser.2021.111509.

Kathirvale, S., Noor, M., Yunus, M. et al. 2003. Energy potential from municipal solid waste in Malaysia. Renewable Energy 29: 559–567. https://doi.org/10.1016/j.renene.2003.09.003.

Kaur, K. and Phutela, U.G. 2016. Enhancement of paddy straw digestibility and biogas production by sodium hydroxide-microwave pretreatment. Renewable Energy 92: 178–184. https://doi.org/10.1016/j.renene.2016.01.083.

Kim, J.K., Oh, B.R., Chun, Y.N. et al. 2006. Effects of temperature and hydraulic retention time on anaerobic digestion of food waste. Journal of Bioscience and Bioengineering 102(4): 328–332. https://doi.org/10.1263/jbb.102.328.

Kim, M. and Speece, R.E. 2002. Reactor configuration-part ii comparative process stability and efficiency of thermophilic anaerobic digestion. Environmental Technology (United Kingdom) 23(6): 643–654. https://doi.org/10.1080/09593332308618380.

Kucharska, K., Rybarczyk, P., Hołowacz, I. et al. 2018. Pretreatment of lignocellulosic materials as substrates for fermentation processes. Molecules 23(11): 1–32. https://doi.org/10.3390/molecules23112937.

Kumar, A. and Samadder, S.R. 2020. Performance evaluation of anaerobic digestion technology for energy recovery from organic fraction of municipal solid waste: A review. Energy 197: 117253. https://doi.org/10.1016/j.energy.2020.117253.

Latha, K., Velraj, R., Shanmugam, P. et al. 2019. Mixing strategies of high solids anaerobic co-digestion using food waste with sewage sludge for enhanced biogas production. Journal of Cleaner Production 210: 388–400. https://doi.org/10.1016/j.jclepro.2018.10.219.

Leung, D.Y.C. and Wang, J. 2016. An overview on biogas generation from anaerobic digestion of food waste. International Journal of Green Energy 13(2): 119–131. https://doi.org/10.1080/15435075.2014.909355.

Li, H., Jin, Y., Mahar, R.B. et al. 2008. Effects and model of alkaline waste activated sludge treatment. Bioresource Technology 99(11): 5140–5144. https://doi.org/10.1016/j.biortech.2007.09.019.

Li, Y., Chen, Y. and Wu, J. 2019. Enhancement of methane production in anaerobic digestion process: A review. Applied Energy 240: 120–137. https://doi.org/10.1016/j.apenergy.2019.01.243.

Lim, J.W. and Wang, J.Y. 2013. Enhanced hydrolysis and methane yield by applying microaeration pretreatment to the anaerobic co-digestion of brown water and food waste. Waste Management, 33(4), 813–819. https://doi.org/10.1016/j.wasman.2012.11.013.

Lin, Y., Wang, D., Wu, S. et al. 2009. Alkali pretreatment enhances biogas production in the anaerobic digestion of pulp and paper sludge. Journal of Hazardous Materials 170(1): 366–373. https://doi.org/10.1016/j.jhazmat.2009.04.086.

Liu, D., Zeng, R.J. and Angelidaki, I. 2008. Effects of pH and hydraulic retention time on hydrogen production versus methanogenesis during anaerobic fermentation of organic household solid waste under extreme-thermophilic temperature (70°C). Biotechnology and Bioengineering 100(6): 1108–1114. https://doi.org/10.1002/bit.21834.

Logan, M. and Visvanathan, C. 2019. Management strategies for anaerobic digestate of organic fraction of municipal solid waste: Current status and future prospects 37(1): 29–37. https://doi.org/10.1177/0734242X18816793.

Long, J.H., Aziz, T.N., Reyes, F.L.D.L. et al. 2012. Anaerobic co-digestion of fat, oil, and grease (FOG): A review of gas production and process limitations. Process Safety and Environmental Protection 90(3): 231–245. https://doi.org/10.1016/j.psep.2011.10.001.

López Torres, M. and Espinosa Lloréns, M. del C. 2008. Effect of alkaline pretreatment on anaerobic digestion of solid wastes. Waste Management 28(11): 2229–2234. https://doi.org/10.1016/j.wasman.2007.10.006.

Lópeza, M., Montserrat Solivaa, F., Martínez-Farréa, X. et al. 2010. Evaluation of MSW organic fraction for composting: Separate collectionor mechanical sorting. Resources, Conservation and Recycling 4(54): 222–228.

Edgar Fernando Castillo M., Diego Edison Cristancho and A. Victor Arellano. 2006. Study of the operational conditions for anaerobicdigestion of urban solid wastes. Waste Management 26: 46–556.

Mahmoodi, P., Karimi, K. and Taherzadeh, M.J. 2018. Hydrothermal processing as pretreatment for efficient production of ethanol and biogas from municipal solid waste. Bioresource Technology 261: 166–175. https://doi.org/10.1016/j.biortech.2018.03.115.

Mancini, G., Papirio, S., Riccardelli, G. et al. 2018. Trace elements dosing and alkaline pretreatment in the anaerobic digestion of rice straw. Bioresource Technology 247(September 2017): 897–903. https://doi.org/10.1016/j.biortech.2017.10.001.

Manser, N.D., Mihelcic, J.R. and Ergas, S.J. 2015. Semi-continuous mesophilic anaerobic digester performance under variations in solids retention time and feeding frequency. In Bioresource Technology 190: 359–366. https://doi.org/10.1016/j.biortech.2015.04.111.

Mao, C., Feng, Y., Wang, X. et al. 2015. Review on research achievements of biogas from anaerobic digestion. Renewable and Sustainable Energy Reviews 45: 540–555. https://doi.org/10.1016/j.rser.2015.02.032.

Mlaik, N., Khoufi, S., Hamza, M. et al. 2019. Enzymatic pre-hydrolysis of organic fraction of municipal solid waste to enhance anaerobic digestion. Biomass and Bioenergy 127: 105286. https://doi.org/10.1016/j.biombioe.2019.105286.

Modenbach, A.A. and Nokes, S.E. 2012. The use of high-solids loadings in biomass pretreatment—A review. Biotechnology and Bioengineering 109(6): 1430–1442. https://doi.org/10.1002/bit.24464.

Mosier, N., Wyman, C., Dale, B. et al. 2005. Features of promising technologies for pretreatment of lignocellulosic biomass. Bioresource Technology 96(6): 673–686. https://doi.org/10.1016/j.biortech.2004.06.025.

Nagao, N., Tajima, N., Kawai, M. et al. 2012. Bioresource technology maximum organic loading rate for the single-stage wet anaerobic digestion of food waste. Bioresource Technology 118: 210–218. https://doi.org/10.1016/j.biortech.2012.05.045.

Nasir, Z., Ahring, B.K. and Uellendahl, H. 2020. Enhancing the hydrolysis process in a dry anaerobic digestion process for the organic fraction of municipal solid waste. Bioresource Technology Reports 11: 100542. https://doi.org/10.1016/j.biteb.2020.100542.

Nayono, S.E., Gallert, C. and Winter, J. 2010. Co-digestion of press water and food waste in a biowaste digester for improvement of biogas production. Bioresource Technology 101(18): 6987–6993. https://doi.org/10.1016/j.biortech.2010.03.123.

Neumann, P., Pesante, S., Venegas, M. et al. 2016. Developments in pre-treatment methods to improve anaerobic digestion of sewage sludge. Reviews in Environmental Science and Biotechnology 15(2): 173–211. https://doi.org/10.1007/s11157-016-9396-8.

Nghiem, L.D., Koch, K., Bolzonella, D. et al. 2017. Full scale co-digestion of wastewater sludge and food waste: Bottlenecks and possibilities. Renewable and Sustainable Energy Reviews 72: 354–362. https://doi.org/https://doi.org/10.1016/j.rser.2017.01.062.

Noxolo, S.T., Edison, M. and Tesfagiorgis, H.B. 2014. Effect of temperature and ph on the anaerobic digestion of grass silage. In Proceedings of the 6th International Conference on Green Technology, Renewable Energy and Environmental Engineering, Cape Town, South Africa, November, 27–28.

Parikh, J., Channiwala, S.A. and Ghosal, G.K. 2005. A correlation for calculating HHV from proximate analysis of solid fuels. Fuel 84(5): 487–494. https://doi.org/10.1016/j.fuel.2004.10.010.

Pavi, S., Kramer, L.E., Gomes, L.P. et al. 2017. Biogas production from co-digestion of organic fraction of municipal solid waste and fruit and vegetable waste. Bioresource Technology 228: 362–367. https://doi.org/10.1016/j.biortech.2017.01.003.

Pérez, M. and Romero, L.I. 2008. Bioresource technology anaerobic digestion of municipal solid wastes: Dry thermophilic performance 99: 8180–8184. https://doi.org/10.1016/j.biortech.2008.03.021.

Planning Commission. 2014. Report of the Task Force on Waste to Energy: Volume I prepared by Planning Commission (PC). Task Force on Waste to Energy I(Volume I): 1–178. http://planningcommission.nic.in/reports/genrep/rep_wte1205.pdf.

Qi, H., Xiao, S., Shi, R. et al. 2018. Impact of anaerobic co-digestion between sewage sludge and carbon-rich organic waste on microbial community resilience. Nature 388: 539–547.

Rajput, A.A. and Sheikh, Z. 2019. Effect of inoculum type and organic loading on biogas production of sunflower meal and wheat straw. Sustainable Environment Research 1(1): 1–10. https://doi.org/10.1186/s42834-019-0003-x.

Ramachandra, T.V., Bharath, H.A., Kulkarni, G. et al. 2018. Municipal solid waste: Generation, composition and GHG emissions in Bangalore, India. Renewable and Sustainable Energy Reviews 82: 1122–1136. https://doi.org/10.1016/j.rser.2017.09.085.

Raposo, F., Banks, C.J., Siegert, I. et al. 2006. Influence of inoculum to substrate ratio on the biochemical methane potential of maize in batch tests. Process Biochemistry 41(6): 1444–1450. https://doi.org/10.1016/j.procbio.2006.01.012.

Rittmann, B.E., Lee, H.S., Zhang, H. et al. 2008. Full-scale application of focused-pulsed pre-treatment for improving biosolids digestion and conversion to methane. Water Science and Technology 58(10): 1895–1901. https://doi.org/10.2166/wst.2008.547.

Rocamora, I., Wagland, S.T., Villa, R. et al. 2020. Dry anaerobic digestion of organic waste: A review of operational parameters and their impact on process performance. Bioresource Technology, 299. https://doi.org/10.1016/j.biortech.2019.122681.

Romero-Güiza, M.S., Peces, M., Astals, S. et al. 2014. Implementation of a prototypal optical sorter as core of the new pre-treatment configuration of a mechanical-biological treatment plant treating OFMSW through anaerobic digestion. Applied Energy 135: 63–70. https://doi.org/10.1016/j.apenergy.2014.08.077.

Romero-Güiza, M.S., Vila, J., Mata-Alvarez, J. et al. 2016. The role of additives on anaerobic digestion: A review. Renewable and Sustainable Energy Reviews 58: 1486–1499. https://doi.org/10.1016/j.rser.2015.12.094.

Rubio, J.A., Fdez-Güelfo, L.A., Romero-García, L.I. et al. 2022. Start-up of the mesophilic anaerobic co-digestion of two-phase olive-mill waste and cattle manure using volatile fatty acids as process control parameter. Fuel 325: 124901.

Sahoo, S.R. and Rao, P.V. 2019. Temperature-phased anaerobic co-digestion of food waste and rice husk using response surface methodology. In Water Resources and Environmental Engineering II (pp. 137–146). Springer Singapore. https://doi.org/10.1007/978-981-13-2038-5.

Sajeena Beevi, B., Madhu, G. and Sahoo, D.K. 2015. Performance and kinetic study of semi-dry thermophilic anaerobic digestion of organic fraction of municipal solid waste. Waste Management 36: 93–97. https://doi.org/10.1016/j.wasman.2014.09.024.

Salwa Khamis, S., Purwanto, H., Naili Rozhan, A. et al. 2019. Characterization of municipal solid waste in malaysia for energy recovery. IOP Conference Series: Earth and Environmental Science 264(1). https://doi.org/10.1088/1755-1315/264/1/012003.

Sambusiti, C., Ficara, E., Malpei, F. et al. 2013. Benefit of sodium hydroxide pretreatment of ensiled sorghum forage on the anaerobic reactor stability and methane production. Bioresource Technology 144: 149–155. https://doi.org/10.1016/j.biortech.2013.06.095.

Saratale, G.D. and Oh, M.K. 2015. Improving alkaline pretreatment method for preparation of whole rice waste biomass feedstock and bioethanol production. RSC Advances 5(118): 97171–97179. https://doi.org/10.1039/c5ra17797a.

Saritha, M., Arora, A. and Lata. 2012. Biological pretreatment of lignocellulosic substrates for enhanced delignification and enzymatic digestibility. Indian Journal of Microbiology 52(2): 122–130. https://doi.org/10.1007/s12088-011-0199-x.

Sarker, S. and Møller, H.B. 2013. Boosting biogas yield of anaerobic digesters by utilizing concentrated molasses from 2nd generation bioethanol plant. International Journal of Energy and Environment 4(2): 199–210.

Sarker, S., Møller, H.B. and Bruhn, A. 2014. Influence of variable feeding on mesophilic and thermophilic co-digestion of Laminaria digitata and cattle manure. Energy Conversion and Management 87: 513–520. https://doi.org/10.1016/j.enconman.2014.07.039.

Schimpf, U., Hanreich, A., Mähnert, P. et al. 2013. Improving the efficiency of large-scale biogas processes: Pectinolytic enzymes accelerate the lignocellulose degradation. Journal of Sustainable Energy and Environment 4(2): 53–60. http://www.forum.tci-thaijo.org/index.php/JSEE/article/view/9830.

Sedighi, A., Karrabi, M., Shahnavaz, B. et al. 2022. Bioenergy production from the organic fraction of municipal solid waste and sewage sludge using mesophilic anaerobic co-digestion: An experimental and kinetic modeling study. Renewable and Sustainable Energy Reviews 153: 111797. https://doi.org/10.1016/j.rser.2021.111797.

Shahriari, H., Warith, M., Hamoda, M. et al. 2012. Anaerobic digestion of organic fraction of municipal solid waste combining two pretreatment modalities, high temperature microwave and hydrogen peroxide. Waste Management 32(1): 41–52. https://doi.org/10.1016/j.wasman.2011.08.012.

Sharma, K.D. and Jain, S. 2019. Overview of municipal solid waste generation, composition, and management in India. Journal of Environmental Engineering 145(3): 04018143. https://doi.org/10.1061/(asce)ee.1943-7870.0001490.

Shi, X.S., Dong, J.J., Yu, J.H. et al. 2017. Effect of hydraulic retention time on anaerobic digestion of wheat straw in the semicontinuous continuous stirred-tank reactors. BioMed Research International, 1–6. https://doi.org/10.1155/2017/2457805.

Siddique, N.I., Munaim, M.S.A. and Wahid, Z.A. 2015. Role of biogas recirculation in enhancing petrochemical wastewater treatment efficiency of continuous stirred tank reactor. Journal of Cleaner Production 91: 229–234. http://dx.doi.org/10.1016/j.jclepro.2014.12.036.

Singh, B., Szamosi, Z. and Siménfalvi, Z. 2019. State of the art on mixing in an anaerobic digester: A review. Renewable Energy 141: 922–936. https://doi.org/10.1016/j.renene.2019.04.072.

Singh, S. 2020. Solid Waste Management in Urban India: Imperatives for improvement. Journal of Contemporary Issues in Business and Government 25(1): 5–38. https://doi.org/10.47750/cibg.2019.25.01.001.

Sivaprakasam, S. and Balaji, K. 2020. A review of upflow anaerobic sludge fixed film (UASFF) reactor for treatment of dairy wastewater. Materials Today: Proceedings 43: 1879–1883. https://doi.org/10.1016/j.matpr.2020.10.822.

Sondh, S., Upadhyay, D.S., Patel, S. et al. 2022. A strategic review on Municipal Solid Waste (living solid waste) management system focusing on policies, selection criteria and techniques for waste-to-value. Journal of Cleaner Production 131908.

Song, Z. lin, Yag, G. he, Feng, Y. zhong et al. 2013. Pretreatment of rice straw by hydrogen peroxide for enhanced methane yield. Journal of Integrative Agriculture 12(7): 1258–1266. https://doi.org/10.1016/S2095-3119(13)60355-X.

Souza, T.S.O., Ferreira, L.C., Sapkaite, I. et al. 2013. Bioresource technology thermal pretreatment and hydraulic retention time in continuous digesters fed with sewage sludge: Assessment using the ADM1. Bioresource Technology 148: 317–324. https://doi.org/10.1016/j.biortech.2013.08.161.

Speece, R.E. 1983. Anaerobic biotechnology treatment for industrial wastewater. Environmental Science and Technology 17(9): 416A–427A. https://doi.org/https://doi.org/10.1021/es00115a001.

Steffen, R., Szolar, O. and Braun, R. 1998. Feedstocks for anaerobic digestion. Institute of Agrobiotechnology Tulin, University of Agricultural Sciences, Vienna, 1–29.

Szaja, A., Golianek, P. and Kamiński, M. 2022. Process performance of thermophilic anaerobic co-digestion of municipal sewage sludge and orange peel. Journal of Ecological Engineering 23(8): 66–76. https://doi.org/10.12911/22998993/150613.

TJayanth, T.A.S., Mamindlapelli, N.K., Begum, S. et al. 2020. Anaerobic mono and co-digestion of organic fraction of municipal solid waste and landfill leachate at industrial scale: Impact of volatile organic loading rate on reaction kinetics, biogas yield and microbial diversity. Science of the Total Environment 748: 142462. https://doi.org/10.1016/j.scitotenv.2020.142462.

Tauseef, S., Premalatha, M., Abbasi, T. et al. 2013. Methane capture from livestock manure. Journal of Environmental Management 117: 187–207.

Teghammar, A. 2013. Biogas production from lignocelluloses: Pretreatment, substrate characterization, co-digestion, and economic evaluation. Chalmers Tekniska Hogskola (Sweden).

Tricase, C. and Lombardi, M. 2009. State of the art and prospects of Italian biogas production from animal sewage: Technical-economic considerations. Renewable Energy 34(3): 477–485. https://doi.org/10.1016/j.renene.2008.06.013.

Visvanathan, C. 2010. Bioenergy production from organic fraction of municipal solid waste (OFMSW) through dry anaerobic digestion. Bioenergy and Biofuel from Biowastes and Biomass, 71–87. https://doi.org/10.1061/9780784410899.ch04.

Wang, X., Lu, X., Li, F. and Yang, G. (2014). Effects of temperature abd Carbon-too-Nitrogen (C/N) ratio on the performance of anaerobic co-digestion of Dairy Manure and Rice Straw: Focusing on Ammonia inhibition. PLOS ONE 9(5): e97265.

Wilkins, D., Rao, S., Lu, X. et al. 2015. Effects of sludge inoculum and organic feedstock on active microbial communities and methane yield during anaerobic digestion. Frontiers in Microbiology 6: 1114. https://doi.org/10.3389/fmicb.2015.01114.

World Energy Council. 2016. World Energy Resources 2016. In World Energy Council 2016 (pp. 6–46). https://www.worldenergy.org/wp-content/uploads/2016/10/World-Energy-Resources_SummaryReport_2016.10.03.pdf.

Wu, C., Huang, Q., Yu, M. et al. 2018. Effects of digestate recirculation on a two-stage anaerobic digestion system, particularly focusing on metabolite correlation analysis. In Bioresource Technology 251: 40–48.https://doi.org/10.1016/j.biortech.2017.12.020.

Wang, X., Lu, X., Li, F. et al. 2014. Effects of temperature and Carbon-Nitrogen (C/N) ratio on the performance of anaerobic co-digestion of DairyManure, Chicken Manure and Rice Straw: Focusing on ammonia inhibition. PLOS ONE, 9(5): e97265. https://doi.org/10.1371/journal.pone.0097265.

Xiao, B., Zhang, W., Yi, H. et al. 2019. Biogas production by two-stage thermophilic anaerobic co-digestion of food waste and paper waste: Effect of paper waste ratio. Renewable Energy 132: 1301–1309. https://doi.org/10.1016/j.renene.2018.09.030.

Xie, R., Xing, Y., Ghani, Y.A. et al. 2007. Full-scale demonstration of an ultrasonic disintegration technology in enhancing anaerobic digestion of mixed primary and thickened secondary sewage sludge. Journal of Environmental Engineering and Science 6(5): 533–541. https://doi.org/10.1139/S07-013.

Yang, Z.H., Xu, R., Zheng, Y. et al. 2016. Characterization of extracellular polymeric substances and microbial diversity in anaerobic co-digestion reactor treated sewage sludge with fat, oil, grease. Bioresource Technology 212: 164–173. https://doi.org/10.1016/j.biortech.2016.04.046.

Yu, D., Liu, J., Sui, Q. et al. 2016. Biogas-pH automation control strategy for optimizing organic loading rate of anaerobic membrane bioreactor treating high COD wastewater. Bioresource Technology 203: 62–70. https://doi.org/10.1016/j.biortech.2015.12.010.

Yuan, X., Wen, B., Ma, X. et al. 2014. Enhancing the anaerobic digestion of lignocellulose of municipal solid waste using a microbial pretreatment method. Bioresource Technology 154: 1–9. https://doi.org/10.1016/j.biortech.2013.11.090.

Zábranská, J., Dohányos, M., Jeníček, P. et al. 2006. Disintegration of excess activated sludge - Evaluation and experience of full-scale applications. Water Science and Technology 53(12): 229–236. https://doi.org/10.2166/wst.2006.425.

Zaher, U., Buffiere, P., Steyer, J.-P. et al. 2009. A procedure to estimate proximate analysis of mixed organic wastes. Water Environment Research, 81(4): 407–415. https://doi.org/10.2175/106143008x370548.

Zamri, M.F.M.A., Hasmady, S., Akhiar, A. et al. 2021. A comprehensive review on anaerobic digestion of organic fraction of municipal solid waste. Renewable and Sustainable Energy Reviews 137: 110637. https://doi.org/10.1016/j.rser.2020.110637.

Zhang, J., Kan, X., Shen, Y. et al. 2018. A hybrid biological and thermal waste-to-energy system with heat energy recovery and utilization for solid organic waste treatment. Energy 152: 214–222. https://doi.org/10.1016/j.energy.2018.03.143.

Zhou, Y., Huang, K., Jiao, X. et al. 2021. Anaerobic co-digestion of organic fractions of municipal solid waste: Synergy study of methane production and microbial community. Biomass and Bioenergy 151: 106137. https://doi.org/10.1016/j.biombioe.2021.106137.

Zhu, D., Asnani, P.U., Zurbrügg, C. et al. 2008. Improving Municipal Solid Waste Management in India: A Sourcebook for Policy Makers. 190. https://openknowledge.worldbank.org/handle/10986/6916.

Ziels, R.M., Karlsson, A., Beck, D.A.C. et al. 2016. Microbial community adaptation influences long-chain fatty acid conversion during anaerobic codigestion of fats, oils, and grease with municipal sludge. Water Research 103: 372–382. https://doi.org/10.1016/j.watres.2016.07.043.

3

KINETICS OF BIOCHEMICAL DEGRADATION OF MUNICIPAL SOLID WASTE IN LANDFILLS

Sameena Begum,[1] *Raashmi Bitra,*[1] *Sudharshan Juntupally*[1,2]
and *Gangagni Rao Anupoju*[1,2,*]

1. INTRODUCTION

Solid waste is defined as the unwanted or useless material generated due to human activities in residential, industrial and commercial areas. Solid waste is classified into four categories, namely, municipal waste, hazardous waste, electronic waste and medical waste. The solid waste that includes household waste, food waste from offices, schools, restaurants, paper waste, green waste, plastic, etc., is called 'municipal solid waste' (MSW), while the waste from factories and industries is called 'industrial waste', and the wastes other than these comprising of oil, medical, industrial and hospital waste are called 'hazardous solid waste'. Although, by definition, the MSW comprises of only domestic waste, in exception in real time situations, it also includes industrial and commercial wastes. Due to rapid industrialization, urbanization and population growth, the exponential growth in MSW generation is inevitable. As per the estimates of the World Bank in 2016, about 2.02 billion metric tonnes (BMT) of MSW is generated globally which is projected to increase by 28.2%, i.e., 2.59 BMT by 2030 and a further increase by 31.3% by the end of 2050, which makes a cumulative increase in MSW generation by 68.3% by 2050 against the current generation rate (Fig. 1) (Begum et al. 2017, Kaza et al. 2018). The typical composition of MSW generated across the globe comprises of food waste, garden waste, plastic waste, wood, textile, metals, glass, paper and cardboard and other wastes. MSW comprises of 40 – 50% of biodegradable organic waste, 17 – 20% recyclables, 11% of hazardous waste, and 21 – 25% of inerts materials (Fig. 2) (Abdel-Shafy and Mansour 2018, Begum et al. 2017). To tackle this huge quantity of MSW generated across the globe, systematic waste management approaches or

---

[1] Bioengineering and Environmental Sciences (BEES) division, Department of Energy and Environmental Engineering (DEEE), CSIR-Indian Institute of Chemical Technology (IICT), Hyderabad – 500007, India.
[2] Academy of Scientific and Innovative Research (AcSIR), Ghaziabad 201002, India.
* Corresponding author: gangagnirao@gmail.com; agrao@iict.res.in

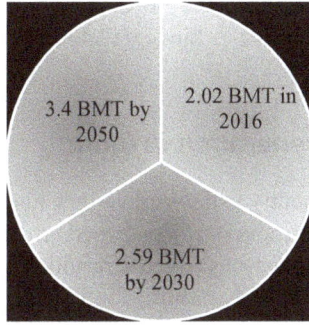

Figure 1. Estimated MSW generation statistics from 2016 to 2050 globally (Kaza et al. 2018).

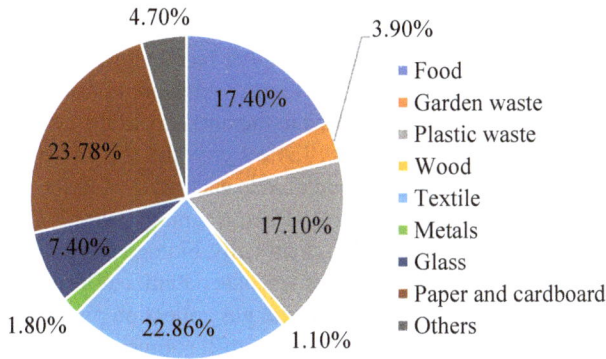

Figure 2. Typical composition of MSW generated globally (Ghosh 2020).

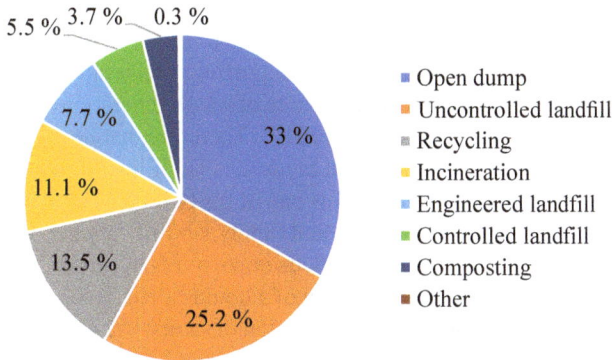

Figure 3. MSW disposal methods in practise globally (Abdel-Shafy and Mansour 2018, Begum et al. 2017).

treatment methods have to be followed to encourage effective waste management. In contrast, landfilling is the most preferred option that is globally practised to get rid of MSW. About 33% of the MSW generated is left unattended while the remaining 67% is collected for processing (HK-MEFCC 2016). About 38.4% of the collected MSW is landfilled either in controlled or uncontrolled or engineered landfills. About 3.7% of MSW is used to produce fertilizer through composting process, 11% is incinerated, 13.5% is recycled and 0.3% of MSW is used for other methods of treatment such as waste to energy (WtE) plants, RDF, etc. (Fig. 3) (Ghosh 2020, MNRE 2018, Statista 2016). These numbers indicate that about 75% of the MSW generated globally is not appropriately treated, i.e., the MSW is left unattended or landfilled in the open lands (Jadhav 2018). Both these approaches of MSW disposal are considered to be an environmental hazard as it results in uncontrolled greenhouse gas (GHG) emissions which directly impacts the global climate and also contributes to human health hazards. The major fraction of MSW that is biodegradable is one of the primary concerns in the present scenario.

1.1 Municipal solid waste generation in India

India is the second most populous country in the world after China and one of the developing nations in terms of its GDP (Gross domestic product). Uncontrolled urbanization, rapid industrialization and population growth are the three major reasons which is resulting in uncontrolled generation of MSW in India. As can be seen from Table 1, Maharashtra, Tamil Nadu, Delhi, Gujarat and Karnataka are the five major states with highest generation of MSW with 22,570 MT, 15,547 MT, 10,500 MT, 10,145 MT and 10,000 MT per day, respectively. The lowest MSW generation is observed in the state of Sikkim with less than 100 MT per day. The 29 states and 7 union territories of India generate about 1,44,241 MT of MSW per day making it an enormous amount of MSW 5,19,26,760 MT per annum (IBEF 2021, Statista 2016). To provide a waste treatment facility for MSW, each state is allocated with specific areas in the outskirts of the cities or towns which are known as MSW dumping sites. Landfills are a discrete area of land usually located in the outskirts of the city or state which receives the MSW generated in a particular city or a state. Table 1 provides the information on the number of landfill sites (open dumps, sanitary landfills) available in each state of India. Landfills are classified into open dump landfills, sanitary or engineered landfills. A sanitary landfill is designed to isolate the MSW from the environment until it is physically, chemically, biologically safe. Sanitary landfilling of MSW prevents environmental hazard through gas or leachate pollution. These dumping sites have been transformed to sanitary landfills that are provided with gas collection pipelines and landfill leachate channels to effectively manage the MSW. A sanitary landfill comprises of a bottom liner, a leachate collection system, a cover and a natural hydrogeologic setting. Unlike the sanitary or engineered landfills, the open dump landfills practise uncontrolled dumping of MSW that results in air, water and soil pollution through GHG emissions and leachate percolation (Kumar and Agrawal 2020). The statistics presented in Table 1 reveal that about 25% of the total MSW generated is being processed using different methods of treatment (recycling, composting, incineration, WtE plants) while the remaining 75% of MSW is left unattended at the source of waste generation and/or is openly dumped (landfilled) in a large area (Jadhav 2018, Statista 2016). Therefore, nationally or internationally, the most widely practised option for MSW disposal method is landfilling it. Notwithstanding the fact that, landfilling of MSW also can be considered as one of the best options if engineered landfills are implemented with landfill gas capturing facility and timely landfill capping. Capture of landfill gas and timely capping of landfills reduces the hazardous impact of GHG emissions to the environment and the bioleaching of the organic waste in the form of a liquid called 'leachate' into the water bodies and soil leading to land and water pollution. Construction of sanitary landfills or engineered landfills for the safe disposal of MSW is essential as these are the landfills that are designed in such a way that it has a facility to capture the GHG emissions comprising mainly of CH_4 and CO_2. These are the two major gases that are a resultant of natural weathering and biodegradation of organic matter in the biodegradable waste that gets piled up into mountains of MSW over a period.

1.1.1 Biodegradable waste in MSW landfills and its impact on the environment

About 40–50% of the MSW comprises of biodegradable organic material which ends up in the landfill and undergoes aerobic and anaerobic digestion due to natural weathering conditions. The top layers of the MSW landfills are exposed to air whereas the inside deep layers are depleted of oxygen resulting in anaerobic environment. Therefore, the top layers of MSW undergo aerobic digestion while the oxygen depleted layers undergo anaerobic degradation over a period of time. The creation

Table 1. State-wise MSW generation rate, number of landfills/dumpsites and the waste processing capacity (Kaza et al. 2018, MNRE 2018, Statista 2016).

S. no.	State	No. of landfills	MSW generation (MT/day)	MSW generation (MT/annum)	Waste processing (%)
1.	Madhya Pradesh	378	6,424	23,12,640	20–30
2.	Maharashtra	320	22,570	81,25,200	39
3.	Karnataka	215	10,000	36,00,000	20–30
4.	Rajasthan	174	6,500	23,40,000	20–30
5.	Punjab	150	4,100	14,76,000	20–30
6.	Bihar	93	1,192	4,29,120	3
7.	Uttar Pradesh	82	15,500	55,80,000	20–30
8.	Assam	76	1,134	4,08,240	20–30
9.	Odisha	54	2,460	8,85,600	2
10.	Gujarat	36	10,145	36,52,200	20–30
11.	Haryana	26	4,514	16,25,040	6
12.	Tripura	17	421	1,51,560	20–30
13.	Uttarakhand	13	1,400	5,04,000	20–30
14.	Jammu and Kashmir	12	1,792	6,45,120	1
15.	Arunachal Pradesh	11	181	65,160	1
16.	Tamil Nadu	4	15,547	55,96,920	10
17.	Telangana	4	7,371	26,53,560	67
18.	Himachal Pradesh	3	342	1,23,120	20–30
19.	Chhattisgarh	2	1,959	7,05,240	74
20.	Sikkim	2	89	32,040	66
21.	Goa	2	240	86,400	62
22.	Delhi	2	10,500	37,80,000	55
23.	Kerala	1	1,576	5,67,360	45
24.	Pondicherry	1	495	1,78,200	3
25.	Chandigarh	1	340	1,22,400	20–30
26.	Mizoram	1	201	72,360	4
27.	Nagaland	1	342	1,23,120	0
28.	Andaman and Nicobar Islands	1	115	41,400	20–30
29.	Daman Diu and Dadra Nagar Haveli	1	81	29,160	0
30.	Meghalaya	1	268	96,480	58
31.	Andhra Pradesh	110	6,063	21,82,680	10
32.	West Bengal	107	7,876	28,35,360	5
33.	Jharkhand		2,327	8,37,720	2
34.	Manipur		176	63,360	50
			1,44,241	5,19,26,760	
25% of MSW is processed and 75% of MSW is openly dumped					

of anaerobic environment inside the landfill results in rapid degradation of organic matter as this environment favours the growth of anaerobic microorganisms and results in the liberation of methane (CH_4) and carbon dioxide (CO_2) in to the atmosphere. CH_4 and CO_2 are the two major GHGs that not only pollute the atmosphere but also result in burning of the MSW in the landfills leading to the release of particulate organic matter into the atmosphere. In addition to the gaseous emissions, the open landfills also contribute to the percolation of leachate from the MSW (Begum et al. 2018). The leachate that is generated from the MSW heaps is known as landfill leachate. Landfill leachate is a thick and complicated liquid that comprises of large amount of organic and inorganic matter that oozes out of the MSW heaps, flows through channels and ends up polluting the water bodies as well as the land. The composition of landfill leachate is dependent on the age of the MSW landfill. If the MSW landfill comprises of waste that is greater than 10 years old, then the leachate percolating from that landfill is called 'legacy leachate'. The leachate generated from the MSW landfill containing waste that is less than 10 years and greater than 5 years is called 'intermediate leachate', while the leachate generated from the landfill which comprises of waste less than 3 years old is called 'young or fresh leachate'. The typical characteristics of all types of leachate generated from an MSW landfill are summarized in Table 2. As can be seen from Table 2, the composition of landfill leachate varies depending on the age of the landfill and accordingly the organic matter accounted in terms of chemical oxygen demand (COD) and other volatile organic acids (VOA), fulvic acid (FA), and humic acid (HA) varies. The concentration of COD and organic matter is high in leachate from a landfill less than 3 years old. The presence of VOA in the landfill leachate can be ascertained from the pH of the sample which ranges between 4 and 6. As the age of the MSW landfill increases, the complexity of the landfill leachate generated from that particular landfill decreases. Therefore, the treatment of young or fresh landfill leachate is highly essential as it is composed of enormous amount of organic and inorganic matter (nitrogen and heavy metals) which proves to be fatal to both human and animal life. Therefore, collection of landfill leachate at the source of generation is necessary which can only be achieved by the transformation of present MSW open dumps to sanitary MSW landfills.

Table 2. Typical composition of MSW leachate in the landfills (Begum et al. 2020b, a, 2018).

Parameter	Young/Fresh	Intermediate	Legacy
Year	< 3	5–10	> 10
pH	4–6	6.5–7.5	> 7.5
COD (g/L)	15–140	10–40	< 10
NH_3-N (g/L)	2–7	1–4	4–7
Heavy metals (g/L)	1–1.2	0.5–0.7	< 0.5
Organic matter (%)	80% VOA	5–30 VOA+HA+FA	< 5 HA+FA

2. FATE OF MSW: FROM GENERATION TO DISPOSAL AND BIODEGRADATION IN THE LANDFILL

The ultimate fate of MSW in the landfills is its biochemical degradation in the presence of microorganisms both aerobes and anaerobes resulting in the release of GHGs along with the generation of leachate which is possible only when the biodegradable fraction of MSW, i.e., organic fraction of MSW (OFMSW) undergoes a series of biochemical reactions in the presence of microorganisms. The fate of MSW from generation to disposal and its biodegradation in the MSW landfill is shown in Fig. 4. The MSW comprising of 40–50% of biodegradable material and the remaining non-biodegradable material ends up in the landfills as composting and incineration are the two methods that share a very less proportion in the entire disposal processing methods (Fig. 3). As can be seen from Fig. 4, the MSW is landfilled either in open dumps yards and/or sanitary or engineered landfills.

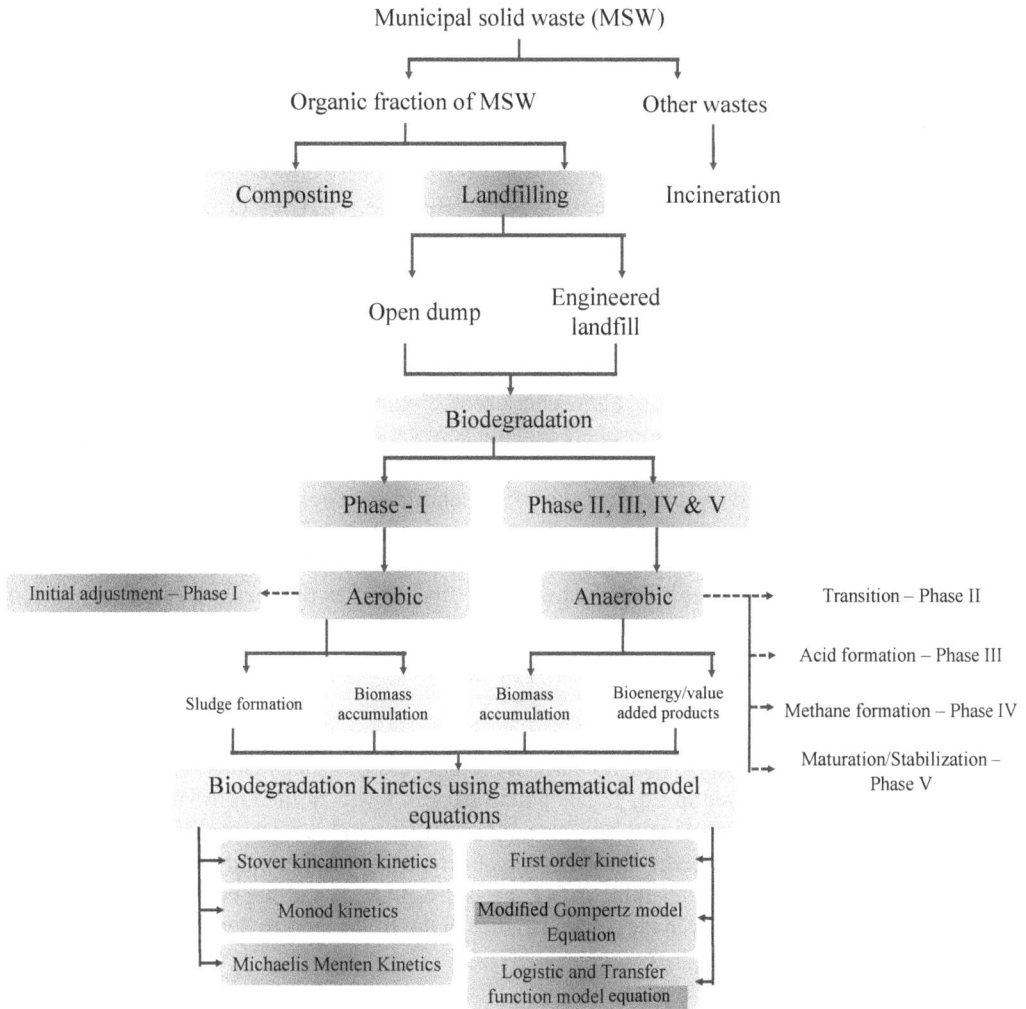

Figure 4. Fate of MSW from generation to disposal and biodegradation in the landfill.

In either case, the OFMSW in the MSW landfill undergoes biodegradation. Biodegradation is defined as the process by which the complex organic matter is decomposed by microorganisms (aerobic and anaerobic) into simpler compounds such as methane, carbon dioxide, water, ammonia, and other volatile organics. As biodegradation occurs in two environments, one is aerobic environment and the other is anaerobic environment. In the aerobic biodegradation process, the organic matter is decomposed by the aerobic bacteria such as *Lactobacillus, Nocardia, Mycobacteria* that survive in the presence of oxygen while in the anaerobic biodegradation process, the organic matter is decomposed by the anaerobic microorganisms belonging to bacterial and archaeal groups such as *Firmicutes, Bacteroidetes* and *Methanogens*. The key organisms responsible for methane generation are *Methanosarcina, Methanosaeta, Methanofollis* and *Methanoculleus* that survive in the oxygen depleted environments. Equations 1 and 2 show the aerobic and anaerobic biochemical reactions, respectively, that take place in the landfill. The main products from aerobic degradation of OFMSW in the landfill are CO_2 and H_2O whereas in anaerobic degradation, CH_4 is also liberated in addition to CO_2 and H_2O. Accumulation of aerobic and anaerobic biomass is the common phenomena in both the biodegradation processes.

Aerobic degradation of organic matter

Organic matter $(C_{polymer})$ + O_2 → Organic matter $(C_{monomers})$ + Biomass + CO_2 + H_2O Eq. 1

Anaerobic degradation of organic matter

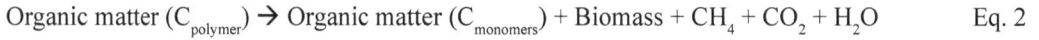

Organic matter $(C_{polymer})$ → Organic matter $(C_{monomers})$ + Biomass + CH_4 + CO_2 + H_2O Eq. 2

2.1 Biochemical reactions of aerobic and anaerobic degradation of OFMSW

The aerobic and anaerobic degradation of OFMSW occurs in the same MSW landfill. The top most layer of the MSW landfill is exposed to air where the OFMSW undergoes aerobic degradation due to natural weathering in the presence of atmospheric temperature and other environmental conditions resulting in the release of CO_2 to atmosphere and the percolation of moisture from top layers to the bottom layers which is defined as the initial adjustment phase (Phase I) of MSW in the landfill. The subsequent layers beneath the top layer of the MSW landfill are depleted of oxygen/air creating a complete anaerobic environment. The anaerobic degradation of OFMSW takes place in four phases as shown in Fig. 4. Transition (Phase II), acid formation (Phase III), methane formation (Phase IV), maturation or stabilization (Phase V) are the four phases that occur in the deeper layers of the landfill (Tamru and Chakma 2015).

2.1.1 Initial adjustment: Phase I

Aerobic degradation of OFMSW is the initial phase in the landfill which results in the generation of CO_2 and moisture as explained in Eq. 1. The bound oxygen levels in the OFMSW increase in this phase. The oxidation of the waste in the presence of facultative aerobic microorganisms belonging to hydrolytic and fermentative bacteria, nitrogen fixing bacteria and sulphur oxidizing bacteria initiates the degradation of organic matter resulting in CO_2, moisture, nitrogen and sulphur compounds. During this phase, the pH and the solids in the organic matter in the MSW start to decrease.

2.1.2 Transition: Phase II

In the transition phase, the OFMSW is exposed to anaerobic environment where the first step of anaerobic degradation or anaerobic digestion begins. Anaerobic digestion (AD) is a complex microbial process where the complex organic compounds are broken down to simpler monomers by the action of a group of microorganisms that are a consortium of hydrolytic and fermentative bacteria, acidogens, acetogens and methanogens belonging to the groups Firmicutes, Bacteroidetes, and Archaea. A few anaerobic microorganisms include *Proteobacteria*, *Actinobacteria*, *Synergistetes*, *Chloroflexi* and so on (Juntupally et al. 2022). These microorganisms act syntrophically, switching from one pathway to another depending on the environmental conditions. In the first step of AD, hydrolysis of organic matter takes place by the action of hydrolytic bacteria such as cellulases, lipases, proteases, etc., resulting in the conversion of complex polymers to soluble glucose units and the liberation of hydrogen as shown in Eq. 3. Due to the impact conversion of complex organic matter to soluble monomer units, the pH of the organic matter during this phase drops down to less than 5 (Begum et al. 2020b, 2018).

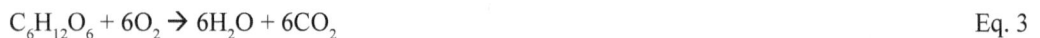

$C_6H_{12}O_6$ + $6O_2$ → $6H_2O$ + $6CO_2$ Eq. 3

2.1.3 Acid formation: Phase III

The second step in the AD process and the third step in the overall biodegradation of OFMSW in the landfill is the fermentation step. In this biochemical degradation process, the soluble monomeric units such as glucose is utilized by the acidogens for the production of short chain carboxylic acids or volatile fatty acids (acetic, propionic, butyric, valeric and lactic acids) along with CO_2, and H_2 through a sequence of biochemical reactions as shown in Eqs. 4 and 5. The pH of the organic matter in this phase further drops to less than 4 as there is a formation of acids and alcohol due to the microbial conversion of organic matter (Begum et al. 2018a, 2020b). The presence of enzymes that are capable of degrading lignocellulosic biomasses results in the formation of sugars such as xylose and fructose that can be utilized by the microorganisms (yeasts) to produce alcohols.

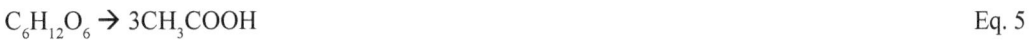

$$C_6H_{12}O_6 + 2H_2 \rightarrow 2CH_3CH_2COOH + 2H_2O \qquad \text{Eq. 4}$$

$$C_6H_{12}O_6 \rightarrow 3CH_3COOH \qquad \text{Eq. 5}$$

Subsequent to the acidogenic phase, acetogenesis begins where in the action of microbes helps in the chain elongation process where all the organic acids are converted to acetate (Eq. 8) for its utilization by the methanogens to produce methane. One of the major indicators of the acid formation phase is the drastic reduction in the pH of the OFMSW (Begum et al. 2016, Kuruti et al. 2017). The VFA to alkalinity ratio during this phase increases rapidly as the acidity in the wastes is predominant than the buffering capacity. Therefore, during this phase of degradation, the generation of methane is on a lower side but the leachate that is generated during this phase is highly complex as it contains numerous intermediate products like acids, and alcohols.

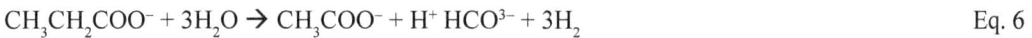

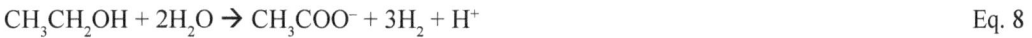

$$CH_3CH_2COO^- + 3H_2O \rightarrow CH_3COO^- + H^+ HCO^{3-} + 3H_2 \qquad \text{Eq. 6}$$

$$C_6H_{12}O_6 + 2H_2O \rightarrow 2CH_3COOH + 2\,CO_2 + 4H_2O \qquad \text{Eq. 7}$$

$$CH_3CH_2OH + 2H_2O \rightarrow CH_3COO^- + 3H_2 + H^+ \qquad \text{Eq. 8}$$

2.1.4 Methane formation: Phase IV

Methane formation phase is the last step in the AD process but the fourth step in the biodegradation of OFMSW in the landfill. In this phase, the acids, H_2, acetate and the alcohols formed are converted to CH_4 and CO_2 by the action of hydrogenotrophic and the acetoclastic methanogens as shown in Eqs. 9, 10 and 11. On the other hand, presence of methylotrophic organisms that utilize methane for their survival, lowers the methane content in the landfill gas. However, the predominance of microbial communities plays a key role in the landfills. Methane generation slows down, if the relative abundance of acid formers is higher than the methanogens, and when the predominance of methane formers is on a higher side, then all the other biochemical pathways slow down. Switching the biochemical pathways in accordance with time makes substantial changes in the microbial ecology in the landfills. The pH of OFMSW during this phase increases as the acids are consumed by the methanogens for the generation of CH_4 and CO_2. However, presence of small amount of organic acids is an exception in this phase as the organisms in the landfill cannot completely consume the available acids (Begum et al., 2021, 2020a, 2019). Equation 10 and Eq. 11 show the reactions where CO_2 and H_2 are converted to CH_4 and CH_3COOH is also converted to CH_4 and CO_2.

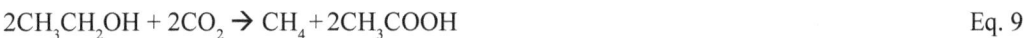

$$2CH_3CH_2OH + 2CO_2 \rightarrow CH_4 + 2CH_3COOH \qquad \text{Eq. 9}$$

$$CO_2 + 4H_2 \rightarrow CH_4 + 2H_2O \qquad\qquad\qquad\qquad\qquad\qquad\qquad \text{Eq. 10}$$

$$CH_3COOH \rightarrow CH_4 + CO_2 \qquad\qquad\qquad\qquad\qquad\qquad\qquad\qquad \text{Eq. 11}$$

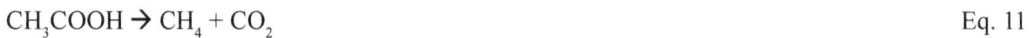

2.1.5 Maturation/Stabilization: Phase V

The final step in the biodegradation of OFMSW in the landfill is the maturation or the stabilization of waste. The organic waste in the landfill shifts to maturation phase where no further degradation takes place. Meaning, the biodegradation phase of the organic matter ends in this phase and the remaining waste becomes recalcitrant towards biodegradation. The stabilized waste is now suitable for capping in the landfill as there is no survival of microbes and has no impact on the environment. However, the stabilized waste has to be efficiently capped by covering the landfill with different layers of clay and other material which does not allow rain or other weathering conditions to penetrate into the landfill (Meyer-Dombard et al. 2020).

All the above explained five phases occur in the landfill over a period of time and as the OFMSW from initial phase reaches the stabilization phase, there is a lot of landfill leachate generation, and gaseous emissions are observed. If sanitary landfilling is practised, all the landfill leachate can be collected at a centralized facility in the landfill and used as a raw material for further processing and value-added products recovery. The landfill gas (CH_4 and CO_2) can be captured and utilized for various applications instead of letting it out to the environment as in the case of open landfills. Therefore, the fate of MSW biodegradation starts from initial adjustment phase and ends at stabilization phase.

2.2 Factors that affect the biodegradation kinetics in a system

It is very well known that the microorganisms, environment and the substrate are the three important factors that affect the biodegradation kinetics in a landfill.

2.2.1 Substrate

Availability of biodegradable material in the landfill such as food waste, green waste, spoiled fruits and vegetables, paper and cardboard is one of the important factors that favours the growth of aerobic and anaerobic microorganisms that are responsible for the biodegradation of organic material through biochemical pathways (Begum et al. 2021). Presence of multiple substrates in the landfill is another factor that influences the overall biodegradation. Waste materials are composed of complex polymers such as the carbohydrates, proteins, lipids and lignocelluloses which are hydrolyzed or solubilized into their simple monomeric units of glucose, fructose, amino acids, glycerol, and other sugars in the suitable environment. The biodegradation of waste material begins with the solubilization of waste and finishes with the conversion of these wastes to environmentally accepted products such as water, carbon dioxide, methane and biomass by the action of suitable microorganisms depending on the biochemical reaction that takes place. Studying the biodegradation kinetics of the biochemical reaction assists in determining the amount of substrate that has been utilized by the microorganisms to produce another product. Availability of substrate itself is not enough for efficient biodegradation to take place but the availability of substrate in adequate amount is necessary to prevent the substrate deficiency or organic shock loading (substrate inhibition) of substrate to the microorganisms. Organic loading rate (OLR) is one of the crucial parameters that plays a key role in promoting or inhibiting the biodegradation process. Substrate to inoculum (S/I), otherwise called food to inoculum (F/I), ratio in terms of COD or volatile solids (VS) should be maintained to prevent substrate inhibition (overload of substrate on the microbes) during the biological process (Gharasoo et al. 2015). Depending on the characteristics of the substrate, the optimum F/I ratio required can be 1:2 and 1:4 for high strength wastes whereas 2:1 for low strength wastes. The metabolic rate of the microorganisms is dependent on the availability of nutrients, adequate substrate and the suitable environmental conditions (Begum et al. 2021, 2020b). Controlling the OLR in the open dumps or uncontrolled landfills is impossible, whereas in the sanitary landfills, OLR can be controlled to ensure efficient biodegradation of the organic material. In the case of bioreactors that are operated either in aerobic or anaerobic environment, OLR can be controlled to promote the specific biochemical conversion. For instance, in an anaerobic digester, an OLR of $1-10$ kg VS/m^3 of digester volume can be given considering the effective maintenance of other operational parameters such as pH, temperature, nutrients, oxidation reduction potential (ORP), and hydraulic residence time (Begum et al. 2018, Jayanth et al. 2021). The OLR can be considered in any of these terms such as the COD, biochemical oxygen demand (BOD), VS, etc., to evaluate the biodegradation kinetics in the aerobic or anaerobic system (Pommier et al. 2007). Availability of micro (Iron, Cobalt, Nickel, Molybdenum) and macro nutrients (Carbon, Nitrogen, Phosphorous and Sulfur) is also one of the critical parameters to be considered to ensure efficient growth of microorganisms. Micro and macronutrients are essential elements in the biological pathways and most importantly in anaerobic digestion process. In the anaerobic digestion process, the methanogenic pathway that involves the activity of both hydrogenotrophic and acetoclastic methanogens for the conversion of acetate to CH_4 and CO_2 requires macro and micro nutrients in suitable quantities to activate the co-enzymes (methyl co-enzyme) that are in turn stimulated by the co-factors (F420). However, extensive research is in progress to optimize the required dosage of nutrient and the type and form of nutrient that is to be added depending on the characteristics of the waste in the landfill (Juntupally et al. 2022).

2.2.2 Environment

Environment is yet another important factor in addition to the availability of substrate that plays a key role in the aerobic or anaerobic degradation of the organic material. Temperature, ORP and pH are the crucial environmental parameters that play a key in the degradation of organic material by stimulating the growth of microorganisms and the rate of biochemical enzymatic pathway. These two parameters affect the microorganisms by stimulating their genomics and shapes the microbial community composition, activity, diversity and alters the overall biochemical pathway. The full range of ORP typically is between 1500 mV to $(-)$1500 mV. As the ORP of the system varies, the biochemical reaction also shifts from one to another. For example, during the aerobic degradation, the ORP of the system moves to positive side of the electrode as the organic material gets oxidized to CO_2, and H_2O whereas in the anaerobic environment, the ORP shifts from positive side of the electrode to negative as the anaerobic degradation is favoured in reduced environments. Meaning, the organic matter is reduced by the microorganisms to CH_4, CO_2 and H_2O. Degradation of organic material can occur at a wide range of temperature from 20°C to as high as 90°C. Depending on the type of biochemical reactions that occur in the waste landfill, the range of temperature is classified into four categories namely, psychrophilic temperature between 15 and 25°C, mesophilic temperature between 30 and 40°C, thermophilic temperature between 50 and 60°C and hyper thermophilic temperature between 80 and 90°C. These temperatures have been classified based on the adaptability and metabolism of microorganisms. For instance, anaerobic degradation of organic material can occur at all these temperature ranges whereas the aerobic degradation significantly decreases at elevated temperature beyond mesophilic temperature as per the reported literature (Zavala et al. 2004). It is reported that the hydrolysis stage has the greatest impact due to temperature. As the temperature increases, the duration of the hydrolytic phase decreases to as short as 18–24 h from 30–36 h. However, in a huge landfill, the hydrolytic state lasts from a few days (Phase I) to months (Phase (II, III) whereas it takes almost a year for the landfill to reach Phase IV. Temperature played a key role in positively influencing the kinetics during the hydrolytic phase of biodegradation (Liwarska-Bizukojc and Ledakowicz 2001). Reports revealed that when the temperature is increased from 20°C to 42°C, the increase in the hydrolysis rate constant from 0.0414 ± 0.0012 and 0.103 ± 0.004 was also observed resulting in the decrease in overall residence time of the process. However, estimating the metabolic activities of microorganisms in the landfills particularly during anaerobic degradation along with the temperature gradient is challenging. In the MSW landfills, depending on the variations in the seasons such as monsoon, spring and rainy seasons, the degree of biodegradation varies and accordingly the growth of specific microbial species is facilitated in the landfill. The microorganisms require moisture to survive and disintegrate the organic matter either in aerobic or anaerobic environments. The degradation of organic matter accelerates when the waste contains notable bound moisture. In the landfills, 'recirculation of leachate' into the deeper layers of the waste is practised to ensure adequate moisture inside the heaps of waste for efficient degradation. In contrast, if the waste material does not contain bound moisture, the biodegradation of that waste is relatively slower compared to the waste that has excess bound moisture. For instance, in the anaerobic digesters, the waste material is diluted with water or wastewater to improve its moisture (60–88%) suitable for the type of anaerobic digestion (wet (10–12% solids)/semi-solid (15–25% solids)/solid state (> 25 %)) process (Begum et al. 2019).

2.2.3 Microorganisms

Microorganisms are the biological catalysts that drive the entire process of biodegradation. In the initial phases of biodegradation in the landfill, the OFMSW remains in aerobic degradation stage but as time progresses, the landfilled waste shifts from aerobic to anaerobic environment due to the predominance of hydrolytic, fermentative, acidogenic, acetogenic and methanogenic microorganisms. Carbon, hydrogen, nitrogen and sulfur are the important macronutrients that are required for the biodegradation of waste in the landfill. Accordingly, depending on the availability of the nutrient, the predominance of the microorganisms and the biochemical pathway shifts from one to another (Juntupally et al. 2017). Aerobic bacteria or the aerobes survive only in the presence of oxygen in the environment as they need molecular oxygen for their growth because aerobic respiration is used to derive energy through oxidative phosphorylation, krebs cycle and glycolysis. In contrast, the anaerobic microorganisms do not require any oxygen for their survival. The anaerobes through complex biochemical pathways derive energy through anaerobic digestion process favouring lactic acid and alcoholic fermentation. Nicotinamide adenine dinucleotide hydrogen acts as the electron carrier which traps the energy from the electron and converts them to "adenosine triphosphate" molecules. For example, Bacteroides, *Firmicutes* predominates in the upper layers of the landfill and facilitates aerobic degradation of the organic matter (Juntupally et al. 2022). In the deeper layers of the MSW landfills, a wide variety of microorganisms belonging to the phylum Proteobacteria, Eubacteria, Actinomycetes, Spirochetes, Methanogens in addition to Yeast, and Fungi can be found. As the biological pathway shifts from aerobic to anaerobic, the microbial population of anaerobes such as the hydrolytic, acetoclastic, hydrogenotrophic microorganisms increases which converts the soluble organic matter to bioenergy (Begum et al. 2018, Pommier et al. 2007, T.A.S et al. 2020).

The aerobic bacteria that are involved in the aerobic biodegradation process are the obligate aerobes, facultative aerobes, microaerophiles and the aerotolerant aerobes. Some of the examples of aerobic bacteria are *Pseudomonas aeruginosa, Nocardia, Mycobacterium tuberculosis, E. coli, Salmonella, Achromobacter, Klebsiella, Citrobacter*, etc. (Online 2021). The anaerobic microorganisms that are involved in biodegradation process are Facultative anaerobes, Obligate anaerobes and Aerotolerant organisms. Some of the examples of anaerobic microorganisms are *Peptostreptococcus, Clostridium, Actinomyces, Propionibacterium*, etc. (Anaerobe 2021).

2.3 Biodegradation kinetics using mathematical model equations

The biodegradation of OFMSW is based on the biochemical reactions taking place in the landfill wherein the transformation of one molecule to another takes place in the cell/by the cell. A biochemical reaction is a conversion of one substance to another mediated by the microorganisms that are regarded as the biological catalysts that can alter the rate and specificity of chemical reactions. Therefore, any chemical reaction mediated by the microbes is called a biochemical reaction. The aerobic and anaerobic biodegradation of OFMSW that occurs in the MSW landfill can be evaluated for its kinetic parameters using various mathematical model equations. The widely used mathematical correlations are Stover Kincannon kinetics, Monod kinetics, and Michaelis Menten kinetics, first order kinetics, Gompertz model equation, Logistic and the Transfer function etc., which are suitable for the determination of growth rate of microorganism, substrate concentration, production formation rate, and lag phase of the process. The correlations are particularly used in anaerobic degradation kinetics in the anaerobic digesters where the same process of transition, acid formation and methane formation take place whereas the earlier mentioned equations are common to both aerobic and anaerobic degradation kinetics. Each of the mathematical model equations and their applicability are described in detail in the below subsections.

2.3.1 Stover Kincannon kinetics

The Stover Kincannon kinetic model is used to determine the substrate utilization rate of the microorganisms in a biochemical pathway. It considers the substrate removal rate as a function of organic loading rate under steady state conditions. The derivative equation from which the final Stover kincannon kinetic model equation is developed is shown in Equations (Eq. 12, Eq. 13, Eq. 14). The final equation is shown in Eq. 15 using which the substrate utilization rate by the microorganisms can be calculated. This model equation is particularly useful in the continuous operating systems in steady state. For example, in an anaerobic bioreactor system operated in a continuous mode, the substrate at a defined organic loading rate is considered as input and the ability of the microorganisms to degrade the substrate at a given loading rate can be determined as the maximum substrate that can be utilized by the microorganisms in the reactor (T.A.S et al. 2020).

$$\frac{ds}{dt} = \frac{Q}{V}(S_i - S_e) \qquad\qquad\text{Eq. 12}$$

$$\frac{ds}{dt} = \frac{U_{max}\left(\dfrac{QS_i}{V}\right)}{K_B + \left(\dfrac{QS_i}{V}\right)} \qquad\qquad\text{Eq. 13}$$

$$\frac{V}{Q(S_i - S_e)} = \frac{K_B}{U_{max}} \times \frac{V}{QS_i} + \frac{1}{U_{max}} \qquad\qquad\text{Eq. 14}$$

$$S_e = S_i - \frac{U_{max}S_i}{K_B + \left(\dfrac{QS_i}{V}\right)} \qquad\qquad\text{Eq. 15}$$

'S_e' is the concentration of organic matter in the substrate (g/L); 'S_i' is the concentration of organic matter in the substrate (g/L); 'U_{max}' is the substrate utilization rate constant (g COD/(L.d)); 'Q' is the flow rate (L/day), 'K_B' is the Saturation constant (g/L.day) and 'V' is the effective volume of the reactor (L). The concentration of organic matter in the substrate can be considered in any form such as COD, BOD, VS, etc. In general, for liquid waste, the organic matter is calculated in terms of COD, whereas in solids, it is VS. The Stover Kincannon kinetic equation has been considered as a useful mathematical model in aerobic and anaerobic treatment processes to determine the growth rate of the biomass against the substrate utilized (Kapdan 2005). Some of the reported literature is shown in Table 3 wherein the organic matter in the waste is considered in terms of COD and VS, respectively. One of the interesting observations that can be made from Table 3 is that the saturation constant increased as the substrate utilization by the microorganisms increased.

Table 3. Variation in kinetic parameters determined as per the reported literature.

Substrate	S_i (g /L)	S_e (g /L)	U_{max} (g COD/L. day)	K_B (g/L.day)	Ref.
Synthetic waste water	0.75–2.25	0.04–0.23	101	106.8	(Borghei et al. 2008)
Domestic waste water	0.28 ± 0.03	0.065 ± 0.003	25.9	24.75	(Nga et al. 2020)
OFMSW (VS)	6.2	3.35	5.24	12.5	(T.A.S et al. 2020)
Landfill leachate (VS)	6.2	3.6	8.69	14.93	
OFMSW + Landfill leachate (VS)	6.2	2.24	11.1	20.11	

2.3.2 Monod Kinetics

The Monod kinetic model equation is useful in determining the growth of microorganisms in a biochemical process based on a single growth controlling substrate. Jacques Monod, a French chemist, proposed this equation to relate the microbial growth rate to nutrient concentration in the aqueous medium. The Monod's kinetic model equation is shown in Eq. 16.

$$\mu = \mu_{max} \left[\frac{S}{S + K_s} \right] \qquad \text{Eq. 16}$$

where, 'μ' is the specific growth rate of the microorganisms (h^{-1}), 'μ_{max}' is the maximum specific growth rate (h^{-1}) that can be achieved, K_s is the half saturation constant (g/L) and S is the concentration of substrate (g/L). These parameters differ between the biochemical reactions that are performed by different microorganisms and the environmental conditions such as temperature, pH and the composition of the waste.

2.3.3 Michaelis Menten kinetics

Michaelis Menten kinetic equation is one of the widely used model equation to determine the kinetic parameters of a biochemical reaction where the microorganisms play a key role in the conversion of reactants to products. In simple words, it is the best-known equation for enzyme kinetics study. This equation has been named after Leonar Michaelis, a German chemist and Maud Menten, a Canadian physician who has defined that this equation is useful in determining the rate of enzyme reaction and the product formed at a given substrate concentration (Schmidt et al. 1985, Zeng and Yang 2020). The Michaelis Menten model equation is shown in Eq. 17.

According to this equation, as the rate of enzyme reaction increases with the input concentration of the substrate, the formation of the product also increases. But at a certain point of the reaction, increase in the substrate concentration doesn't increase the rate of the reaction.

$$r = \frac{r_{max}[S]}{[S]+K_m}$$

Eq. 17

where S is the substrate concentration (g/L); K_m is the Michaelis-Menten constant; r is the rate of product formation (g/L) and r_{max} is the maximum rate of product formation (g/(L.d)).

2.3.4 First order kinetics

The first order kinetics equation shown in Eq. 18 is generally used to determine the maximum methane generation potential per unit of substrate decayed, and specific growth of biomass per unit of substrate utilized in biochemical reactions, particularly in anaerobic degradation process. The first order kinetic equation can be used to calculate the time required to solubilize the complex organic matter to simple monomeric sugars that can be in the form of soluble COD, glucose or fructose, volatile fatty acids (VFA), etc. (Bekins et al. 1998, Reddy et al. 2022, Schmidt et al. 1985).

$$P = S \times (1 - e^{-kt})$$

Eq. 18

Equation 19 shows the first order kinetic equation where 'P' is the product formed at a given time 't' (g/(L.d)), 'S' is the concentration of the substrate (g/(L)), e is the mathematical constant (2.718), k is the first-order decay rate (1/d).

2.3.5 Gompertz model

Most of the modelling equations which describes a sigmoidal growth curve contain mathematical parameters such as a, b, c, etc., instead of parameters S, R_m, λ with a biological meaning. All the growth models containing mathematical parameters were rewritten and substituted to obtain a modified expression of the biological parameters. Equation 19 and Eq. 26 shows the Gompertz and the modified Gompertz model equations, respectively (Reddy et al. 2022).

$$P(t) = s.e^{-e^{(b-ct)}}$$

Eq. 19

where 's' is an asymptote, 'b' shows the displacement along x-axis, 'c' sets the growth rate on y-axis.

In order to obtain the inflection point of the growth curve, the 2nd derivative of the Gompertz model with respect to 't' is to be calculated as shown from Eq. 21 to Eq. 27.

$$\frac{d^2p}{dt^2} = sc^2 e[-e(b-ct)].\ e(b-ct)\ [\exp(b-ct)-1]$$

Eq. 20

At the inflection point, where $t = t_i$

$$\frac{d^2p}{dt^2} = 0 \rightarrow t_i = \frac{b}{c}$$

Eq. 21

The maximum specific growth rate can be derived by the first derivative at the inflection point

$$R_m = \frac{sc}{e}$$

Eq. 22

The parameter 'c' in the Gompertz model equation can be substituted by the above expression

$$c = \frac{R_m e}{s}$$ Eq. 23

The lag phase, λ can be defined as the tangent line through the inflection point which intercepts the x-axis

$$\lambda = \frac{(b-1)}{c}$$ Eq. 24

The parameter 'b' in the gompertz model equation can be substituted by the above expression

$$b = \frac{R_m e}{s}\lambda + 1$$ Eq. 25

The parameter 's' in the gompertz equation, substituted as "S", yields the modified gompertz equation

$$P = S \times e\left\{-e\left[\frac{R_m e}{s}(\lambda - t) + 1\right]\right\}$$ Eq. 26

Both the gompertz model and the modified gompertz model expressions were developed in order to describe the biological and bacterial growth in preference to their formation of the product, for, e.g., in case of methane production from wastes (solid and liquid), there are two situations having different initial conditions, i.e., the initial microbial mass is not zero due to the additional seeding of anaerobic activated sludge to the reactor whereas the product volume is zero. Hence, the gompertz and modified gompertz models can be used only with the error correction to simulate the product formation and microbial growth.

2.3.6 Logistic model

The logistic function shown in Eq. 27 is simply designed to avoid unidentifiable parameters. The curve itself explains the sigmoidal growth of microbes in its environment, which increases exponentially at an initial stage and finally reaches stabilization state when maximum product is formed. It predicts the maximum production rate and maximum capacity of product formed (Reddy et al. 2022).

$$P = \frac{S}{\left\{1 + e\left[4 \times \left(\frac{R_m}{S}\right) \times (\lambda - t) + 2\right]\right\}}$$ Eq. 27

2.3.7 Transference function model

The transfer function model shown in Eq. 28, which is mainly used for control purposes and considers that any systematic process can be analyzed by acquiring inputs and generating outputs, was evaluated (Reddy et al. 2022).

$$P = S\left\{1 - exp\left[\frac{-R_m}{S} \cdot (t - \lambda)\right]\right\}$$ Eq. 28

In all the above equations, Eq. 19, Eq. 20, Eq. 21, Eq. 27 and Eq. 28, 'P' is the maximum product formed (g/L) with respect to time t (days); 'S' is the product yield of substrate concentration (g/L); 'R_m' is the maximum specific growth rate (g/L.d); 'λ' is the lag phase of the process (d); 'e' is Euler's function equal to 2.7183.

The modified gompertz model, logistic model and the transfer function model equations shown in Eq. 19, Eq. 27 and Eq. 28 have wide range of applications and are extensively used to estimate the lag phase of the anaerobic digestion process and to estimate the maximum rate of product (methane) formed. Among the three widely accepted mathematical models to estimate the product formed from a given substrate. One of the assumptions that can be made in these equations is that the rate of product (biogas/methane) production is proportional to the microbial activity; however, the proportionality parameter decreases with time.

3. CONCLUSIONS

Kinetic study of any process is helpful in predicting the rate of the reaction under certain environmental conditions. The aerobic and anaerobic biodegradation of MSW in the landfills occurs in five phases resulting in different products such as water, carbon dioxide, biomass, methane and other trace gases. Generation and liberation of landfill gas into the atmosphere and landfill leachate on to the land is the result of all the biodegradation phases that occur in the MSW landfills. Controlled or sanitary landfills gives us an opportunity to estimate the products formed from a given amount of input or substrate unlike the open dumps or uncontrolled landfills. The action of different group of microorganisms that shift the chemical reaction from one phase to another in the MSW landfills can be determined using certain kinetic equations such as Monod Kinetic equation, Michaelis Menten equation, Stover Kincannon kinetics, Gompertz kinetics, first order kinetics, Gompertz model equation, logistic and transfer function models, which consider that the rate of product formation is directly proportional to the rate of substrate utilization. These equations allow us to get the information on the mechanism and the important kinetic parameters of the biochemical reactions that occur in the landfills under suitable environmental conditions.

ACKNOWLEDGEMENT

The authors are grateful to the Director-CSIR-Indian Institute of Chemical Technology (IICT) for his encouragement (IICT/Pubs./2022/085) in carrying out this work. The authors are thankful to Council of Scientific and Industrial Research (CSIR) for providing research.

REFERENCES

Abdel-Shafy, H.I. and Mansour, M.S.M. 2018. Solid waste issue: Sources, composition, disposal, recycling, and valorization. Egypt. J. Pet. 27: 1275–1290.

Anaerobe. 2021. Anaerobes [WWW Document]. URL https://www.microscopemaster.com/anaerobes.html.

Begum, S., Ahuja, S., Rao, A. et al. 2017. Significance of decentralized biomethanation systems in the framework of municipal solid waste treatment in India. Curr. Biochem. Eng. 4: 2–8. https://doi.org/10.2174/221271190299915100 1134836.

Begum, S., Anupoju, G.R., Sridhar, S. et al. 2018. Evaluation of single and two stage anaerobic digestion of landfill leachate: Effect of pH and initial organic loading rate on volatile fatty acid (VFA) and biogas production. Bioresour. Technol. 251: 364–373.

Begum, S., Anupoju, G.R. and Eshtiaghi, N. 2021. Anaerobic co-digestion of food waste and cardboard in different mixing ratios: Impact of ultrasound pre-treatment on soluble organic matter and biogas generation potential at varying food to inoculum ratios. Biochem. Eng. J. 166: 107853. https://doi.org/10.1016/j.bej.2020.107853.

Begum, S., Arelli, V., Anupoju, G.R. et al. 2020a. Optimization of feed and extractant concentration for the liquid–liquid extraction of volatile fatty acids from synthetic solution and landfill leachate. J. Ind. Eng. Chem. 90: 190–202.

Begum, S., Golluri, K., Anupoju, G.R. et al. 2016. Cooked and uncooked food waste: A viable feedstock for generation of value added products through biorefinery approach. Chem. Eng. Res. Des. 107: 43–51. https://doi.org/10.1016/j.cherd.2015.10.032.

Begum, S., Juntupally, S., Anupoju, G.R. et al. 2020b. Comparison of mesophilic and thermophilic methane production potential of acids rich and high-strength landfill leachate at different initial organic loadings and food to inoculum ratios. Sci. Total Environ. 715: 136658. https://doi.org/10.1016/j.scitotenv.2020.136658.

Begum, S., Rao, A.G., Bhargava, S.K. et al. 2019. Waste-to-Energy Production through a Biorefinery System. WASTE-TO-ENERGY (WTE) 85.

Bekins, B.A., Warren, E. and Godsy, E.M. 1998. A comparison of zero-order, first-order, and monod biotransformation models. Groundwater 36: 261–268.

Borghei, S.M., Sharbatmaleki, M., Pourrezaie, P. et al. 2008. Kinetics of organic removal in fixed-bed aerobic biological reactor. Bioresour. Technol. 99: 1118–1124.

Gharasoo, M., Centler, F., Van Cappellen, P. et al. 2015. Kinetics of substrate biodegradation under the cumulative effects of bioavailability and self-inhibition. Environ. Sci. Technol. 49: 5529–5537.

Ghosh, S.K. 2020. Sustainable Waste Management: Policies and Case Studies: 7th IconSWM–ISWMAW 2017, Volume 1. Springer.

HK-MEFCC. 2016. Solid Waste Management Rules Revised After 16 Years; Rules Now Extend to Urban and Industrial Areas.

IBEF. 2021. 70 Percent Municipal Solid Waste (MSW) processing achieved under Swachh Bharat Mission – Urban.

Jadhav, R. 2018. 75% of Municipal Garbage in India Dumped without Processing.

Juntupally, S., Begum, S., Allu, S.K. et al. 2017. Relative evaluation of micronutrients (MN) and its respective nanoparticles (NPs) as additives for the enhanced methane generation. Bioresour. Technol. 238: 290–295. https://doi.org/10.1016/j.biortech.2017.04.049.

Juntupally, S., Begum, S., Arelli, V. et al. 2022. Evaluating the impact of Iron Oxide nanoparticles (IO-NPs) and IO-NPs doped granular activated carbon on the anaerobic digestion of food waste at mesophilic and thermophilic temperature. J. Environ. Chem. Eng. 10: 107388. https://doi.org/10.1016/j.jece.2022.107388.

Kapdan, I.K. 2005. Kinetic analysis of dyestuff and COD removal from synthetic wastewater in an anaerobic packed column reactor. Process Biochem. 40: 2545–2550.

Kaza, S., Yao, L., Bhada-Tata, P. et al. 2018. What a waste 2.0: A global snapshot of solid waste management to 2050. World Bank Publications.

Kumar, A. and Agrawal, A. 2020. Recent trends in solid waste management status, challenges, and potential for the future Indian cities—A review. Curr. Res. Environ. Sustain. 2: 100011. https://doi.org/10.1016/j.crsust.2020.100011.

Kuruti, K., Nakkasunchi, S., Begum, S. et al. 2017. Rapid generation of volatile fatty acids (VFA) through anaerobic acidification of livestock organic waste at low hydraulic residence time (HRT). Bioresour. Technol. 238: 188–193. https://doi.org/10.1016/j.biortech.2017.04.005.

Liwarska-Bizukojc, E. and Ledakowicz, S. 2001. RNA assay as a method of viable biomass determination in the organic fraction of municipal solid waste suspension. Biotechnol. Lett. 23: 1057–1060.

Meyer-Dombard, D.R., Bogner, J.E. and Malas, J. 2020. A review of landfill microbiology and ecology: A call for modernization with 'next generation' technology. Front. Microbiol. 11: 1127.

MNRE. 2018. Government of India, Ministry of New and Renewable Energy.

Nga, D.T., Hiep, N.T. and Hung, N.T.Q. 2020. Kinetic modeling of organic and nitrogen removal from domestic wastewater in a down-flow hanging sponge bioreactor. Environ. Eng. Res. 25: 243–250.

Online, B. 2021. Aerobic bacteria [WWW Document]. URL https://www.biologyonline.com/dictionary/aerobic-bacteria.

Pommier, S., Chenu, D., Quintard, M. et al. 2007. A logistic model for the prediction of the influence of water on the solid waste methanization in landfills. Biotechnol. Bioeng. 97: 473–482.

Reddy, A., Begum, S., Juntupally, S. et al. 2022. Silica extraction followed by biogas generation from rice straw: Investigating the impact of pretreatment on purity of silica, biogas yield and microbial diversity along with insights on techno-economic analysis. J. Environ. Chem. Eng. 108274.

Schmidt, S.K., Simkins, S. and Alexander, M. 1985. Models for the kinetics of biodegradation of organic compounds not supporting growth. Appl. Environ. Microbiol. 50: 323–331.

Statista. 2016. Distribution of Municipal Solid Waste Treatment and Disposal Worldwide.

T.A.S, J., Mamindlapelli, N.K., Begum, S. et al. 2020. Anaerobic mono and co-digestion of organic fraction of municipal solid waste and landfill leachate at industrial scale: Impact of volatile organic loading rate on reaction kinetics, biogas yield and microbial diversity. Sci. Total Environ. 748: 142462. https://doi.org/10.1016/j.scitotenv.2020.142462.

Tamru, A.T. and Chakma, S. 2015. Effects of landfilled MSW stabilization stages on composition of landfill leachate: A review. Age 1: 0–5.

Zavala, M.A.L., Funamizu, N. and Takakuwa, T. 2004. Temperature effect on aerobic biodegradation of feces using sawdust as a matrix. Water Res. 38: 2406–2416.

Zeng, H. and Yang, A. 2020. Bridging substrate intake kinetics and bacterial growth phenotypes with flux balance analysis incorporating proteome allocation. Sci. Rep. 10: 1–10.

4

BIOGAS GENERATION FROM SOLID STATE ANAEROBIC DIGESTION OF WASTE MATERIAL

Sadaf Shakeel,[1] Rajkumar Joshi[2] and Mohammad Zain Khan[1,]*

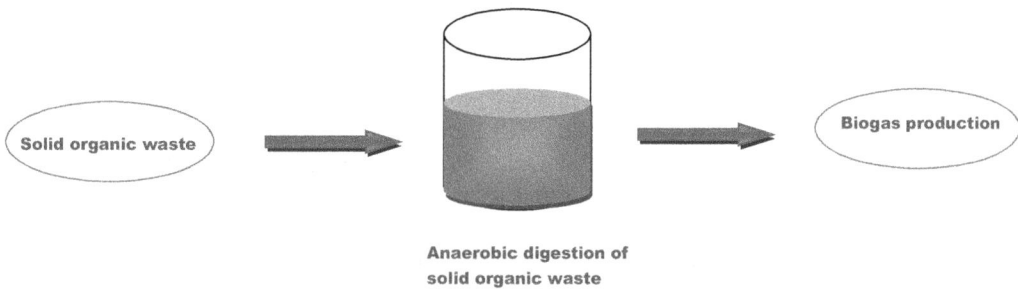

Anaerobic digestion of
solid organic waste

1. INTRODUCTION

World is looking for alternative energy sources due to the rising prices of fossil fuels, limited reserves and increasing climate change issues (Zamri et al. 2021). Renewable energy is the best solution to all such problems (Zamri et al. 2021). It can reduce the harmful impact of global warming caused by green house gas emissions (Khan et al. 2016). Biogas production is a promising alternative source of energy that has been given tremendous attention in the past years (Paritosh et al. 2021a). Global production of biogas is rising at a tremendous rate with Europe leading the production, accountable for 50% global production (Harsha and Singh 2021). As per the World Bioenergy Association report, a total of 1.33 EJ biogas was produced in 2017, and is expected to increase in coming years (Paritosh et al. 2021b). Other nations are also focusing on this platform to achieve future energy needs. It is highly advantageous since its environmentally friendly, checks green house gas emissions and also

---

[1] Industrial Chemistry Research Laboratory, Department of Chemistry, Faculty of Sciences, Aligarh Muslim University, Aligarh 202002, India.
[2] Department of Science and Technology New Delhi, India.
* Corresponding author: zn.khan1@gmail.com

produces fertilizers of high quality (Liu et al. 2021). It has been mainly produced from anaerobic digestion (AD) conditions. It is composed of carbon dioxide (CO_2) (20%–30% v/v), methane (CH_4) (60%–70% v/v), hydrogen sulfide (H_2S) (0.1%–3% v/v), moisture and other trace contaminants (Paritosh et al. 2021a). It is mainly produced from waste; therefore, countries like India which hold surplus amount of waste can easily adopt this technique for waste handling and energy demands. Major cause of large-scale environmental pollution is inappropriate disposal and decomposition of biodegradable waste generated from various sources like agricultural, municipal sources and industrial sources (Khan et al. 2016). Solid waste production is tremendously rising with the world population and economic growth. Municipal solid wastes (MSW) and biodegradable agricultural residues constitute a large part of solid waste production (Zhou and Wen 2019). Utilization of these solid wastes in AD will provide renewable energy and solve waste disposal problem simultaneously. Solid organic waste includes food waste (FW), waste activated sludge (WAS), yard waste (YW), agricultural waste (AW) and animal manure (AM) (Zhang et al. 2019). This solid organic waste carries huge amount of renewable energy which could be utilized via AD technology (Zhang et al. 2016). AD is a platform where such organic waste is converted to renewable energy. It was first observed 3000 years ago by a group of ancient people (Assyrian Empire) for water bath heating (Harsha and Singh 2021). After that, people made matured fertilizer via pile digestion of cattle manure which resulted in the formation of gases during digging process (Abbasi et al. 2012). In 1776, an Italian scientist Volta figured out that flammable gas generated during this process and organic matter degraded are related to each other (McCarty 2001). Methane presence was recognized by Sir Humphry Davy (Harsha and Singh 2021). Pasteur, a scholar, realized that release of biogas generated is related to the microorganism involved in the process (Abbasi et al. 2012). Leper colony of economical capital of India, Bombay, in 1859 started the modern form of biogas plant which was mostly applied to waste treatment, agriculture and sewage water management (Harsha and Singh 2021). Later technique evolved as one of the important renewable energy technology for producing carbon neutral fuels. Europe is leading in this technology with 17,783 biogas plants operating there to generate 10,532 MW of electricity (Harsha and Singh 2021). India has installed 49.57 lakhs household type biogas plants in 2018 (CSO, Energy statistics 2019). China is generating 9000 million m^3 biomethane from 105 modern biogas plants and 43 million residential level plants (Harsha and Singh 2021).

AD is a biochemical process where organic materials undergo degradation in anaerobic conditions via microorganisms to produce biogas (Li et al. 2011). The AD process involves four steps: hydrolysis, acidogenesis, acetogenesis, and methanogenesis (Zhang et al. 2019). During hydrolysis, fermentative bacteria convert high molecular weight substrates to low molecular weight water soluble substrate like amino acids, volatile fatty acids and glucose (Fig. 1). This step is the slowest step of whole process and rate determining step also (Zhou and Wen 2019). In acidogenesis step, VFAs production takes place along with H_2S, NH_3, and CO_2 as by-products. VFAs are converted to acetate in acetogenesis. This acetate, hydrogen and carbon dioxide is finally converted to methane in the final step of methanogenesis catalysed by methanogens (Zhou and Wen 2019). The biogas generated carries 17–25 MJ m^{-3} calorific value (Zhang et al. 2019). This energy can be used as electrical/ heat energy after burning (Zhang et al. 2019). Major part of this energy could be sold to the local grid while some part can be sued for running biogas plants (Axelsson et al. 2012). Digestate left behind can be processed further to biochar or can be recycled back as fertilizer since it is rich in phosphorus and nitrogen nutrients (Inyang et al. 2010). Methane productivity is very high from AD; therefore, it seems best suited for the same. Microbial catalysts play a very important role in whole AD process. These microbes are quite sensitive towards change in the temperature variations and pH. Therefore, optimization of process is highly essential for enhancing biogas productions (Zhou and Wen 2019). Other factors like biomass composition, digester designs, pretreatment and addition of enzymes, and microbes are also very important in determining the AD performance (Zhou and Wen

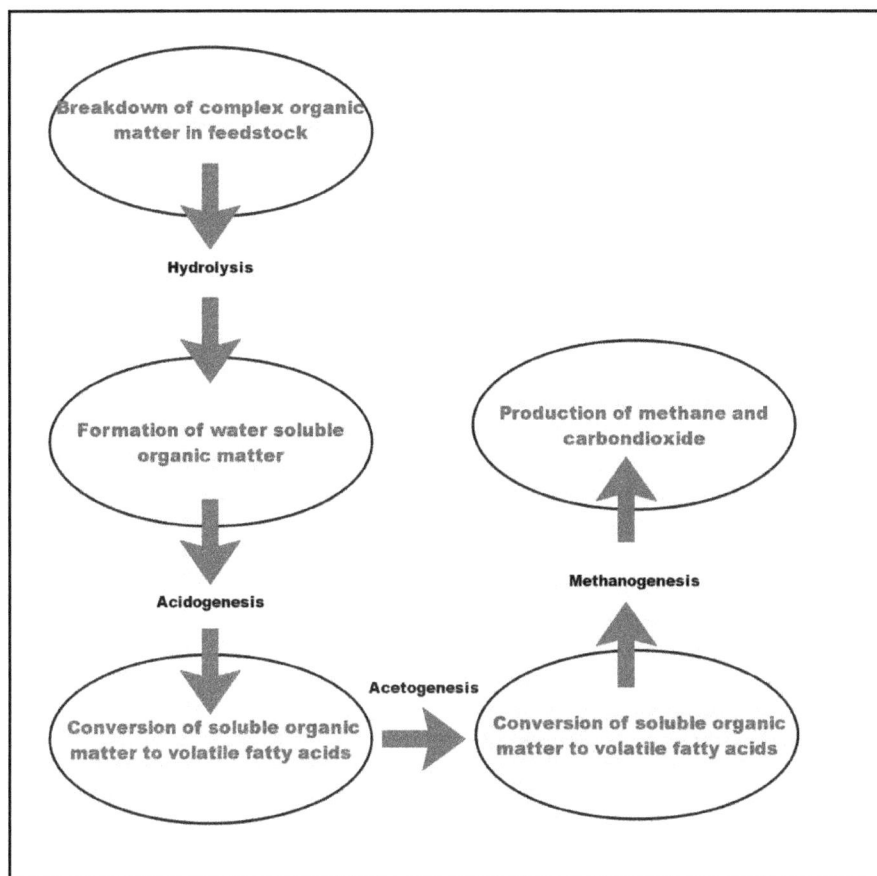

Figure 1. Schematic of solid state anaerobic digestion of organic waste material.

2019). AD process can be classified into solid state anaerobic digestion (SS-AD) (TS content is > 15%) and liquid state anaerobic digestion (LSAD) (TS content < 15%). SS-AD has advantages over LSAD since it requires lower energy for stirring or heating; it can tolerate high organic loading, less material handling, small volume is required, and total parasitic energy loss is very low (Zhou and Wen 2019, Paritosh et al. 2021b, Jenkins et al. 2008). Biogas obtained from SS-AD is comparable to the output of LS-AD (Saady and Massé 2015). Digestate of SS-AD contains low moisture content and handling of this digestate is much easier than the LS-AD effluent. Digestate of SS-AD can be used as pelletized fuel or fertilizer (Li et al. 2011). Europe's 54% of the total installed AD plants use SS-AD technology (De Baere 2008). North America and Canada also use SS-AD technology (De Baere 2008). Other countries are also looking forward to this technology due to its environmental and economical benefits over landfill, incineration and composting which are used for waste disposal methods (Jenkins et al. 2008). Number of factors affect SS-AD performance like temperature, microbial community, inoculation, pre-treatment, reactor types, reactor design and co-digestion (Li et al. 2011). SS-AD organic loading requires a very low water demand; therefore, higher volumetric production of methane is obtained from it (Zhou and Wen 2019). Wastewater generated is also very low during the SS-AD process (Zhou and Wen 2019). LS-AD treats substrates that are high in moisture like sewage and animal manure and also requires a large amount of water during the process which causes methane generation from it (Zhou and Wen 2019). Disposing of waste digestate is also a big problem associated with LS-AD (Zhou and Wen 2019). Therefore, more research interests developed in SS-AD in past decades as more advantages are associated with it than LS-AD.

2. FUNDAMENTAL ASPECTS OF SS-AD AND ITS PROCESS OPERATIONS

2.1 Inoculums

Inoculums constitute nutrients, microbes and water required in SS-AD reactors. Main inoculums include ruminant cultures, digested manure and sewage sludge (Karthikeyan and Visvanathan 2013). Microbes play the most vital role in each state of SS-AD; they are mainly categorized as hydrolytic, fermentative, acetogenic, and methanogenic. First hydrolytic bacteria works on complex particulate compounds and convert it to water soluble dimeric or monomeric substrates (Li et al. 2011). Here most of organic material is converted to VFAs via fermentation which later converted into biogas; therefore, hydrolysis is a rate determining step of SS-AD and also determines the biomass conversion efficiency (Li et al. 2011). Some solid wastes, mainly municipal and agricultural wastes, contain cellulose which is an insoluble compound. This cellulose undergoes enzymatic hydrolysis via cellulolytic bacteria such as Acetovibrio Streptomyces Erwinia, Clostridium, Ruthese into minococcus, Thermomonospora Cellulomonas Microbispora, Baceriodes, and Bacillus (Lo et al. 2009). Fermentative bacteria consume these water soluble compounds generated during hydrolysis and convert into VFAs, hydrogen gas, carbon dioxide, and alcohols (Li et al. 2011). Acetogenic bacteria consume CO_2 and reduce it to acetate via Wood-Ljungdahl pathway. These bacteria include acetogenic: Sporomusa, Acetobacterium and acetogenic, and nonacetogenic: Clostridium, Eubacterium, Ruminococcus (Schmidt et al. 2009). Methanogens utilize carbon dioxide, hydrogen and acetate in methane production. Methanogens and acetogens follow a symbiotic relation with each other, where acetogens cannot tolerate high hydrogen pressures and methanogens act as hydrogen consumer (Gerardi 2003). Inoculums rich in methanogens are essential for SS-AD since most of the feedstock lack methanogens. Microbial community of inoculums needs to be characterized for better understanding the role of each microbial community and its functioning. Shi et al. (2014) studied the microbial community of corn stover in SS-AD and reported the enriched communities of archae and bacteria (Shi et al. 2014). Yan et al. (2015) have studied the rice straw SS-AD and reported the dominance of Bacteroidia, Betaproteobacteria, Clostridia, Gammaproteobacteria and Methanobacteria (Yan et al. 2015). Psychrobacter was found in the mixed kitchen waste, sludge and pig manure (Li et al. 2013). Hydrolysis of carbohydrate into biomic acids and lactate has shown the dominance of Lactobacillaceae and Pseudomonadaceae species (Zhou and Wen 2019). Porphyromonadaceae and Enterobacteriaceae were dominant in acidogenesis stage and methanogens species were found to be dominant in methanogenic stage (Zhang et al. 2017). Syntrophic relation encourages the methanogens to consume products obtained from acidogenesis like acetate, butyrate, ethanol and other intermediates and produce methane. Therefore, improvement in the syntrophic relation is required for enhancing SS-AD process. Carbon-based conducting materials were found to improve this syntrophic relation by enhancing the interspecies electron transfer. Methanogens use IIET pathway via consumption of H_2 and HCOOH for producing methane in the presence of a favorable hydrogen partial pressure (Paritosh et al. 2021a). A high hydrogen partial pressure may affect the syntrophic relation negatively, which could lead to the accumulation of VFAs. DIET pathway involves the release of electrons by the microbes, which is later captured by electron seeking microorganisms. Conductive materials' addition helps in enhancing the DIET (Paritosh et al. 2021b).

2.2 Feedstocks

Different types of organic solids are used as SS-AD feedstock (Table 1). The characteristics and composition of organic wastes influence greatly the performance of SS-AD like its startup, biogas yield, retention time, volatile solids and conversion ratio (Li et al. 2011, Paritosh et al. 2021a). High

Table 1. Different feed stocks used in solid state anaerobic digestion of organic waste material.

S. no.	Feedstock	TS (%)	T (°C)	CH$_4$ yield	References
1.	Rice straw	20	37	263 L/kg VS (120)	Mustafa et al. (2016)
2.	Palm fruit bunches	20	40	73.3 m^3/tonne (135)	Mamimin et al. (2021)
3.	Rice husk	21		18 L/kg TS (55)	Nugraha and Matin (2017)
4.	Rice straw	21	37	190 L/kg VS (26)	Momayez et al. (2018)
5.	Rose stalk	12.1	55	117 L/kg VS (144)	Liang et al. (2016)
6.	Rice straw	20	35	240 L/kg VS (20.2)	Wang et al. (2017)
7.	Sugarcane bagasse	15	35	143 L/kg VS (67)	Lima et al. (2018)
8.	Rice husk	21		45 and 21 ml/g TS	Nugraha et al. (2018)
9.	Food waste and sewage sludge	> 15%	35	0.35 L CH4/(g VS reduced)	Arelli et al. (2021)
10.	Solid sago waste		30	27.91 mL/g TS	Sumardiono et al. (2021)
11.	Empty fruit bunches	11.5–79.0 g·L^{-1}	28–32	281 mL CH4·g^{-1} VS	Mamimin et al. (2021)
12.	Organic residues (cornstalk, cattle manure and tomato residues)		35	283.3 L/kg VS	Wang et al. (2021)

solid content wastes like industrial wastes, municipal solid wastes, and agricultural residues (wheat straw, corn stover, rice straw) are used as feed stocks in SS-AD process (Zhou and Wen 2019). The organic municipal solid wastes like yard and food waste also constitute an important part of feedstock for SS-AD and are used commonly in different parts of the world. 1.3 billion tons food waste was generated per year worldwide (Zhou and Wen 2019). Food wastes carries very high amount of water soluble organic compounds which can be easily gets converted to VFAs (Li et al. 2011). However, an excess of VFAs may lead to the drop in the pH and inhibit the process of methanogenesis (Zhou and Wen 2019).

To reduce the inhibition impact of VFAs, co-digestion is used where feedstocks rich in carbohydrates are co-digested with other feedstocks (Paritosh et al. 2021b). Lu et al. (2007) studied the co-digestion of woodchips and sludge with dog food at meso- and thermophilic conditions with 43% (TS) content (LU et al. 2007). Lissens et al. (2004) co-digested 10% green algae, 70% food waste, and 20% fecal matter and obtained 90% biogas yield (Lissens et al. 2004). Yard waste includes leaves and grasses, and constitutes a large part of municipal solid waste. Such type of wastes is affected by its collection methods. Collection strategy affects the biogas yield and disposal of left digestate (Bolzonella et al. 2006). Variations in the season and regional practices change the organic composition of the collected municipal solid wastes which greatly affects the biogas yield (Forster-Carneiro et al. 2007). Agricultural wastes are most promising feedstocks for SS-AD because of their low costs, high potential biogas yields and plentiful supply (Li et al. 2011). It mainly constitutes wheat straw, corn stover and rice straw. Carbohydrates' presence in the crop residues is not easily available for the fermentation process since it is protected by hemicellulose, cellulose, and lignin which hamper the degradation of carbohydrates (Li et al. 2011). Therefore, pretreatment of such lignocellulosic biomass is done using steam explosion, dilute acid, ammonia and lime (Lissens et al. 2004). Pretreatment removes the hemicellulose or lignin, increases the surface area and deceases the cellulose crystallinity (Mosier et al. 2005). Corn stover is the most widely used agricultural waste feedstock, reported to yield much higher biogas production in SS-AD than LS-AD (Zhou and Wen 2019). High sugar yields (90% glucose yield and almost 80% xylose yield) are found to be obtained from corn stover steam pretreatment using SO$_2$ acid catalyst at 190°

C for 5 min after enzymatic hydrolysis of 72 h (Öhgren et al. 2005). Similarly, sodium hydroxide treated rice straw and corn stover increased the biogas yield by 64.5% and 48.5%, respectively, compared to their untreated forms (He et al. 2008). Wheat straw also widely used agricultural waste feedstock and produced higher methane in SS-AD as compared to LS-AD and spent wheat straws were found to be more digestible than raw wheat straw since it contains rough fiber and serrations at the edge whereas raw wheat straw contains long and smooth intact fibers (Zhou and Wen 2019). Methane yields were increased when food wastes percentages were increased to 10% and 20% at 2 and 1 F/I ratios, respectively (Paritosh et al. 2021a). Food waste and distiller's grain co-digestion gave better synergistic effects (Paritosh et al. 2021a). Co-digestion of hay and soybean processing waste at F/I ratio of 3 gave higher methane yield than mono processing of both wastes separately (Zhu et al. 2014).

2.3 Batch vs. continuous operations

SS-AD uses two operation modes, namely, batch and continuous operations. Maintenance of batch operation is very easy since it requires very less operating costs, capital, and process control requirements (Zhou and Wen 2019). There are some limitations associated with batch SS-AD as well, like production of biogas varies with time and majority is obtained at a peak production time only. Large amount of inoculums' requirement is another limitation of SS-AD process (Zhou and Wen 2019). Sources of inoculums also affect the SS-AD process significantly. Continuous SS-AD produces methane at a consistent rate, which makes it better than batch SS-AD process (Zhou and Wen 2019). Continuous SS-AD involves the addition of waste feedstock to the anaerobic reactor regularly and removal of product in an equal amount. Continuous SS-AD designs involve plug-flow reactor in which complete mixing of digester contents is not done. Rather, digestate is passed through feed port and travels to the exit. This process includes heavy equipments which can handle materials which are dry and viscous in nature and cannot easily flow. These reactors are maintained at 20 % solids; if solid content falls lower than this, it may lead to the accumulation of sediments in the anaerobic reactor (Li et al. 2011). Solid retention time (SRT), Organic loading rate (OLR) and CH_4 production are three important factors that determine the interactions between the substrates and microorganisms (Zhou and Wen 2019). These parameters are used in the evaluation and designing of the performance of continuous SS-AD process. Conversion capacity of an AD system depends upon the OLR. Maximum level of OLR depends upon factors like as feedstock characteristics, microbial activity, reactor design, toxicity level, pH and temperature (Amani et al. 2010). Preference is usually given to a high OLR level for improving utilization efficiency and reducing digester size. But there is a drawback associated with high OLR too; it could lead to overproductions of VFAs which can disturb the balance between methanogens and acidogens (Zhou et al. 2017). Nguyen et al. (2016) reported the slowdown of the acclimatization of bacteria in the new environment and caused an increase in the adaptation time from 2 days to 31 days when they increased OLR from 2.3 kg VS/m³ to 9.2 kg VS/m³ (Nguyen et al. 2016). Solid retention time (SRT), which is the time up to which the organic compounds remain in the digester, is another important operational parameter that determines continuous SS-AD performance. Optimal SRT is needed during the continuous SS-AD process which depends upon OLR, feedstock and TS. Decrease in the SRT may lead to insufficient substrate utilization and washing out of microorganisms. On the other hand, a longer SRT would cost higher for its maintenance and will require larger reactor volumes as well, thus not economical (Zhou and Wen 2019). Methane productions also get influenced from SRT. An increase in SRT will elevate methane productions as well for green, fruit and vegetables containing organic waste (Zhou and Wen 2019). Continuous mode operations are not suitable for energy crops or lignocellulosic biomass (Li et al. 2011).

2.4 Single and multi-stage operations

Operations of SS-AD can take place in multiple stages or single stage. Single stage operation takes place in only one reactor vessel where organic substrates are converted to biogas in multiple steps (Zhou and Wen 2019), whereas in multistage operation, more than one reactor vessels are used and different conversion steps take place in different reactors. Diverse growth of different microbes takes place in different compartments of multi-stage digesters. Optimization and operation of different microbial communities occurs independently at each stage of multi stage digester which supports higher biogas production (Zhang et al. 2019). A two stage AD operation is the most common multistage operation where hydrolysis/acidogenesis reaction occurs in first reactor and methanogenesis occurs in second reactor. Single stage reactor consumes lesser cost in its designing and building. OLR is limited in single stage digester so that the overproductions of VFAs could be avoided and rapid pH drop as well (Jenkins et al. 2008). Two stage digester can hold each conversion step at their own optimum conditions. In terms of performance also, two stage digester is better than single stage digester. Panjičko et al. (2015) reported that single stage digester of brewery spent grain (BSG) was limited by inhibitors, like phenolic substances, furan derivatives, and weak acids which were generated during the lignocelluloses degradation in BSG (Panjičko et al. 2015). In case of multi stage digester, separation of reactors for hydrolysis and acidogenesis and methanogenesis improves the production of biogas and biodegradation of feedstock. There are three stage AD digesters, which are used in co-digestion of horse manure and food waste (Zhou and Wen 2019).

First-stage of three stage digesters consists of hydrolysis, second-stage consists of acidogenesis (operated in solid state) and third stage consists of methanogenesis (operated in liquid state). Such a hybrid system enhanced methane yield by 11.2%–22.7%, where methanogenic archaea were abundant compared to single stage reactor (Zhou and Wen 2019). Despite the improvements in the biogas production rates, multistage reactors are limited to be applied at a commercial level due to its operating costs and capital (Zhou and Wen 2019).

3. FACTORS INVOLVED IN THE ANAEROBIC DIGESTION PROCESS

3.1 Nutrients

Balanced nutrients are required for the growth of anaerobic microbes; these nutrients mainly constitute of nitrogen, carbon, minerals and phosphorous. Carbon is mostly obtained from organic materials as they are rich in it. Nitrogen and phosphorous are needed for protein and nucleic acid synthesis, respectively (Takashima et al. 1990). An optimum C/N ratio is needed for bacterial growth in AD-SS systems. Therefore, choice of feedstock should be made by carefully choosing its C/N ratio as it affects it digestion process. Previous studies reported 20:1 to 30:1 as operating C/N ratio range with 25:1 as optimum ratio for bacterial growth of anaerobic microbes in SS-AD systems (Liu et al. 2018). Improper C/N ratio could lead to the release of high VFA and total ammonia nitrogen (TAN) which will cause accumulation of VFA which will inhibit the AD process and biogas production (Parkin and Owen 1986). Optimum C/N ratio differs with the feedstock types. Different studies used different type of waste feedstock and reported different optimum C/N ratio. Too low C/N ratio leads to the insufficient utilization of carbon and elevates the chances of ammonium inhibition, whereas a very high C/N ratio leads to the less availability of nitrogen for bacterial growth (Zhou and Wen 2019). Iron, nickel, cobalt, and sulfur are trace elements essential for fermentation of methane. 15% phosphorus is required for bacterial growth (Zhou and Wen 2019). Iron is used as a supplementation for activation of enzymes like PEP carboxylase, ATPase, and serine transhydroxymethylase. Nickel is required in coenzymes like hydrogenase, and CO dehydrogenase in methanogenic microorganisms.

Cobalt is used in the CO dehydrogenase activity (CODH) and methyl transferase. Molybdenum is a part in molybdoprotein called CO_2 reductase which reduces CO_2 to formate and then to CH_4 (Zhou and Wen 2019). Jin et al. (2022) performed the enzymatic pre-treatment and SS-AD of wheat straw and found lignocellulose degradation and enhanced yield of biogas (77.59 L/kg volatile solid) at C/N ratio of 25.

3.2 Feedstock-to-inoculum ratio

Another very important factor is SS-AD, which is the ratio between feedstock and inoculum ratio (F/I). A very high F/I ratio leads to VFAs overproduction that will build up an acidic environment and will cause methanogens inhibitions. Few studies found that starting phase of AD is highly influenced by the F/I ratio (Paritosh et al. 2021a). Zhou and Wen (2019) reported methane yield from rice straw feedstock to be inversely proportional to ratio of feedstock and inoculum (F/I ratio) (Zhou and Wen 2019). Highest methane production rates were obtained at lowest F/I ratio ranging from 2:1–5:1 for palm oil mill residues (Zhou and Wen 2019). Corn stover gave the highest methane generation (81.2 L/kg VS) than wheat straw, leaves and yard waste at F/I ratio of 2 (Liew et al. 2012). Cui et al. (2011) used spent wheat straw and raw wheat straw at 2, 4 and 6 F/I ratios and found the maximum yield at F/I ratio 4 and 2. Modification of F/I ratio can increase the alkalinity of SS-AD system (Ward et al. 2008). Requirement of a low F/I ratio is one of the limitation in batch SS-AD. Capson-Tojo et al. (2017) has shown that a very low F/I ratio (lower than 0.25) can produce biogas from food waste and cardboard mixture in batch SS-AD. Similarly in a study where yard trimmings were used in SS-AD operation, lowest F/I ratio (0.2 to 2) gave highest methane production (Veeken et al. 2000). Kassongo et al. (2022) studied the SS-AD of lignocellulosic biomass and reported highest methane yield (6.45 L CH4kg-1VS) at 7/3 S/I (substrate to inoculum) ratio.

3.3 pH

pH also affects the digestion of a SS-AD system greatly. Changes in pH significantly affect the stability of the SS-AD process and methanogenic and acidogenic microbial activates. It also helps in regulation and dynamic detection of the AD process (Zhang et al. 2019). The ideal pH of an SS-AD system for methanogenesis step is between 6.8–7.2 which ensures very high activity of acidifying and methanogenic bacteria (Garcia-Peña et al. 2011). According to Boe (2006), suitable pH range for methanogenic archaea growth is 5.5–8.5 and optimal range is 6.5–8.0 (Boe 2006). Hwang et al. (2004) have shown that pH 4.0–8.5 is suitable range for fermentative bacteria functioning and optimal range is 5.0–6.0 (Hwang et al. 2004). Different microbes need different optimal pH (Ward et al. 2008). pH maintenance specially during high content of solids in AD is important since mixing is essential for homogeneity (Zhang et al. 2019). Acidogens take place between pH range of 5.5 and 6.5, whereas methanogens are highly active at pH 6.5–8.2 (Mao et al. 2015). pH is influenced by various parameters during the AD process. pH was low (< 6.5) initially in SS-AD of organic municipal solid wastes (OMSW) due to the high concentration of VFAs and later increased to 8 once VFAs decreased (Di Maria et al. 2017). Alkalinity of the system affects the pH of the SS-AD process. A drop in pH was observed for a system with low level of alkalinity (1,036 mg $CaCO_3$/kg). On increasing alkalinity (> 1,700 mg $CaCO_3$/kg) via adjusting F/I ratio, pH decreases only slightly which shows the buffer capacity of the system to resist pH fluctuations (Shi et al. 2014). A balance in the concentration of VFAs and bicarbonate is needed for maintaining a stable pH during the SS-AD process which could be done by reduction in the organic loading, addition of bases/bicarbonates and modification of F/I ratio to increase alkalinity (Ward et al. 2008).

3.4 Temperature

SS-AD commonly occurs under mesophilic or thermophilic or psychrophilic condition. Mesophilic occurs under 37°C temperature, thermophilic under 55°C and psychrophilic condition occurs below 20°C (Chiu and Lo 2016). With the increase in the temperature, production of biogas, growth of microbes and metabolic rate increases (Kim et al. 2006). Some authors conducted AD under both thermophilic and mesophilic ranges and reported higher yield at high temperature.

Biogas production is twice under thermophilic than in psychrophilic condition. Abdelgadir et al. (2014) has shown that diffusion rates improve with the increasing temperature. Some studies have shown pathogen destruction at high temperatures and digestate with less pathogen is produced in this process (Smith et al. 2005). Thermophilic AD has shorter hydraulic retention time (HRT), shorter startup time, and higher methane yield compared to mesophilic AD (Zhou and Wen 2019). There are some limitations in thermophilic AD-like poor stability and reliability, microbes' sensitivity towards environmental changes, less diversity in microbial community, rapid VFAs production during hydrolysis process creating an imbalance between acidogenesis and methanogenesis, high heating energy, and shifting in NH_3/NH_4^+ equilibrium toward the cytotoxic ammonia due to higher temperature (Mao et al. 2015). Therefore, theomorphic digesters are not used commercially in SS-AD process. Many authors have indicated that mesophilic conditions are the more feasible choice for AD (Guo et al. 2010). In mesophilic conditions, organic material is stabilized, and fermented sludge is produced in less quantity and can be used as dung (Gebreeyessus et al. 2016). Process in mesophilic conditions is stable and requires less energy as compared to thermophilic AD (Gebreeyessus et al. 2016). Few studies have shown that thermophilic AD has shown worse sludge dewaterability than mesophilic AD (Chi et al. 2010). There are drawbacks also like in mesophilic conditions, SS-AD has shown poor start up performance and generates lower methane yield compared to thermophilic condition (Li et al. 2011). Some studies also conducted AD in psychrophilic conditions so that cold zones where temperature is less due to their topographic and climatic characteristics can also adopt AD for biogas production (Rajagopal et al. 2017).

3.5 Inhibitors

There are various compounds found to inhibit SS-AD process affecting its stability, decrease methane yield and cause microbial population shift (Chen et al. 2008). Easily digestible feedstock can cause rapid hydrolysis followed by acidification. This generates very high amount of VFAs that eventually inhibit methanogens. Hydrogen also acts as an inhibitor of VFAs' degradation due to its high partial pressure (Zhou and Wen 2019). A balanced pressure of CO_2/H_2 is needed in the headspace for production of acetate/CH_4 (Abbassi-Guendouz et al. 2012). Degradation of nitrogen containing compounds produce ammonia during AD process. This ammonia is required for bacterial grow and neutralizing VFAs, but excess amount of ammonia inhibits methanogenesis (Li et al. 2016). Free ammonia is also present in the AD reactor since ammonia exists in equilibrium between free ammonia (NH_3) and ammonium ion (NH_4^+). This free ammonia can cause proton imbalance in microbial cells as it can penetrate cell membranes easily (Zhou and Wen 2019). Metal ions like Mg^{2+}, Na^+ and Ca^{2+} are found to be antagonistic to the inhibition of ammonia where presence of an ion decreases the toxicity of another ion (Braun et al. 1981). A 10% increase in the yield of methane was obtained using combination of Na^+ and Mg^{2+} or Na^+ and K^+ as compared to using Na^+ alone (Kugelman and McCarty 1964). Acclimation also affects the degree of inhibition caused by ammonia. Melbinger and Donnellon (1971) were the first ones who exposed methanogens slowly to ammonia increasing concentrations so that methanogens can adopt themselves in such conditions. After that, many studies have reported adaptation to wide range of inhibitors (Parkin and Miller 1983, Speece and Parkin 1983). Few studies show the toxicity caused by heavy metals leading to the failure of AD digester

(Swanwick et al. 1969). In brewery spent grain feedstock, a degradation product called *p*-cresol is produced which inhibits the SS-AD process (Paritosh et al. 2021a). This was reported to be overcome by use of granular biomass in SS-AD process (Panjičko et al. 2017). D-limonene which is found in citrus fruit waste can damage the cell membrane of microbes (Paritosh et al. 2021a). This could be removed by steam distillation and solvent extraction (Paritosh et al. 2021a). 5-hydroxyl methyl furfural and furfurals produced during the pretreatment of hemicellulose-rich substrates may affect the SS-AD process adversely (Paritosh et al. 2021a).

3.6 Mixing

Mixing in certain degree is required in SS-AD for enhancing the organic substrate transfer to microorganisms and to maintain the feed homogeneity. This prevents the sedimentation of floating lighter materials, and dense particles, facilitating the release of gas bubbles which are trapped in the feedstock (Zhou et al. 2017). Different methods can be adopted for mixing like liquid (leachate) recirculation, biogas recirculation and solid mixing using augers. Leachate recirculation reduces the inoculum amount and facilitates the diffusion of nutrients (André et al. 2018). There are some drawbacks also associated with the leachate recirculation like it leads to the VFAs accumulation and other compounds that are inhibitors. Leachate dilution with fresh water can solve the problem (André et al. 2018). Solid content affects the mixing - feed with less than 5% TS doesn't get much affected by mixing, while feed with higher TS (more than 10%) showed improvement in the yield of biogas with mixing compared to its unmixed feed (Harsha and Singh 2021). 50–1500 rpm mixing range (not optimum) is used by most studies depending on the type of feed and OLR (Harsha and Singh 2021).

4. FURTHER IMPROVEMENT IN THE PERFORMANCE

4.1 Feedstock pre-treatment

Feedstock pretreatment is required in hydrolysis (rate-limiting step); different methods are summarized below which have been used for pretreatment of feedstock to increase the biogas production. Pre-treatment is done to speed up the biodegradation rate and also increase the substrate solubility as well as methane production (Mirmohamadsadeghi et al. 2019). It also increases availability of the substrates to the anaerobes which gives high biogas yield. It increases the uptake of organic compounds from cells which will increase the biogas yield (Mirmohamadsadeghi et al. 2019).

4.2 Physical treatment

Methods like milling and grinding are used as physical treatment methods. These methods reduce the size of the feedstock particles and increase the surface area for microbial utilization (Zhou and Wen 2019). Mechanical pretreatments increase the surface area improving bacteria and substrate contact (Ren et al. 2018). But too fine particles also affect the performance of AD negatively. Milling or comminution decreased the size of particles and increased surface area in feedstock. Thermal treatment is also very effective treatment method; it enhances the reaction rate, removes pathogens, decreases the digestion system viscosity and improves dewater ability (Jain et al. 2015). Thermal treatments require a suitable temperature and time combination. Ultrasound treatment has also been used and generated mechanical effects and chemical effects (Zhou and Wen 2019). However, prolonged exposure to UV waves can cause changes in the substrates via physical and chemical impacts and breakdown of cell walls, cell lysis, micro-bubble production, partial pressure reduction, free radical production, and solid organic substrates solubilization (Khoo et al. 2020). Liquid shearing (Nah et al. 2000) and milling (Motte et al. 2014) were also used in few studies as pre-treatment method. Microwave treatment of feedstock modified structure and thermal effects which enhanced the sludge solubility, shortened initial lag phase and enhanced the biogas production (Zhou and Wen 2019). This technique has been used in various fields successfully due to its intense heating technique. Steam explosion was also used as an effective pretreatment technique. Though it didn't enhance much methane yield, it is widely used due to various advantages associated with it like low energy input, commercially available equipment and low pollution propensity. Due to these advantages, there are more installations of steam explosion as a pretreatment technique at an industrial level. There are some drawbacks associated as well, like safety issues, furfural formation, and scalding risks (Zhang et al. 2019). Hydrothermal pretreatment is another important technique used in SS-AD process. It has advantages over other methods since it doesn't require any chemicals (Zhang et al. 2019). It holds better advantages than biological pretreatment methods - it can be easily commercialized in the short term, and is economical and environmentally less challenging, and therefore various waste management industries and fields are considering it. Its major drawback is the requirement of very high energy for heating liquid water (Zhang et al. 2019). Liu et al. (2022) used freeze vacuum drying pre-treatment technique on corn straw and reported improved methane yield during SS-AD process.

4.3 Chemical treatment

Chemical pretreatment methods are very essential since they break down the recalcitrant structures of feedstock. There are various chemical pretreatment methods like alkaline, acids or oxidants. Type of feedstock and reagents used also influence the chemical treatment methods' effectiveness. Easily digestible carbohydrates containing feedstock like starch are not suitable for chemical pretreatment methods since it can accelerate the degradation of starch which will accumulate VFAs overproduction and accumulation. Potassium hydroxide, lime, sodium hydroxide and ammonium hydroxide are used in alkaline treatment and usually take place at ambient temperature. Alkaline treatment of lignocellulose removes lignin, and improves the cellulose and hemicelluloses accessibility to enzymes and microbes. It also prevents the drop in pH since the alkali presence neutralizes the carboxylic acids produced from the lignocellulose degradation in the stage of acidogenesis (Liao et al. 2016). Acid treatment has more effect as compared to alkali treatment. There are some limitations of acid treatment like production of hydroxymethylfurfural (HMF) and furfural during the acid treatment which inhibits AD process. Additional bases are also needed in SS-AD for neutralizing the pH. Therefore, alkali treatment method is more favorable than acid treatment. Ketsub et al. (2022) performed SS-AD of lignocelluloses using different pre-treatment methods and reported maximum yield of methane (6.9 L/kg VS/day) from KOH-pretreatment. Ozonation is also used as chemical treatment method. Ozone is a strong oxidant which decomposes into radicals and reacts with the insoluble and soluble substrates fractions (Braguglia et al. 2012). Ariunbaatar et al. (2014) obtained the optimum ozone dosage of 0.05–0.5 g O^3/g TS (Ariunbaatar et al. 2014). Organic solvents have been sued as well as chemical treatment methods in lignocellulose-based feedstock and higher biogas productions were reported (Mirmohamadsadeghi et al. 2014). Xie et al. (2022) have used urea pre-treatment in SS-AD of corn straw and reported enhanced methane yield.

4.4 Biological treatment

Biological treatment methods like use of microorganisms and/or enzymes are used widely for AD process and helps in the breaking down of the recalcitrant structure present in the feedstock. Peptidase, carbohydrates, and lipase are enzymes which are commonly used and added for speeding up the digestion process in the LS-AD system (Ariunbaatar et al. 2014). However, this enzyme addition approach is reported in SS-AD system. Microorganisms like white-rot fungi have been used in SS-AD since it can decompose lignin, altering the linkage between polysaccharides and lignin (Wan and Li 2011). The fungi *Trichoderma reesei* and *Pleurotus ostreatus* are used for decomposition of rice straw to enhance biogas production in SS-AD process (Mustafa et al. 2016). *Ceriporiopsis subvermispora*, a white-rot fungus, is very effective in degrading liginin while preserving cellulose (Ge et al. 2015). Zhao et al. (2014) reported enhanced biogas production by 154% using *C. subvermispora*-treated SS-AD (Zhao et al. 2014). Ge et al. (2015) reported a 3.7-fold increase in biogas yield using *C. subvermispora* treated albizia chips (Ge et al. 2015). Composting is another way of biological treatment of feedstock in SS-AD. It is an aerobic process facilitated by fungi and bacteria. Pre-aeration is found to improve the efficiency of composting via reducing the excessive organic compounds of feed stocks reducing the risk of VFA acidification and overproduction in SS-AD process (Charles et al. 2009). Pre-aeration should be done carefully since excess may lead to the introduction of oxygen to methanogens and can be toxic to methanogens.

4.5 Co-digestion

It is a beneficial and sustainable strategy, first used in Europe to improve the performance of SS-AD process (Lacovidou et al. 2012). Simultaneous digestion of more than one substrate to create a synergistic effect to improve biogas production is done in co-digestion (Cárdenas-Cleves et al. 2018).

Sewage sludge obtained from wastewater treatment plants (WWTP) is considered as a good co-substrate based on its total organic carbon concentration, contribution of alkalinity and macro and micronutrients' presence (Lin et al. 2018). Hagos et al. (2017) and Siddique and Wahid (2018) have shown that the nutrient requirement affects the process of co-digestion which is related to the C/N ratio and eventually affects the microbes involved; therefore, choice of co-substrate should be done carefully. Lu et al. (2017) found that the best C/N ratio is between 17 and 24 when they used cattle manure and corn straw as co-substrates. Ren et al. (2018) have used lignocellulosic co-substrates with RA like macro and micro algae, which increased the methane production (Cogan et al. 2016). Their research found that the ideal ratios for thermophilic and mesophilic conditions in co-digesting food waste and sewage sludge are 2:1 and 3:1, respectively (Arelli et al. 2021). Co-digestion using different feed stocks is needed for adjusting the carbon-nitrogen (C/N) ratio of the substrates. Co-digestion improves the nutrient profiles and gives a microbial community which is balanced; it provides a desirable moisture content and possesses great economic advantages (Zhou and Wen 2019). There are also some limitations associated with co-digestion like extra cost, premixing requirement, and increased effluent COD (Kangle et al. 2012). Ratio in co-digested substrate is important aspect as well for successful SS-AD. Li et al. (2016) achieved the highest CH_4 yield using a mixed feedstock of three different wastes consisting of tomato residues, corn stover, and dairy manure 13:33:54 (TS based) (Li et al. 2016). Zhang et al. (2022) co-digested sorghum vinegar residue and livestock manures and reported effective dilution of inhibitory substances like sodium ion, potassium ion, ammonium ion and improved methane production by 10.1–58.2%.

4.6 Additives

Different additives have been used as supplementation in AD system. These additives improve the digestion performance greatly **(Liu et al. 2021)**. Biochar, a charcoal-like product, has been used as an additive in AD, which enhances the stability of process and increases the alkalinity (Liu et al. 2021). Liang et al. (2017) found that addition of biochar increased methane production rate by 18% (Liang et al. 2017). Shen et al. (2016) has shown that biochar addition increased biogas production by 92.3% (Shen et al. 2016). Some studied have reported that addition of conductive materials was also found to enhance the biogas production. Nanoparticles' (NPs) addition improved biogas production from SS-AD like addition of metal oxides, zero-valent metals, nano-ash and carbon based substances. Activated carbon also enhanced the biogas yield since it acts as an acceptor of electron, works as a redox mediator, promotes direct interspecies electron transfer between microbes, and acts as a suitable immobilization matrix for anaerobic microbes due to its porous and amorphous nature (Zhang et al. 2019).

5. LATEST STRATEGIES FOR ENHANCING BIOGAS PRODUCTION FROM AD PROCESS

Additives, optimization, genetic technology, pretreatments, operating parameters optimization, bioreactor design and co-digestion are major strategies for improving biogas production from AD. Optimization of operating parameters includes the temperature, investigations on the pH, OLR, particle size, C/N ratio and hydraulic retention time (HRT) (Harsha and Singh 2021). Optimization of the bioreactor involves enhancing its performance by modifying specific features and characteristics of the bioreactor setup (Zhang et al. 2019). Genetic engineering a pivotal role in this process through the targeted manipulation of enzyme microbial strains and the extraction of valuable insights from metagenomics-based mining of microbial communities (Zhang et al. 2019). Usage of genetically modified microorganisms and enzymes that were genetically engineered improved AD stability of biogas production. Use of highly efficient biomolecular tools for manipulations and genetic analyses were found to improve the production of biogas from SS-AD process (Zhang et al. 2019). Metagenomics engineering can be coupled with artificial neural networks to form a precise platform for fermentation which will enhance the performance of the AD process (Zhang et al. 2019). Some additional tools are also available now for careful study of bacterial community called microbial genomics resources. Genomic information possesses the relevant physiological properties of the bacteria which will help in determining the performance of the bacteria under various temperature conditions (Zhang et al. 2019). Molecular biological approaches will also allow non-cultivable bacteria identification from consortia. Therefore, reliance on culturing will be reduced. Discovery of cultivable microbes will be done using molecular biological methods. Genetic engineering tools can manipulate either the particular enzyme or cells for improving the biogas productions from AD process (Zhang et al. 2019). Genetic engineering tools used to be high in cost but nowadays faster development of genetic engineering approaches and synthetic biology reduces the higher cost of these methods (Zhang et al. 2019). Use of nanoparticles as additives is recently found to improve the biogas production from AD process (Liu et al. 2021). Use of nanoparticles is associated with many advantages like low cost, and high surface area to volume which improves the surface biochemical reaction in the SS-AD process. Organic loading rate (OLR) is the amount of raw material continuously charged into the anaerobic reactor per unit volume per day (Zhang et al. 2019). OLR varies with the variations in the feedstocks' characteristics, hydraulic retention time and operating temperature. A very high OLR leads to the VFAs accumulation while a very low OLR could affect the nutrients of the fermentation microbes (Zhang et al. 2019). For evaluating the maximum endurable OLR, type of feed, operation cost and operation conditions should be considered. Menon et al. (2017) reported that higher OLR is required in thermophilic system (Menon et al. 2017). HRT is the time required for complete degradation of the organic matters. It is expressed in hours or days. It is an important parameter that influences the microbial community, and therefore its optimization is essential (Zhang et al. 2019). Mao et al. (2015) reported 15-30 days HRT for treating solid organic wastes (SOWs waste) (Mao et al. 2015). Too low HRT results in the VFAs accumulation and a very high HRT will cause insufficient usage of the feedstock (Mao et al. 2015). Optimization of anaerobic bioreactor is important for enhancing the performance of the AD process. One-stage digester is simple in operation and construction, low cost, and easy to design. However, due to limitations of one stage digester, it is not used as much (Zhang et al. 2019). Two stage and multi stage digestive systems ensure a more stable AD system as these provide well suitable environment for various microbial communities. Multi stage systems are found to be better in performance as compared to the single stage system (Zhang et al. 2019).

6. CONCLUSION

AD is highly sustainable and environmentally friendly and has advantages over other biological conversion methods like composting and landfill. It reduces the greenhouse gas emission (GHG), generates the renewable energy and nutrient rich sludge which can be used as a fertilizer. Generation of electricity form methane is highly economical. A zero waste producing system could be developed. Area occupied in AD is also less compared to that of composting and landfilling. There are some challenges that need to be addressed like characteristics of feedstocks that highly influence the process. Substrate quality should be standardized which includes the collection, sorting and pretreatment before AD process. Microbial community's behavior greatly impacts the production of biogas and its quality. Production of valuable intermediate products should be given serious attention for maintaining the stability of the AD process. A high solid hydrolysis stage is good for multistage process since it provides good conditions for diverse microbial growth. Utilization of genetic engineering technologies like metagenome technology, gene sequencing technologies, and synthetic biology will likely improve the microbial community involved in the digestion process. Outcome and efficiency of AD process depends upon the composition and behavior of inoculum used; therefore, a careful study of the microbial community, its members, and number of species should be done for enhancing the performance of biogas production from AD process. Methane produced during AD process can be converted further to other value added products.

REFERENCES

Abbasi, T., Tauseef, S.M. and Abbasi, S.A. 2012. Biogas Energy, 1–169.

Abbassi-Guendouz, A., Brockmann, D., Trably, E. et al. 2012. Total solids content drives high solid anaerobic digestion via mass transfer limitation. Bioresour. Technol. 111: 55–61.

Abdelgadir, A., Chen, X., Liu, J. et al. 2014. Characteristics, process parameters, and inner components of anaerobic bioreactors. BioMed Res Int.

Amani, T., Nosrati, M. and Sreekrishnan, T.R. 2010. Anaerobic digestion from the viewpoint of microbiological, chemical, and operational aspects—A review. Environ. Rev. 18: 255–278.

André, L., Pauss, A. and Ribeiro, T. 2018. Solid anaerobic digestion: State-of-art, scientific and technological hurdles. Bioresour. Technol. 247: 1027–1037.

Arelli, V., Mamindlapelli, N.K., Begum, S. et al. 2021. Solid state anaerobic digestion of food waste and sewage sludge: Impact of mixing ratios and temperature on microbial diversity, reactor stability and methane yield. Sci. Total Environ. 793: 148586.

Ariunbaatar, J., Panico, A., Esposito, G. et al. 2014. Pretreatment methods to enhance anaerobic digestion of organic solid waste. Appl. Energy 123: 143–156.

Axelsson, L., Franzén, M., Ostwald, M. et al. 2012. Perspective: Jatropha cultivation in southern India: Assessing farmers' experiences. Biofuels, Bioprod Biorefining 6: 246–256.

Boe, K. 2006. Online monitoring and control of the biogas process. Dissertation, Technical University of Denmark, see, http:// www. risoe. dk/ rispu bl/ NEI/ nei- dk- 4757. pdf.

Bolzonella, D., Pavan, P., Mace, S. et al. 2006. Dry anaerobic digestion of differently sorted organic municipal solid waste: A full-scale experience. Water Sci. Technol. 53: 23–32.

Braguglia, C.M., Gianico, A. and Mininni, G. 2012. Comparison between ozone and ultrasound disintegration on sludge anaerobic digestion. J. Environ. Manage. 95: S139–S143.

Braun, B., Huber, P. and Meyrath, J. 1981. Ammonia toxicity in liquid piggery manure digestion. Biotechnol. Lett. 3: 159–164.

Capson-Tojo, G., Trably, E., Rouez, M. et al. 2017. Dry anaerobic digestion of food waste and cardboard at different substrate loads, solid contents and co-digestion proportions. Bioresource Technology 233: 166–175.

Cárdenas-Cleves, L.M., Marmolejo-Rebellón, L.F. and Torres-Lozada, P. 2018. Anaerobic codigestion of sugarcane press mud with food waste: Effects on hydrolysis stage, methane yield, and synergistic effects. Int. J. Chem. Eng. pp. 1–8.

Charles, W., Walker, L. and Cord-Ruwisch, R. 2009. Effect of pre-aeration and inoculum on the start-up of batch thermophilic anaerobic digestion of municipal solid waste. Bioresour. Technol. 100: 2329–2335.

Chen, Y., Cheng, J.J. and Creamer, K.S. 2008. Inhibition of anaerobic digestion process: A review. Bioresour. Technol. 99: 4044–4064.

Chi, Y.Z., Li, Y.Y., Ji, M. et al. 2010. Mesophilic and thermophilic digestion of thickened waste activated sludge: A comparative study. Adv. Mater. Res. 113-116: 450–458.

Chiu, S.L. and Lo, I.M. 2016. Reviewing the anaerobic digestion and codigestion process of food waste from the perspectives on biogas production performance and environmental impacts. Environ. Sci. Pollut. Res. 23: 24435–24450.

CSO. 2019. Energy Statistics, 26: 1–123.

Cogan, M/ and Antizar-Ladislao. B. 2016. The ability of macroalgae to stabilise and optimise the anaerobic digestion of household food waste. Biomass Bioenergy 86: 146–155.

Cui, Z., Shi, J. and Li, Y. 2011. Solid-state anaerobic digestion of spent wheat straw from horse stall. Bioresour. Technol. 102: 9432–9437.

De Baere, L. 2008. Partial stream digestion of residual municipal solid waste. Water Sci. Technol. 57: 1073–1077.

Di Maria, F., Barratta, M., Bianconi, F. et al. 2017. Solid anaerobic digestion batch with liquid digestate recirculation and wet anaerobic digestion of organic waste: Comparison of system performances and identification of microbial guilds. Waste Manag. 59: 172–180.

Forster-Carneiro, T., Pérez, M., Romero, L.I. et al. 2007. Dry-thermophilic anaerobic digestion of organic fraction of the municipal solid waste: Focusing on the inoculum sources. Bioresour. Technol. 98: 3195–3203.

Garcia-Peña, E.I., Parameswaran, P., Kang, D. et al. 2011. Anaerobic digestion and co-digestion processes of vegetable and fruit residues: Process and microbial ecology. Bioresour. Technol. 102: 9447–9455.

Ge, X., Matsumoto, T., Keith, L. et al. 2015. Fungal pretreatment of albizia chips for enhanced biogas production by solid-state anaerobic digestion. Energy and Fuels 29: 200–204.

Gebreeyessus, G.D. and Jenicek, P. 2016. Thermophilic versus mesophilic anaerobic digestion of sewage sludge: A comparative review. Bioengineering 3: 15.

Gerardi, M.H. 2003. The Microbiology of Anaerobic Digesters. John Wiley & Sons; DOI:10.1002/0471468967.

Guo, X., Wang, C., Sun, F. et al. 2014. A comparison of microbial characteristics between the thermophilic and mesophilic anaerobic digesters exposed to elevated food waste loadings. Bioresour. Technol. 152: 420–428.

Hagos, K., Zong, J., Li, D. et al. 2017. Anaerobic co-digestion process for biogas production: Progress, challenges and perspectives. Renew. Sustain. Energy Rev. 76: 1485–1496.

Harsha, G. and Singh, M.N. 2021. Art of anaerobic digestion: An overview. Res. J. Chem. Environ. 25: 168–176.

He, Y., Pang, Y., Liu, Y. et al. 2008. Physicochemical characterization of rice straw pretreated with sodium hydroxide in the solid state for enhancing biogas production. Energy and Fuels 22: 2775–2781.

Hwang, M.H., Jang, N.J., Hyun, S.H. et al. 2004. Anaerobic biohydrogen production from ethanol fermentation: The role of pH. J. Biotechnol. 111: 297–309.

Inyang, M., Gao, B., Pullammanappallil, P. et al. 2010. Biochar from anaerobically digested sugarcane bagasse. Bioresour. Technol. 101: 8868–8872.

Jenkins, B.M., Williams, R.B., Adams, L.S. et al. 2008. Current anaerobic digestion technologies used for treatment of municipal organic solid waste. California Environmental Protection Agency Report.

Jin, X., Ai, W. and Dong, W. 2022. Lignocellulose degradation, biogas production and characteristics of the microbial community in solid-state anaerobic digestion of wheat straw waste. Life Sci. Space Res. 32: 1–7.

Kangle, K.M., Kore, S.V., Kore, V.S. et al. 2012. Recent trends in anaerobic codigestion: A review. Univers. J. Environ. Res. Technol. 2: 210–219.

Kassongo, J., Shahsavari, E. and Ball, A.S. 2022. Substrate-to-inoculum ratio drives solid-state anaerobic digestion of unamended grape marc and cheese whey. Plos One 17: e0262940.

Karthikeyan, O.P. and Visvanathan, C. 2013. Bio-energy recovery from high-solid organic substrates by dry anaerobic bio-conversion processes: A review. Rev. Environ. Sci. Biotechnol. 12: 257–284.

Ketsub, N., Whatmore, P., Abbasabadi, M. et al. 2022. Effects of pretreatment methods on biomethane production kinetics and microbial community by solid state anaerobic digestion of sugarcane trash. Bioresour. Technol. 352: 127112.

Khan, M.Z., Nizami, A.S., Rehan, M. et al. 2016. Microbial electrolysis cells for hydrogen production and urban wastewater treatment: A case study of Saudi Arabia. Appl. Energy, 1–11.

Khoo, K.S., Chew, K.W., Yew, G.Y. et al. 2020. Integrated ultrasonic assisted liquid biphasic flotation for efficient extraction of astaxanthin from Haematococcuspluvialis. Ultrason. Sonochem, 67: 105052.

Kim, J.K., Oh, B.R., Chun, Y.N. et al. 2006. Effects of temperature and hydraulic retention time on anaerobic digestion of food waste. J. Biosci. Bioeng. 102: 328–332.

Kugelman, I.J. and McCarty, P.L. 1964. Cation toxicity and stimulation in anaerobic waste treatment. J. Water Pollut. Control Fed. 37: 97–116.

Lacovidou, E., Ohandja, D.G. and Voulvoulis, N. 2012. Food waste codigestion with sewage sludge–realising its potential in the UK. J. Environ. Manag. 112: 267–274.

Li, A., Chu, Y., Wang, X. et al. 2013. A pyrosequencing-based metagenomic study of methane-producing microbial community in solid-state biogas reactor. Biotechnol. Biofuels 6: 1–17.

Li, Y., Park, S.Y. and Zhu, J. 2011. Solid-state anaerobic digestion for methane production from organic waste. Renew. Sustain. Energy Rev. 15: 821–826.

Li Yangyang, Li Yu, Zhang, D. et al. 2016. Solid state anaerobic co-digestion of tomato residues with dairy manure and corn stover for biogas production. Bioresour. Technol. 217: 50–55.

Liang, Y.G., Cheng, B., Si, Y.B. et al. 2016. Effect of solid-state NaOH pretreatment on methane production from thermophilic semi-dry anaerobic digestion of rose stalk. Water Sci. Technol. 73: 2913–2920.

Liang, Y., Qiu, L., Guo, X. et al. 2017. Start-up performance of chicken manure anaerobic digesters amended with biochar and operated at different temperatures. Nat. Environ. Pollut. Technol. 16: 615–621.

Liao, X., Li, H., Zhang, Y. et al. 2016. Accelerated high-solids anaerobic digestion of sewage sludge using low-temperature thermal pretreatment. Int. Biodeterior. Biodegrad. 106: 141–149.

Liew, L.N., Shi, J. and Li, Y. 2012. Methane production from solid-state anaerobic digestion of lignocellulosic biomass. Biomass and Bioenergy 30: 125–132.

Lima, D.R.S., Adarme, O.F.H., Baêta, B.E.L. et al. 2018. Influence of different thermal pretreatments and inoculum selection on the biomethanation of sugarcane bagasse by solid-state anaerobic digestion: A kinetic analysis. Ind. Crops Prod. 111: 684–693.

Lin, L., Xu, F., Ge, X. et al. 2018. Improving the sustainability of organic waste management practices in the food-energy-water nexus: A comparative review of anaerobic digestion and composting. Renew. Sustain. Energy Rev. 89: 151–167.

Lissens, G., Verstraete, W., Albrecht, T. et al. 2004. Advanced anaerobic bioconversion of lignocellulosic waste for bioregenerative life support following thermal water treatment and biodegradation by Fibrobacter succinogenes. Biodegradation 15: 173–183.

Liu, C.M., Wachemo, A.C., Yuan, H.R. et al. 2018. Evaluation of methane yield using acidogenic effluent of NaOH pretreated corn stover in anaerobic digestion. Renew. Energy 116: 224–233.

Liu, M., Wei, Y. and Leng, X. 2021. Improving biogas production using additives in anaerobic digestion: A review. J. Clean. Prod. 297: 126666.

Liu, Z., Huang, J., Yao, Y. et al. 2022. Effect of pretreatment by freeze vacuum drying on solid-state anaerobic digestion of corn straw. Fermentation 8: 259.

Lo, Y.C., Saratale, G.D., Chen, W.M. et al. 2009. Isolation of cellulose-hydrolytic bacteria and applications of the cellulolytic enzymes for cellulosic biohydrogen production. Enzyme Microb. Technol. 44: 417–425.

Lu S. Guang, Imai, T., Ukita, M. et al. 2007. Start-up performances of dry anaerobic mesophilic and thermophilic digestions of organic solid wastes. J. Environ. Sci. 19: 416–420.

Lu, X., Jin, W., Xue, S. et al. 2017. Effects of waste sources on performance of anaerobic co-digestion of complex organic wastes: taking food waste as an example. Sci. Rep. 7: 1–9.

Mamimin, C., Chanthong, S., Leamdum, C. et al. 2021. Improvement of empty palm fruit bunches biodegradability and biogas production by integrating the straw mushroom cultivation as a pretreatment in the solid-state anaerobic digestion. Bioresour. Technol. 319: 124227.

Mao, C., Feng, Y., Wang, X. et al. 2015. Review on research achievements of biogas from anaerobic digestion. Renew. Sustain. Energy Rev. 45: 540–555.

McCarty, P.L. 2001. The development of anaerobic filter and its future. Water Sci. Technol. 44: 149–156.

Melbinger, N.R. and Donnellon, J. 1971. Toxic effects of ammonia-nitrogen in high rate digestion. J. Water Pollut. Control. Fed. 43: 1658–1670.

Menon, A., Wang, J.Y. and Giannis, A. 2017. Optimization of micronutrient supplement for enhancing biogas production from food waste in two-phase thermophilic anaerobic digestion. Waste Manag. 59: 465–475.

Mirmohamadsadeghi, S., Karimi, K., Zamani, A. et al. 2014. Enhanced solid-state biogas production from lignocellulosic biomass by organosolv pretreatment. Biomed. Res. Int., 1–6.

Mirmohamadsadeghi, S., Karimi, K., Tabatabaei, M. et al. 2019. Biogas production from food wastes: A review on recent developments and future perspectives. Bioresour. Technol. Rep. 7: 100202.

Momayez, F., Karimi, K. and Horváth, I.S. 2018. Enhancing ethanol and methane production from rice straw by pretreatment with liquid waste from biogas plant. Energy Convers. Manage. 178: 290–298.

Mosier, N., Hendrickson, R., Ho, N. et al. 2005. Optimization of pH controlled liquid hot water pretreatment of corn stover. Bioresour. Technol. 96: 1986–1993.

Motte, J.-C., Escudié, R., Hamelin, J. et al. 2014. Substrate milling pretreatment as a key parameter for solid-state anaerobic digestion optimization. Bioresour. Technol. 173: 185–192.

Mustafa, A.M., Poulsen, T.G. and Sheng, K. 2016. Fungal pretreatment of rice straw with Pleurotus ostreatus and Trichoderma reesei to enhance methane production under solid-state anaerobic digestion. Appl. Energy 180: 661–671.

Nah, I.W., Kang, Y.W., Hwang, K.-Y. et al. 2000. Mechanical pretreatment of waste activated sludge for anaerobic digestion process. Water Res. 34: 2362–2368.

Nguyen, D.D., Chang, S.W., Jeong, S.Y. et al. 2016. Dry thermophilic semi-continuous anaerobic digestion of food waste: Performance evaluation, modified Gompertz model analysis, and energy balance. Energy Convers. Manag. 128: 203–210.

Nugraha, W.D. and Matin, H.H.A. 2017. The effect of enzymatic pretreatment and c/n ratio to biogas production from rice husk waste during solid state anaerobic digestion (SS- AD). In MATEC Web of Conferences Vol. 101: 02016.

Nugraha, W.D., Keumala, C.F. and Matin, H.H.A. 2018. The effect of acid pretreatment using acetic acid and nitric acid in the production of biogas from rice husk during solid state anaerobic digestion (SS-AD) Vol. 31: 01006.

Öhgren, K., Galbe, M. and Zacchi, G. 2005. Optimization of steam pretreatment of SO_2-impregnated corn stover for fuel ethanol production. Appl. Biochem. Biotechnol. - Part A Enzym. Eng. Biotechnol. 124: 1055–1067.

Panjičko, M., Zupančič, G.D. and Zelić, B. 2015. Anaerobic biodegradation of raw and pre-treated brewery spent grain utilizing solid state anaerobic digestion. Acta Chim. Slov. 62: 818–827.

Paritosh, K., Kumar, V., Pareek, N. et al. 2021a. Solid state anaerobic digestion of water poor feedstock for methane yield: An overview of process characteristics and challenges. Waste Dispos. Sustain. Energy 3: 227–245.

Paritosh, K., Yadav, M., Kesharwani, N. et al. 2021b. Strategies to improve solid state anaerobic bioconversion of lignocellulosic biomass: An overview. Bioresour. Technol. 331: 125036.

Parkin, G.F. and Owen, W.F. 1986. Fundamentals of anaerobic digestion of wastewater sludges. J. Environ. Eng. 112: 867–920.

Parkin, G.F. and Miller, S.W. 1983. Response of methane fermentation to continuous addition of selected industrial toxicants. In: Proceedings of the 37th Purdue Industrial Waste Conference, West Lafayette, Ind.

Rajagopal, R., Bellavance, D. and Rahaman, M.S. 2017. Psychrophilic anaerobic digestion of semi-dry mixed municipal food waste: for North American context. Process Saf. Environ. Prot. 105: 101–108.

Ren, Y., Yu, M., Wu, C. et al. 2018. A comprehensive review on food waste anaerobic digestion: Research updates and tendencies. Bioresour. Technol. 247: 1069–1076.

Saady, N.M.C. and Massé, D.I. 2015. A start-up of psychrophilic anaerobic sequence batch reactor digesting a 35% total solids feed of dairy manure and wheat straw. AMB Express. 5: 1–10.

Schmidt, S., Biegel, E. and Müller, V. 2009. The ins and outs of Na+ bioenergetics in Acetobacterium woodii. Biochim. Biophys. Acta - Bioenerg. 1787: 691–696.

Shen, Y., Linville, J.L., Ignacio-de Leon, P.A.A. et al. 2016. Towards a sustainable paradigm of waste-to-energy process: Enhanced anaerobic digestion of sludge with woody biochar. J. Clean Prod. 135: 1054–1064.

Shi, J., Xu, F., Wang, Z. et al. 2014. Effects of microbial and non-microbial factors of liquid anaerobic digestion effluent as inoculum on solid-state anaerobic digestion of corn stover. Bioresour. Technol. 157: 188–196.

Siddharth, J., Shivani, J., Wolf, I.T. et al. 2015. A comprehensive review on operating parameters and different pretreatment methodologies for anaerobic digestion of municipal solid waste. Renew. Sustain. Energy Rev. 52: 142–154.

Siddique, M.N.I. and Wahid, Z.A. 2018. Achievements and perspectives of anaerobic co-digestion: A review. J. Clean. Prod. 194: 359–371.

Smith, S., Lang, N., Cheung, K. et al. 2005. Factors controlling pathogen destruction during anaerobic digestion of biowastes. Waste Manag. 25: 417–425.

Speece, R.E. and Parkin, G.F. 1983. The response of methane bacteria to toxicity. In: Proceedings of the 3rd International Symposium on Anaerobic Digestion, Boston, MA.

Sumardiono, S., Adisukmo, G., Hanif, M. et al. 2021. Effects of pretreatment and ratio of solid sago waste to rumen on biogas production through solid-state anaerobic digestion. Sustainability 13: 7491.

Swanwick, J.D., Shurben, D.G. and Jackson, S. 1969. A survey of the performance of sewage sludge digesters in Great Britain. J. Water Pollut. Control. Fed. 68: 639–653.

Takashima, M., Speece, R.E. and Parkin, GF. 1990. Mineral requirements for methane fermentation. Crit. Rev. Environ. Control. 19: 465–479.

Veeken, A.H.M. and Hamelers, B.V.M. 2000. Effect of substrate-seed mixing and leachate recirculation on solid state digestion of biowaste. Water Sci. Technol. 41: 255–262.

Wan, C. and Li, Y. 2011. Effectiveness of microbial pretreatment by Ceriporiopsis subvermispora on different biomass feedstocks. Bioresour. Technol. 102: 7507–7512.

Wang, X., Li, Q., Zhou, Z. et al. 2017. Research on effect of combination of steam explosion and calcium oxide pretreatment on rice straw solid-state anaerobic digestion. J. Agro-Environ. Sci. 36: 394–400.

Wang, R., Zhang, Y., Jia, S. et al. 2021. Comparison of batch and fed-batch solid-state anaerobic digestion of on-farm organic residues: Reactor performance and economic evaluation. Environ. Technol. Innov. 24: 101977.

Ward, A.J., Hobbs, P.J., Holliman, P.J. et al. 2008. Optimisation of the anaerobic digestion of agricultural resources. Bioresour. Technol. 99: 7928–7940.

Xie, Z., Zou, H., Zheng, Y. et al. 2022. Improving anaerobic digestion of corn straw by using solid-state urea pretreatment. Chemosphere 293: 133559.

Yan, Z., Song, Z., Li, D. et al. 2015. The effects of initial substrate concentration, C/N ratio, and temperature on solid-state anaerobic digestion from composting rice straw. Bioresour. Technol. 177: 266–273.

Zamri, M.F.M.A., Hasmady, S., Akhiar, A. et al. 2021. A comprehensive review on anaerobic digestion of organic fraction of municipal solid waste. Renew. Sustain Energy Rev. 137: 110637.

Zhang, J., Loh, K.C., Lee, J. et al. 2017. Three-stage anaerobic co-digestion of food waste and horse manure. Sci. Rep. 7: 1–10.

Zhang, K., Pei, Z. and Wang, D. 2016. Organic solvent pretreatment of lignocellulosic biomass for biofuels and biochemicals: A review. Bioresour. Technol. 199: 21–33.

Zhang, L., Loh, K.C. and Zhang, J. 2019. Enhanced biogas production from anaerobic digestion of solid organic wastes: Current status and prospects. Bioresour. Technol. Rep. 5: 280–296.

Zhang, J., Qi, C., Wang, Y. et al. 2022. Enhancing biogas production from livestock manure in solid-state anaerobic digestion by sorghum-vinegar residues. Environ. Technol. Innov. 26: 102276.

Zhao, J., Zheng, Y. and Li, Y. 2014. Fungal pretreatment of yard trimmings for enhancement of methane yield from solid-state anaerobic digestion. Bioresour. Technol. 156: 176–181.

Zhou, H. and Wen, Z. 2019. Solid-state anaerobic digestion for waste management and biogas production. Adv. Biochem. Eng. Biotechnol. 169: 147–168.

Zhou, Y., Li, C., Nges, I.A. et al. 2017. The effects of pre-aeration and inoculation on solid-state anaerobic digestion of rice straw. Bioresour. Technol. 224: 78–86.

5

MUNICIPAL SOLID WASTE MANAGEMENT AND EFFECT OF PROCESS PARAMETERS ON THE TREATMENT METHODS

Shraddha Shukla, Shweta Rai, Senthamizh R.,
Shubhadarshini Sahoo, Prashant Yadav and *Joyabrata Mal**

1. INTRODUCTION

With a drastic increase in municipal solid waste (MSW) production due to rapid industrialization and urbanization, MSW management (MSWM) has become a critical issue which needs urgent attention. Various human activities are responsible for the yield of MSW which comprise mainly food waste, plastic, textile, wood, paper, etc. (Zhang et al. 2021). The global yield of MSW is estimated to be 2.01 billion tonnes annually and is expected to reach up to 3.40 billion tonnes by 2050 (Pujara et al. 2019, Zhang et al. 2021). In China, it is growing at a rate of 6% annually and was estimated at 212 million tonnes of MSW in 2018 (Du et al. 2021). Similarly in India, it is growing at 5% yearly and is expected to produce more than 56 million tonnes of MSW every year (CPCB Report 2017–18, EAI Report 2019, Pujara et al. 2019). The total production and disposal of solid waste by the major countries around the globe is shown in Table 1.

The management of enormous volume of MSW due to the rapid and unplanned urbanization is, thus, becoming a serious challenges for all the developed and developing countries to have a safer and sustainable disposal of MSW. Unfortunately, almost 80%–90% of the toal MSW are simply dumped improperly in open and is the most commonly employed strategy to manage these MSW (Saja et al. 2021). The improper dumping and burning of MSW without proper segregation, thus, cause severe soil, water and air pollution and the effect of these practices is shown in Fig. 1. In India too, without

Department of Biotechnology, MNNIT Allahabad, Prayagraj-21104, India.
* Corresponding author: joyabrata@mnnit.ac.in, joyabrata2006@gmail.com

Table 1. Major contributor of MSW generation, treatment methods and energy recovery (Kundariya et al. 2021).

Country	Solid waste management (MT/D)	Treatment process	Capacity of electricity generation (MW)
USA	6,24,700	Composting, AD, Landfilling, Recycling and Resource Recovery, Mechanical biological treatment (MBT)	2254
China	5,20,548	Composting, Thermal (Incineration, Pyrolysis, Conventional and Plasma Arc gasification)	–
Brazil	1,49,096	Composting, Recycling and Resource Recovery, Sanitary Landfilling, Incineration	–
Japan	1,44,466	Recycling and Resource Recovery, Landfilling	1501
India	1,09,589	Composting and Vermicomposting, Landfilling, AD	274
Germany	1,27,816	Recycling and Resource Recovery, Composting	1888
Russia	1,00,027	Composting, AD, Recycling and Resource Recovery, Incineration, Landfilling	-
UK	97,342	Recycling, Resource recovery, AD, MBT, Composting, Incineration, Landfilling	781
Spain	72,137	Landfilling, Recycling and Resource Recovery, Composting	251
South Africa	53,425	Disposal, Incineration, Recycling and Resource Recovery	-
South Korea	48,397	Recycling and Resource Recovery	184
Thailand	39,452	Recycling and Resource Recovery	75
Switzerland	14,329	Composting, Landfilling, Recycling and Resource Recovery	398
Sweden	12,329	Composting, Landfilling, Recycling and Resource Recovery	459
Denmark	10,959	Composting, Landfilling, Recycling and Resource Recovery	325
Singapore	7,205	Recycling and Resource Recovery	128

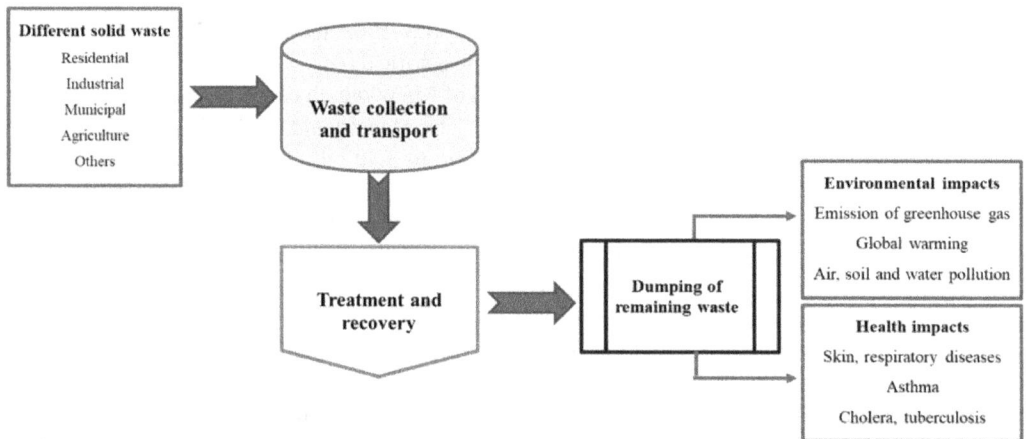

Figure 1. Schematic Diagram on Types of Solid Wastes and Environmental and Health Impacts of Dumped Solid Wastes (Kundariya et al. 2021).

proper segregation, improper and open disposal of most of MSW poses one of the biggest challenges in MSWM. Besides increasing population, rapid urbanization and high cost of MSWM, lack of awareness is also one of the major issues which can significantly affect the overall MSWM system.

1.1 Sources and types of solid waste

Municipal, industrial, and hazardous wastes are the different types of solid waste (Table 2). The majority of MSW in developing countries comes from households (55%–80%), followed by markets or commercial sectors (10%–30%). Solid waste from these sources are generally quite diverse in nature. Humans generate municipal garbage through their daily activities. Hazardous wastes are substances that pose a threat to plants, animals, and humans, and industrial waste is the result of industrial activity. Radioactive waste, chemicals, biological wastes, flammable wastes, and explosives are only a few examples of common hazardous waste. Some major causes of solid waste generation are:

- Overpopulation: The rise of pollution is influenced by a number of factors, including overpopulation.
- Urbanization: Solid waste is a problem in cities because people have the habit of utilizing a range of products and then abandoning them. The majority of urban solid wastes are dumped into bodies of water and land without adequate treatment, which is the principal source of much environmental contaminants (Varjani et al. 2021b).
- Affluence: In affluent societies, per capita consumption is relatively high, and people discard many products on a daily basis, resulting in a considerable amount of solid trash.
- Technology: The way people use things has changed as a result of technological advancements. It is readily obvious in the packaging business for the majority of low-cost goods. For example, plastic bottles and containers, and cans are replacing glass bottles and containers and the non-biodegradable plastic materials used in packaging are one of the major issues in solid waste pollution (Sharholy et al. 2008).

Table 2. Sources and types of solid wastes.

Source	Types of solid wastes
Residential	Food wastes, plastics, leather and textiles, cardboards and papers, wood, ashes, glass, metals and other wastes (e.g., oil, consumer electronics, batteries, tyres), etc.
Industrial (refineries, chemical and power plants, mineral extraction and processing light and heavy manufacturing, construction, etc.)	Cleaning wastes, packaging materials, food wastes, construction and destruction materials (e.g., steel, wood, concrete, etc.), ashes, various other industrial wastes and scrap materials, etc.
Institutional and commercial (hospitals, schools and offices, markets, shopping malls, hotels and restaurants)	Papers and cardboards, plasticss, food wastes, glass, metals, etc.
Municipal services	Cleaning materials, landscape and park maintenance materials, general wastes from recreational areas like parks and beaches, sludge
Agriculture	Agricultural wastes including pesticides, food wastes, etc.

1.2 Composition and characterization of solid waste

Composition of solid waste depends largely on the demographic areas: population and its lifestyle, waste management awareness and rules, economic situation and industrial development. Hence, it differs not only from country to country, but also within municipalities in the same country, while the strategy for MSWM depends on its amount and type. Based on its calorific value, elemental composition and estimated gasesous emission, MSW is treated via gasification and incineration. However, the very existence of toxic compounds in the ash depending on the MSW compositions can be a potential threat and can not be ignored. Hence, it is important to have clear idea about the waste composition which will be vital to decide the strategy for MSW between composting or biogas production via biological conversion. On the other hand, depending on its initial sources, the composition and the physical and chemical properties of MSW also varies (Nabegu 2010). For example, the presence and amount of organic materials in MSW depends the initial sources of MSW which largely control the rate of MSW biodegradation (Abdel-Shafy et al. 2018).

Moreover, different solid wastes like vegetative waste or food waste, plastics, wood, papers, rubbers, leather, textiles, and building debris (construction and deconstruction materials) are difficult to segregate or categorize and are generally included all in their composition. However, the presence of such waste is a major limitation in segregating and propoer management and reuse of MSW. So, it is important to segregate them properly prior to any major MSW treatment process.

1.3 Harmful effects of solid waste on environment

Direct combustion of MSW causes release of gaseous compounds, particulate and volatile matter into air (Fig. 1). Emission of toxic gases into the atmosphere during the burning of waste like rubbers and plastics is another serious issue. Similarly, degradation of MSW from open dumping can cause leaching of chemical and biological toxic contaminatants and cause severe soil and groundwater pollution (Cremiato et al. 2018, Fernandez-Nava et al. 2014). Moreover, organic waste decomposition also releases a large volume of methane gas which is responsible for global warming and climate change (Arafat et al. 2015). Furthermore, presence of high organic waste in MSW also facilitates microbial infections causing infectious diseases in the workers involved in MSWM and people living in/nearby MSW dumping site (Arafat et al. 2015). Lack of awareness and various inappropriate transportation and treatment strategies of solid waste can thus pose severe environmental and health risk. Stray animals looking for food on the roadside trashes also litter it which causes aesthetics pollution in environment. Thus, proper management of solid waste is critical to reduce pollution and to create a pollution-free, healthy environment. Composting, recycling, incineration, pyrolysis, disposal, and landfills are only few of the strategies generally used to manage solid waste pollution.

2. SOLID WASTE MANAGEMENT AND RESOURCE RECOVERY

2.1 Pyrolysis

Pyrolysis refers to the thermal degradation of solid waste at high temperature by breaking of the chemical bonds in inert atmospheric condition (Zhang et al. 2020). It requires external energy input, which provides the amount of heat requires for thermal decomposition of MSW (Du et al. 2021). In contrast to more conventional thermochemical strategies, which involve sequential gasification and subsequent catalytic up-grading, pyrolysis offers a less time-consuming and less expensive pathway to produce chemicals and fuels from biomass. So, various researcher have investigated the possibility of producing various fuels, solvents, chemicals, and other valuable products through the pyrolysis by using different biomass as feedstocks (Bridgwater et al. 2007, Chen et al. 2003, Yaman 2004). Typically, when organic materials are pyrolyzed, char - solid residue rich in carbon - is left behind along with volatile chemicals. Carbonization is the term used to describe pyrolysis that primarily produces carbon as the byproduct. Pyrolysis is known to be the initial stage of the combustion or gasification processes (Zhou et al. 2013, 2017). Dioxins formation is prevented and NO_x formation is reduced by it because of the inert atmospheric condition and low temperature. Pyrolysis process undergoes a series of chemical reaction and after the process, pyrolysis oil and char are produced as end products (Massaro et al. 2014). Pyrolysis oil can be used as an alternative of fuel. It has a very high calorific value and it contains various oxygenated compounds, benzene derivatives, and various aromatic hydrocarbons (Quesada et al. 2019). Typically, the ideal range for pyrolysis temperature in the process of making biochar is 500–800°C. The three main categories of pyrolysis processes are: (i) slow pyrolysis, which involves slow heat transfer rates over a long time with temperature below

300°C, (ii) moderate pyrolysis, which involves moderate heat transfer rates between 300°C and 500°C, and (iii) fast pyrolysis, which involves rapid heat transfer over a lesser time, but at higher temperatures above 500°C (Dhyani et al. 2018). 500°C is thought to be the ideal temperature for the start of pyrolysis and the production of char (Liang et al. 2016).

2.1.1 Pyrolysis of plastics

Plastics are of very high strength, are resistant to corrosion, and have insulating property. Therefore, they are widely used in packaging of products, household chores and other appliances medical appliances (Miandad et al. 2016, Al-Salem et al. 2017). But plastic plays a major role in environmental pollution. The most commonly found polymers in MSW includes acrylonitrile butadiene styrene (ABS), polystyrene (PS), nylon, polyethylene (PE), Kevlar, polyvinyl chloride (PVC), vomex, epoxy, low density polyethylene (LDPE), Polypropylene (PP) and high-density polyethylene (HDPE). Pyrolysis of plastics leads to generation of recoverable energy and heat (Miandad et al. 2016). In pyrolysis, H_2 and hydrocarbon are, respectively, produced from carbon and hydrogen present in plastic. The pyrolysis of polysterene and polyethylene is generally done at 300–500°C temperature and it was found that the yield of hydrocarbon is high at higher temperature (Sophonrat et al. 2018). In co-pyrolysis of LDPE and cellulose, it was found that there is an increase in tar yield and reduction in gas yield in comparison to pyrolysis (Gunasee et al. 2017).

2.1.2 Pyrolysis of biomass waste

During the pyrolysis of biomass waste, CO_2/CO is produced as it contains high amount of oxygen and carbon content. The H/C ratio is increased by co-pyrolysis of biomass and fossil feedstocks which lead to production of more hydrocarbon gas (Du et al. 2021). The MSW component, which can be pyrolyzed, includes furnitures, clothing, bottles, newspapers, leather, rubber, glasses, food waste, etc. The co-pyrolysis of other components like tires (Chen et al. 2019) and biomass (Chen et al. 2019) along with food waste can increase the ratio of gas and olefins produced as the end products of pyrolysis. If the food waste contains high moisture contents, then it may hinder the process of pyrolysis. The pyrolysis of agricultural biomass in a N_2 atmosphere at 300, 400, 500 and 600°C was recently demonstrated by Mlonka-Medrala et al. (2021) and suggested that with an increase in temperature, the concentration of hydrogen, methane and the pyrolysis gas yield increases. Mohammad et al. (2020) explored four different agricultural wastes including willow, rice straw, UGU plant and bagasse to assess the two-stage pyrolytic-catalytic hydrogenation procedure. The Ni/Al_2O_3 was used as catalyst for the second-stage reaction which helped to enhance the yield and rate of methane production. The amount of cellulose and hemicellulose in biomass can also affect the methane generation (Xu et al. 2021). The synergistic impact between paper and PVC pipes were also reported recently in a co-pyrolysis study regarding the hydrogen rich syngas production and reported to obtain highest H_2 yield under 900°C during the co-pyrolysis (Wang et al. 2021).

2.1.3 Pyrolysis of rubber waste

Rubber is used in manufacturing of tires, gaskets, roll coverings, hoses, and crap tubes and in medicine sectors (Xiang et al. 2019). By changing the conditions of pyrolysis, products like any gas, charcoal and oil can be produced. The main components of pyrolysis gas are CH_4, C_2H_8 and CO. Monomers called 1, 3-butadiene and styrene are produced by pyrolysis of styrene-butadiene rubber (SBR) in the initial stage (Yang et al. 2021). H_2 is produced from the pyrolysis of natural rubber. Leung et al. (2002) pyrolyzed tire powder by using a rapidly heating device and the temperature reached 1200°C/

min, resulting in 5% to 23% increase in gas yield and 57% to 43% reduction in the yield of pyrolytic oil (Leung et al. 2002). If rubber is heated rapidly, it can lead to the rubber breaking into small molecular gas, improving gas yield. Catalyst can be used to increase the gas yield and proportion of hydrogen. Kordoghli et al. (2017) studied the catalytic pyrolysis of waste tires by mixing of oyster shells with Al_2O_3, MgO, $CaCO_3$, and ZSM-5 and found that there is 45% increase in gas yield as compared to non-catalyzed pyrolysis (Kordoghli et al. 2017). The efficiency of pyrolysis can be increased by the co-pyrolysis of rubber with other MSW (Du et al. 2021).

2.1.4 Resource recovery

Pyrolysis is carried out when there is no air, causing the formation of products in three states, namely solid carbon residues (biochar), liquid like (bio-oil), and non-condensable gases (Sipra et al. 2018). The product of pyrolysis varies due to variation in composition because of different parameters, i.e., type of reactor, form of feed stock, composition, etc. He et al. (2010) demonstrated that when pyrolysis of MSW is done with dolomite, syngas is produced. Higher the temperature, higher will be the yield (He et al. 2010). The production nitrogen oxides (NOx) and sulfur dioxides (SOx) are low in pyrolysis than that of incineration which leads into lower air pollution and higher syngas production. So, the energy generation and chance of energy recovery is also high in case of pyrolysis (Sipra et al. 2018, Veses et al. 2020).

2.1.5 Factors affecting treatment and resource recovery

Various factors can affect the pyrolysis process and resource recovery (Bamboriya et al. 2019) which has been described below:

- *Pyrolysis temperature* – At different pyrolysis temperature, different end products are produced for the same feedstock. The rate of production of gas in pyrolysis process is directly proportional to the pyrolysis temperature.
- *Heating rate* - The heating rate and heating temperature affect the ratio of the components that will form in pyrolysis process. If the temperature and heating speed is low, then the organic molecules get enough time for break down and then they recombine to form a thermosetting compound which is very difficult to degrade. If the temperature and heating speed is high, then the organic compounds break down and form many organic molecules of low molecular weight and there is increase in gas production.
- *Moisture content of garbage* – If the waste contains large amount of moisture, then to reach the balanced heat, it needs more amount of auxiliary fuel. The steam produced from it affects the proportion of various products formed in gasification process. Water vapor and carbon dioxide formed in the gasiffication reaction reduces the carbon content of the residue and increases the hydrogen and carbon monoxide content.
- *Oxygen* – Oxygen acts as an oxidizing agent in pyrolysis. It undergoes combustion and provides thermal energy and ensures the pyrolysis reaction. Therefore, the provision of oxygen is very important and requires strict control.

2.2 Incineration

Incineration is a thermal treatment process which includes combustion of waste substances (Shah et al. 2021). The main purpose of incineration is to decrease the volume of solid waste by thermal treatment of waste. Heat produced from this process is used to generate electricity (Materazzi et al. 2019). Many other substances like fly ash, slag and gases like sulfur dioxide, and nitrous oxides are also produced in this process which causes damage to the environment. This is one of the most feasible method for combustion of the mass solid waste in a cost-effective way without any pre-treatment and to generate energy from them (Kundariya et al. 2021). The most important benefit is the complete elimination of all organisms and mitigation of hazardous pollutant via mineralization. Incinerators protect the environment and people from toxic pollutants (Makarichi et al. 2018, Brunner et al. 2015) and is one of the most common methods for the disposal because of stringent landfill disposal legislation in many conutries (Scarlat et al. 2015, Kumar et al. 2017). The ash generated from incinerated plants can be used for the construction of highways and manufacturing of cement (Wang et al. 2018). Incineration is the most reliable and economical process in case of mass combustion and energy can be generated without pretreatment as it is an unprocessed method (Joseph et al. 2020). The pollutants released from incineration process can be controlled by air control system. For example, fabric filters or electrostatic precipitators are used for the removal of particulate matter, while flue-gas control systems is used to track NOx emissions. Selective catalytic reduction (SCR) can be used to remove nitrogen oxides (NOx) from flue gas as well (Kundariya et al. 2021).

2.2.1 Treatment method

The materials which can be reused are treated by incineration process. There are three types of thermal treatment, i.e., vitrification, melting and sintering (Zhang et al. 2021). In vitrification process, there is mixing of residues with precursor materials of glass and a single-phase amorphous, glassy product is added at 1000–1500°C. In retention mechanisms, there is formation of chemical bonds between the inorganic species present in the residue and glass forming materials, such as silica, and the residue constituents are encapsulated by a layer of glassy material. Melting process is similar to the vitrifying process which results in a multiphase product like molten metal, but there is no addition of glass in this process. Metals can be separate from the product and can be recycled. In sintering process, the residues are heated to a level at which bonding of particles occurs and reconfiguration of residue takes place, which form a denser, less porosity and higher strength product at a temperature around 900°C. At the starting phase of incineration, sintering takes place to some extent. Thermal treatment by incineration residues form homogeneous, denser product which has a good leaching property. The major drawbacks of the above processes are that all of them require high amount of energy.

Pollutants like heavy metals, dioxin, furan, etc., are released from incinerator, which are present in gas, water or ash and can be released during the process. So special landfill for disposal is needed. These pollutants cause threat to the health of humans and environment. Incineration of plastic like polyvinyl chloride (PVC) produce highly toxic pollutants. No safe way is there to prevent their production; however, they can be trapped in high tech filters or in ash which is very expensive. The temperature required by the heat exchangers to attempt energy recovery also increase the dioxin production. As a consequence, the pollutants spread all over the environment and then enter into the food chain. The use of combustion-based technologies to treat municipal solid waste (MSW) is a controversial topic while it offers a viable technical solution for waste management. Except the environmental and health issues, some economical issues are also associated with incineration. The equipment required for pollution control takes up most of the incinerator's size and cost. To enable the capture of particle and dangerous materials, the majority of incinerators incorporate a bag-house or electrostatic precipitator which makes it economically less feasible.

2.2.2 Resource recovery

Incineration process is done in presence of oxygen in which oxygen reacts with combustibles component at 800°C or above resulting in production of ash, thermal energy and gases (Shah et al. 2021). Primarily, it controls the combustion of waste with heat recovery for the production of steam, which further produces electricity (Materazzi et al. 2019). This method requires very small area for installation and there is greater reduction in waste volume. So this is the most preferable method (Luet al. 2017, Wang et al. 2018). The amount energy produced can be determined by the density of waste material, the amount of moisture present in it and inert material percentage present in it (Pavlas et al. 2011). Ash is another by-product of incineration of MSW that can be used to make various building materials like brick, cement, etc. (Shah et al. 2021). However, the plants necessitate constant monitoring to treat the gaseous pollutant in exhaust which is costly and limits its application.

2.2.3 Factors affecting treatment and resource recovery

- *Moisture content of waste* – If the percentage of moisture content present inside the waste is high, then to destroy it more amount of fuel is required.
- *Inorganic salts* – If the waste contains more amount of alkaline or inorganic salt, then some trouble may arise in disposal of waste in incineration process. A slag or cake is created which reduces the incineration process and on the furnace surface a fraction of salt gets collected.
- *High sulphur or halogen content* – The waste contains chlorides and sulfides which may form compounds that produce acids in the off gas.

2.3 Composting

Composting is a process in which organic waste is biologically broken down or degraded in the presence of oxygen to create manure or stabilized waste (Thomas et al. 2020). Organic waste that is used in composting is mainly municipal solid waste (MSW). In composting, the organic waste is degraded by microorganisms such as bacteria, fungi, algae, and protozoa under controlled environmental conditions. Resultant compost matter can be used as soil conditioner. MSW consist of different types of wastes such as farm waste, urban waste, residential waste, administrative waste, industrial waste, etc. (Abdel-Shafy et al 2018). In India, MSW contains 70–75% organic wastes (Ramachandra et al. 2018). The chemical composition of Indian MSW contains 0.64% nitrogen, 0.67% Phosphorus, 0.68% potassium and 26% C/N ratio (Malav et al. 2020). For the proper management practices, a proper understanding of the composition and the volume of the municipal solid waste must be known. This also helps to give an idea about how much energy that can be recovered from those collected municipal solid waste. There are many different advanced technologies for composting apart from the basic method of composting. They are vermi-composting, windrow composting, Indore method of composting, Bangalore method of composting, etc.

2.3.1 Treatment method

2.3.1.1 Indian Bangalore composting

This type of composting is mainly used for night soil and town refuses (SBMG 2015). This method was invented in 1939 (SBMG 2015). This is also known as hot fermentation mechanism. Plant leaf, garbage, animal dung, urine are mainly used as raw material in this type of composting method. It is an anaerobic method of composting process. In this method, a pit of dimension 10 m*1.5 m*1.5 m is filled with the layers of compost material and then it is covered with 15 cm of thick soil layer. After 2 to 3 days, the anaerobic microorganisms degrade the composed material which results in an increase in the temperature. The refuse gets fully stabilized after 4 to 5 months and the color of the compost changes into brown color which is odorless and in the powdered form which is known as manure (SBMG 2015).

2.3.1.2 Vessel composting

Any composting which takes place in an enclosed area that can be a building or a container or a vessel.This method relies on different varieties of forced aeration and different mechanical turning techniques which increases the composting process at a greater extent. It is an expensive and labor intensive process or method. This is a new and advanced technology which is different from other composting techniques. The complete system is enclosed in a tank or container or a vessel which has an outlet exhaust system for emission of gases that are harmful and odor is filtered by biofilters that are fitted at the exhaust unit. Rotation of the container and aeration pumps generally provides the aeration that is used to maintain steady air flow rate. The ideal moisture content (40–60%) can easily be maintained as well since the entire system is enclosed which reduces the water dependency (Rynk et al. 1992).

2.3.1.3 Windrow composting

In this type of composting process, the compost organic materials are separated and placed in a long narrow pile in the trapezoidal or triangular shape (CPHEEO 2016). The efficiency or the effectiveness of the composting method or process depends upon the size of the pile of the windrow. If the size of the windrow pile is too small, then it cannot withstand the weather conditions. If the size of the windrow pile is too large, then it affects the aeration of the composed material. The average size of the windrow pile should be 2–5 m in width at the bottom and 1–3 m in height (Kuhlman 1990).

2.3.1.4 Vermicomposting

This is the type of composting method in which earthworm is used to digest and breakdown the organic materials. All the materials are converted into granular form; mainly, they are rich in nitrogen content (SBMG 2015). In this type of composting, the temperature remains low as compared to the composting method due to low microbial action (Angima et al. 2011). In vermi-composting, C/N ratio decreases rapidly due to the mineralisation of the organic compound in the presence of earthworms (Bansal and Kapoor 2000). If there is a presence of fungi in the composting process, then it is beneficial for the earthworms because earthworms also take it as a supplementary diet and can grow at faster rate (Adi and Noor 2009).

2.3.2 Resource recovery

As compared to the recycling of the inorganic materials, the organic waste can be compost which can be used as organic fertilizers or manure in the agriculture fields. According to CRINIRDPR (2016), appropriate composting techniques may be used to compost the separated organic waste and government authorized recyclers' seller might be contacted to sell the inorganic recyclable waste materials. The other non-recyclable waste materials are generally sent to the closely situated municipality for proper landfill. This compost is rich in plant macronutrients like nitrogen, phosphorus, potassium and other essential nutrients which can be recovered by the composting process (Varjani et al. 2021a, Shah et al. 2021). Composts can be applied to improve field crops, seedlings, growing nursery flowers, plants, and herbs. Although it helps to improve the physico-chemical and biological attributes of soils, presence of heavy metals and toxic contaminants in compost can limit its application and neccessiate further research (Wei et al. 2000).

2.3.3 Factors affecting treatment and resource recovery

Temperature: The temperature around 65°C is the most appropriate temperature for the composting process because high microbial activity can be seen at this temperature which results in the acceleration of the degradation of organic materials. In vermi-composting treatment method, the temperature requirement is 30–35°C for the initial 15 days; after that, the temperature gets reduced to 25 to 30°C. This temperature is less than composting (https://compost-turner.net/composting-technologies/factors-affect-composting-process-and-compost-quality.html).

Aeration: The aerobic composting is much faster than anaerobic composting because of the aerobic conditions which boosts the degradation of the organic material and also there is no foul smell of the waste (CPHEEO 2016). There are three methods of aeration. They are natural, passive and forced aeration. Out of these three methods of aeration, passive aeration is the best aeration method. In this method of aeration, perforated pipes are used for aeration.

Moisture Content: In composting, moisture content affects the rate of biodegradation of the compost material through water potential, microbial growth rate and water activity. In the compost material, free air space should also be maintained. Free air space depends on the density and the moisture content. If the free air space is low, then it will decrease the efficiency of the composting process. Low moisture content can also lead to decrease in the efficiency of the composting process.

Carbon to Nitrogen (C: N) Ratio: In India, 30:1 is the optimum C/N ratio as per solid waste manual (CPHEEO 2016). Decrease in the C/N ratio below 25:1 leads to the production of foul smell, while increase in the C/N ratio above 30:1 may enhance the quality of the compost. But, if the C/N ratio is more than 35:1, then degradation of the compost will be hampered. According to Kapetanios et al. (1993), optimum C/N ratio for composting of the kitchen waste must be between 25 and 35. If C/N ratio falls below 25, total nitrogen concentration reduces, but excess of nitrogen is present in the form of nitrite and ammonia when C/N ratio is above 35 which helps in better degradation of the compost material.

2.4 Anaerobic digestion

Anaerobic Digestion (AD) is an organic matter decomposting method using anaerobic microorganisms and have been implemented successfully in the treatment of food wastes, agricultural wastes, and wastewater. AD can reduce chemical oxygen demand (COD) and Biological oxygen demand (BOD) from waste streams and produce renewable energy. Hence, it is widely used in treating food waste, agricultural wastes and wastewater (Jain et al. 2015). AD is an adaptable technology for handling various organic wastes and odor emissions; also, it offers an advantage of utilizing less energy for operational purposes and also has potential for recovering electricity (Shah et al. 2021). Anaerobic digestion is done in 4 stages as follows (Fig. 2): hydrolysis, acidogenesis, acetogenesis and methanogenesis using hydrogenotrophic, acidogenic, acetogenic and methanogenic bacteria (Kundariya et al. 2021).

2.4.1 Treatment method

First step in the AD is enzymatic hydrolysis where the complex organic polymers in the substrate are hydrolyzed using extracellular enzymes into soluble molecules (Kundariya et al. 2021). This is followed by fermentation where the compounds that are reduced by hydrolysis are further converted intocombination of volatile shortchain fatty acids, hydrogen, acetic acids using fermentative bacteria. The next step is acidogenesis, where the organic acids are converted to acetate, hydrogen and carbon

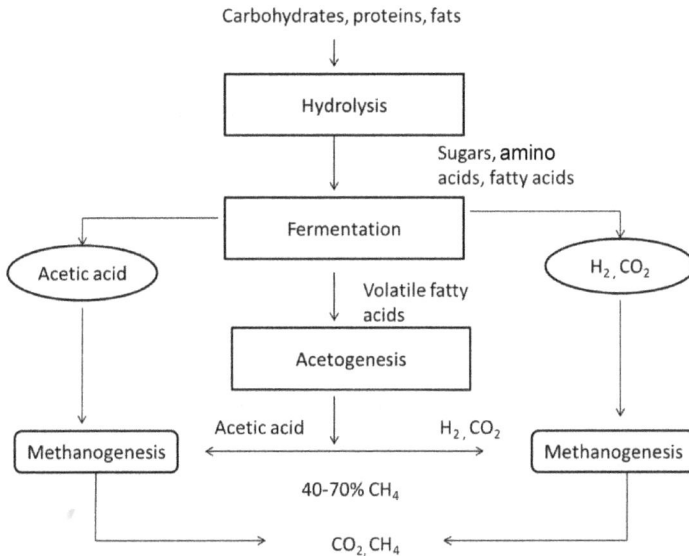

Figure 2. Schematic representation of anaerobic digestion process (Jain et al. 2015).

dioxide by the acetogenic bacteria. The last step is methanogenesis where the compounds obtained in the previous step are utilized by methanogenic bacteria for producing methane.

Hydrolytic bacteria are capable of converting complex compounds into soluble forms and also play a significant role in controlling reaction rate, thereby determining the biomass conversion potency (Jain et al. 2015). Cellulolytic microorganisms like Bacillus, Clostridium, Erwinia, Streptomyces, etc., are known to be capable of hydrolyzing cellulosic biomass found in various municipal and agricultural wastes. Fermentative bacteria consume the products produced by hydrolysis that are soluble and convert them to other intermediates of which acetate and carbon dioxide are primarily necessary for producing methane. Acetogenic bacteria are obligate producers of hydrogen, hence they cannot survive high partial pressure of hydrogen; a symbiotic relationship exists between the acetogens and the methanogens as methanogens consume hydrogen that is produced by acetogens. The methanogenic bacteria produce 30% of methane by carbon dioxide and hydrogen redox reaction and 70% by convertingacetic acid (Jain et al. 2015).

The types of anaerobic digesters and approaches used for anaerobic digestion process are listed below:

Continuous digester: Single stage continuous process involves the conversion of complex organic compounds to biogas, taking place in single chamber which is fed with raw materials regularly as consumed residue is moved out. The two-stage process includes a setup where the acidogenic and methanogenic chambers are separated physically. The acid production takes place in first stage and the diluted form of acids are used as feed for the next chamber in which bio methanation happens, after which collection of biogases can be done (Abu-Dahrieha et al. 2011, Jain et al. 2015).

Batch digester: The digester is fed between intervals when the digestion process is complete after which the reactor is emptied. The biocel reactor was developed to simplify handling of materials and eliminate need for mixing, thereby reducing the cost of process as it facilitates achievement of high loading and conversion rate (Rapport et al. 2008). Sequential batch anaerobic composting has two to three batch reactors in sequence where leachate is transferred between the reactors. Anaerobic phased solid digester is a double stage digester that uses batch loading for the stimulation of rapid production of organic acid. It combines the high solids reactors in initial stage and low solids mixed biofilm reactor in next stage. Bio-converter digester is a sequential batch system consisting of single stage digester (Jain et al. 2015).

Dome and drum digester: Floating gas holder plants are constructed either above or underneath the ground. Its design can be spherical, rectangular, or cylindrical, etc., and can be either horizontal or vertical. Fixed dome digester is constructed as an underground fermentation chamber with dome on top for storage of gas and outlet for gas from the fixed dome and digested slurry which leads to an outlet tank (Abu-Dahrieha et al. 2011, Jain et al. 2015).

2.4.2 Resource recovery

AD is a technique used for recovering energy from biodegradable waste, such as food waste and sludge from animals (Fig. 3). Organic fraction of MSW is commonly used as substrate for AD because of its ease of accessibility and high moisture content. By providing energy to replace fossil fuels, AD has a lower impact on air quality and aids in reducing carbon dioxide emissions. Microorganisms are used in AD to convert biomass into biogas (Kundariya et al. 2021). Biogas rich in methane and farmable digestate are primary outputs of AD process (Shah et al. 2021). Biogas enrichment is carried out in order to remove CO_2, H_2S and water vapors. Impurities in biogas can be removed by using pressurized water as an adsorbent liquid and counter current absorption to remove CO_2 and H_2S (Jain et al. 2015). It is estimated that 20 to 80 m^3 of biogas and 474 kWh of energy can be generated from 1 tons of organic waste (Pujara et al. 2019). Biogas produced by anaerobic digestion process is colorless and odorless and can be used as a substitute to Liquefied Petroleum Gas (LPG). The digested sludge of anaerobic digestion can be used as manure as it contains nitrogen, potassium and phosphorous, thereby having nutritive value. It comprises of less levels of cadmium and chromium compared to mineral fertilizers (Patwa et al. 2020).

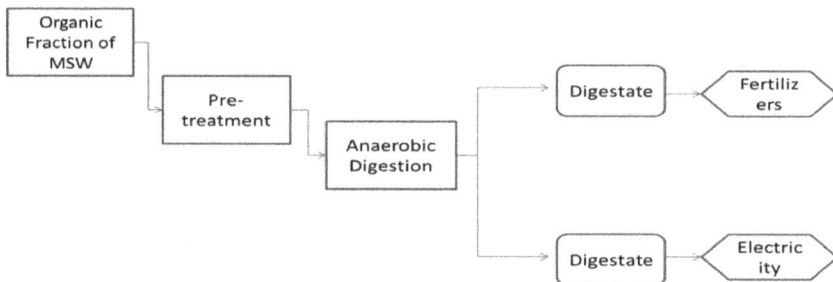

Figure 3. Schematic representation of resource recovery from anaerobic digestion process (Shah et al. 2021).

2.4.3 Factors affecting treatment and resource recovery

Temperature: Temperature plays a significant role in AD as it has an effect on the thermodynamic equilibrium of biochemical reaction, stability, microbial growth rate, kinetics, and mainly the yield of methane (Panigrahi and Dubey 2019). Various researches have reported that digestion at higher temperatures occurs more rapidly compared to lesser temperatures. Favorable temperatures depend on the microorganism type. The optimum temperature for psychrophilic is 20°C, for mesophilic it is 35°C and in case of thermophilic it is 55°C (Jain et al. 2015).

pH: Formation of organic acids at faster rate is unfavorable for methane forming microbes as the pH drops. For normal fermentation and production of gas, an ideal pH of 6.5 to 8 must be maintained (Jain et al. 2015). For maintaining ideal methanogen populations, a pH range of 7 to 7.5 has to be maintained in the anaerobic digester (Dahiya and Joseph 2015).

C/N ratio: Suitable ratio of C/N ratio of is importance for metabolic processes of microorganisms (Panigrahi and Dubey 2019). Anaerobic bacteria grow at C/N ratio ranging from 20/1 to 30/1 in AD system, having an optimum ratio of 25/1. High ammonia-nitrogen release and high volatile fatty acids' (VFA) accumulation can occur due to improper ratio of C/N which may result in inhibition of the anaerobic digestion process by decreasing activity of methanogens (Jain et al. 2015).

Solid content: Increase in solid content decreases the methane yield and COD removal (Jain et al. 2015). Higher the reduction in total solids, higher is the efficiency of the anaerobic digester. The organic matter of feed consists of volatile solids which are consumed during AD for production of biogas (Dahiya and Joseph 2015).

Digester design: Digester must be provided with feed in same time, quality and quantity everyday to maintain the microbes in constant concentrations of organic solids. Gas production was found to be optimum when the ratio of diameter to depth was in range of 0.66 to 1 and satisfactory when diameter of depth 16 ft and diameter 4 to 5 ft was used as difference in depths varies with the temperature (Jain et al. 2015).

Mixing: Mixing is an important step for making food available to the bacteria. Slight mixing is found to improve fermentation process and vigorous mixing may retard the process of digestion (Jain et al. 2015). Mixing is generally done by pumped circulation, mechanical agitation and gas circulation. It plays a crucial role in achievement of high performance of AD.

Retention time: Hydraulic retention time and Solid retention time (SRT), if low, can affect the overall stability of process by posing a risk of biomass washout from the reactor, whereas the SRT being high can lead to inadequate utilization of components in digester; hence, SRT is to be maintained for minimum 4 days and maximum 14 weeks (Panigrahi and Dubey 2019). Retention time must not be less than 2 to 4 days because the methane organisms may come out along with the slurry, which may affect biogas production. Feeding rate depends on the retention period and feedstock type (Jain et al. 2015).

Pressure: It is reported that fermentation is affected by pressure on slurry surface; hence, it is better to maintain lower pressure (Jain et al. 2015). Pressurization is also found to improve the efficiency of biogas collection. Operational pressure of 20 bars was found to be adequate for enrichment of methane (Panigrahi and Dubey 2019).

Toxicants and inhibition: Removal of oxygen present in headspace can be carried out by head space flushing, thereby mitigating headspace inhibition. Methanogens require strong reducing environment for their optimum growth. Hence, reducing agents like cysteine, titanium III, sulphide, etc., are added to reactor. It is ideal to maintain oxidation-reduction potential of 200 mV to 300 mV at pH 7 (Panigrahi and Dubey 2019). Process of AD is affected by unwanted components like wood shavings, sand, glass, metals, straw or polymeric components. Food wastes contain organic acids in high amounts which can be converted easily into VFA, presence of which in excess during early stages of process can lead to extreme drop in pH, thereby inhibiting methanogenesis. This can be reduced by using two phase digestion system or carrying out co-digestion of carbohydrate rich feedstock with others. Accumulation of compounds such as acetic acid, propionic acid, and butyric acid that are intermediates produced during bio-digestion causes drop in pH, thereby leading to reduced methane formation rate (Jain et al. 2015). Nutrients can be toxic depending on their dissolved concentration and chemical forms. Inorganic components like heavy metals, sulphide and free ammonia are found to be inhibitory compounds. Acetogens are less sensitive to toxicity of trace elements compared to methanogens in process of anaerobic digestion (Panigrahi and Dubey 2019).

2.5 Landfilling

A large amount of MSW produced and discarded on daily basis into a large area of land is considered as a landfill (Varjani et al. 2021b, Shah et al. 2021). Landfill is considered to be the method of final disposal; it consists of different layers and has soil layer on top of it (Ozbay et al. 2021). It is also considered to be the most economical and common method. It is designed for disposing of solid wastes in a safe manner; here the top and bottom layers play a crucial role as they prevent the leachate from penetrating into the soil. Penetration of leachate can cause emission of volatile organic compounds and greenhouse gases, contamination of ground water, rotten odor and various other environmental problems. 70% of the collected municipal solid waste is preferred to be landfilled (Pujara et al. 2019). The important components required for landfill designing include landfill foundation, leachate collection tank, barrier cap, liner, station for gas collection, soil cover and gas fare station. Landfills are classified into sanitary landfills, open dump landfills, and semi-controlled landfills (Ozbay et al. 2021). An open dump landfill is where the solid waste is disposed in open environment, while the semi- controlled landfill is one located in designated sites for dumping of solid wastes after carrying out sorting, shredding and compacting. The sanitary or engineered landfill is done by digging dumpsites for storage of solid wastes (Nanda and Berruti 2021). Development, implementation and operational plans are required for ensuring the optimal conditioning and effective functioning of the landfill system (Warith 2003).

2.5.1 Treatment method

Bioreactor landfills (Fig. 4) are different from conventional landfills by their usage of hydrolytic, putrefying, fermentative and methanogenic microorganisms for decomposing solid waste by creating favorable conditions in the landfill. By increasing the anaerobic decomposition of biodegradable components, bioreactor landfills can accelerate the mineralization and stabilisation of solid waste. It uses leachate recirculation to increase the microbial activity for quicker organic waste processing and degradation (Nanda and Berruti 2021). Bioreactor landfills are designed for minimizing rainwater infiltration, leachate migration and maximizing rate of generation of landfill gases (Warith 2003). They make use of microbial processes for transforming and stabilizing decomposable organic constituents of waste within duration of 5 to 10 years by implementing bioreactor processes. It increases the effectiveness of the process, conversion rates and extent of decomposition of organic waste (Pacey et al. 1999). Microbial processes tend to reduce solid waste

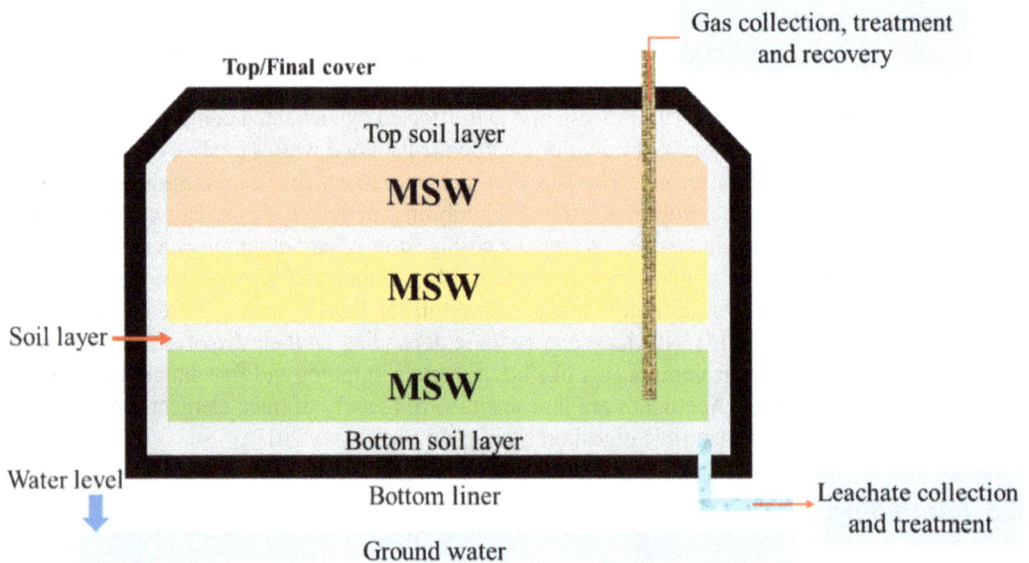

Figure 4. Schematic representation of bioreactor landfill (Nanda and Berruti 2021).

and the leachate circulation system enhances biodegradation of waste, which leads to acceleration of waste breakdown rate, thereby decreasing the stabilization time required. Hence, they minimize the probability of environmental issues as longer the landfill is un-stabilized, the longer it remains as a causative source for environmental issues (Warith 2003). Specific management methods and modifications in operation are required to enhance the microbial decomposition, which can be done by liquid addition and management strategy. Apart from that, the bioreactor process is optimized by methods such as waste shredding, nutrient addition, pH adjustment, management of temperature, and pre and post disposal conditioning of waste (Pacey et al. 1999).

2.5.2 Resource recovery

When methanogens breakdown complex molecules, landfill gases are produced. They include about 90% of methane and other gases and leachate including carbon monoxide, alcohols, nitrogen, organosulfur compounds, heavy metals and hydrocarbons (Fig. 5). These gases are further subjected to treatment and used for production of energy (Pujara et al. 2019). Waste that is degraded by microbes in landfill is stable; thereby, conversion of it into energy and secondary materials can be beneficial (Du et al. 2021). Landfill gases formation occurs either spontaneously or takes up to 0.5 to 3 years and reduces as the landfill matures. Hence, recovery of landfill gases within a period of three years is essential. Landfill mining can be classified into *in situ* and *ex situ* mining. *In situ* mining involves the recovering of resources such as landfill gases and leachate. It also involves removing of contaminants present in water and soil without excavating stored waste. On the other hand, *ex situ* mining involves the recovering of resources through excavation of stored stream of waste for subjecting it to further on-site treatment. The formation of landfill gas is depicted in Fig. 5 through a number of chemical processes, including hydrolysis, anaerobic oxidation, fermentation, acidogenesis, acetogenesis, and methanogenesis. Hydrolytic fermentative bacteria, hydrogenotrophic methanogens, proton-reducing acetogens and acetoclastic methanogens are among the bacterial communities that participate in the anaerobic digestion of organic waste in a landfill and the consequent gas production (Nanda and Berruti 2021).

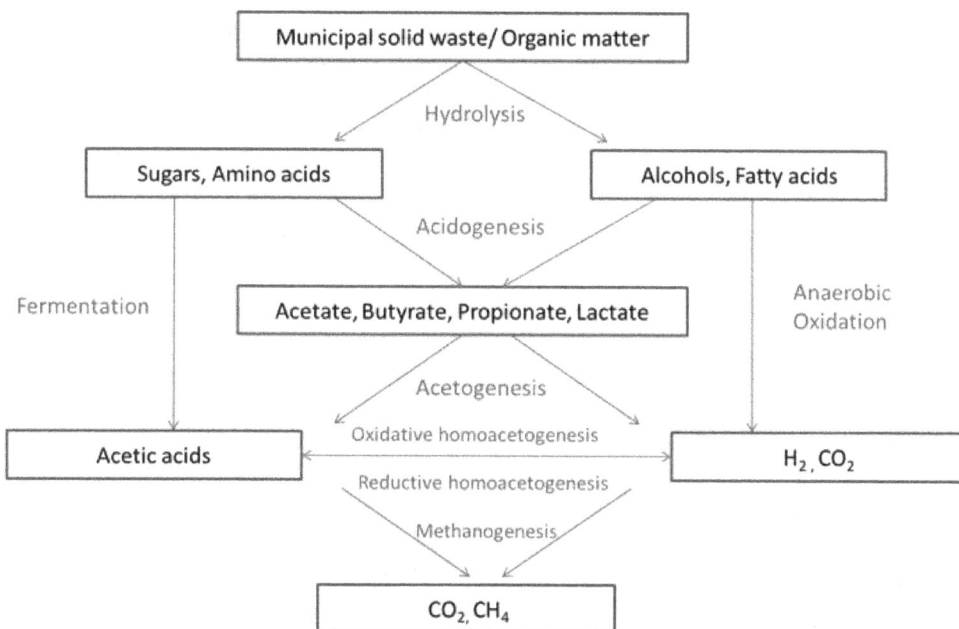

Figure 5. Mechanism of landfill gas formation (Nanda and Berruti 2021).

The landfill gases can be generated by the following five stages:

- **Initial adjustment stage:** The organic components in solid wastes undergo decomposition by microorganisms in presence of oxygen, after which it is filled in a landfill where the biological composition by aerobic organisms takes place. This generates heat and higher temperature (Warith 2003).

- **Transition phase:** Exhaustion of oxygen takes place and anaerobic condition is developed. Electron acceptors such as nitrate and sulfate are converted into hydrogen sulfide and nitrogen gas. Methane generation happens when values of oxidation reduction potential ranges from 150 mV to 300 mV. Due to high concentration of carbon dioxide and organic acids being present, the pH of leachate generated drops (Warith 2003).

- **Acid phase:** Compounds with high molecular mass are enzymatically hydrolyzed into compounds that can serve as source of energy for microorganisms, followed by acidogenesis, where the microbial compounds from enzymatic hydrolysis are converted into compounds of low molecular mass using acidogens. Carbon dioxide and small quantity of hydrogen gas are generated during acid phase. As various inorganic constituents are solubilized, the pH of leachate drops and also the BOD, COD and conductivity of leachate tends to increase (Warith 2003).

- **Methane fermentation phase:** Hydrogen gas and acetic acid are converted to CO_2 and methane by methanogens. This is followed by a rise in pH and decrease in BOD, COD and heavy metal concentration of the leachate (Warith 2003).

- **Maturation phase:** The landfill gas generation is reduced during this phase due to less availability of the nutrients after removal of leachate. Gases primarily produced during this phase include methane and carbondioxide, apart from which nitrogen is also present in small quantity (Warith 2003).

2.5.3 Factors affecting treatment and resource recovery

Factors that influence landfill mining include composition of waste dumped, extent of degradation of waste, treatment and operating procedures, leachate and gas collection facilities, and markets available for recycling of materials that are recovered (Nanda and Berruti 2021). Critical parameters that influence the landfill gas generation include nutrient content, pH, bacterial content, oxygen content, temperature, and moisture level. These parameters may also influence the other parameters that are involved in controlling of the solid waste degradation rate and other related activities (Warith 2003).

- **Moisture content:** Aqueous environment is required for production of gas, decomposition of solid waste, transportation of nutrients and bacteria inside the landfill. Moisture content increases gas production by 25 to 50% when compared to that produced in presence of less moisture content. Hence, it's required to maintain moisture content of 15 to 40% for proper functioning of landfill (Warith 2003).

- **pH level:** Acidic pH causes increase in solubility, ion exchange between leachate and organic compounds and decrease in adsorption. Hence, it is optimum to maintain a pH of 6.5 to 7.5, within which methane production is maximum as the growth rate of methanogens is higher. A pH less than 6 and higher than 8 lowers the methane production. Presence of alkalinity, industrial waste and groundwater infiltration can affect the pH level (Warith 2003).

- **Nutrient content:** Nutrients required for anaerobic degradation include hydrogen, oxygen, nitrogen, phosphorous, calcium, magnesium and other trace compounds. These play a crucial role as the gas generation is greater when the nutrient digestion is more. Gas production can be retarded by presence of heavy metals as they are toxic (Warith 2003).

- **Temperature:** The heat generated from anaerobic decomposition process can lead to rise in temperature. Methane formation rate is affected by rise in temperature. Landfill temperature of 30 to 45°C is ideal for a deep landfill with moderate water flux (Warith 2003).
- **Non-technical barriers:** Some of the non-technical barriers faced by the bioreactor landfills include limitation of regulatory awareness, insufficient experience, lack of financing, time extensions for planning and licensing, and increased regulatory constraints (Pacey et al. 1999).

3. CHALLENGES AND FUTURE PERSPECTIVE

Developed and developing countries are the major manufacturer of the globe, thus generating huge amount of solid waste. Solid waste has been found to have harmful environmental effect and hazardous health impact. Solid waste adversely affects the environment in many ways: it enters water sources, thus contaminating the consumable water, resulting in water-borne diseases like Minmata and Blue-Baby syndrome in consumers (Kundariya et al. 2021, Varjani et al. 2021b, Shah et al. 2021). Small particles have the ability to enter air and pollute it by increasing PM2.5 and PM10 values. It can be observed that capital city of developing countries such as Delhi (capital of India) and Seoul (capital of South Korea) have toxic air. They have been found to have five times more toxicity than the scale set up by the World Health Organisation (WHO). Sustainable solid waste management is therefore the need of the hour. The treatment methods currently available are very efficient, but the cheap methodology for solid waste reduction from environment have it its own advantages and harmful impacts. It is important to assess different solid waste composition and management strategies and then select the most appropriate approach. Pyrolysis has limits: when heavy metals exceeds certain levels, it may have adverse effect on soil and under-ground water. Incineration generates green-house gases. Composting requires proper management and emits very unpleasant smell. Anaerobic digestion requires manure which is not available in many parts of the globe such as Shanghai and Singapore. Landfilling needs vast region of land and is time consuming. These are the challenges particular to techniques but management of solid waste and its treatment is equally challenging. As in many parts of the country, waste segregation in not appropriate which results in collection of mixture of waste which is not suitable for any type of treatment plan. Some techniques are specific with type of material they degrade (Varjani et al. 2021b, Shah et al. 2021). Hence, it is required to maintain the quality of solid waste. Different parts of the country generates different composition of waste, thus there is need to study and follow appropriate process. Even the co-ordination between union and state waste management institute is a problem.

There are studies and researches in progress to reduce hazardous impact of waste management techniques (Shah et al. 2021). Solid waste management can be practiced at various levels such as at home, within society, in city/town/village and among states. Wet waste can be used for bio-composting by institutes; similarly, kitchen waste can be treated at home with lawns and large garden. Segregation of waste at source is very crucial and its importance and awareness need to be spread among citizens (Varjani et al. 2021, Shah et al. 2021). Industries need to elute out waste only after waste treatment. Laws and policies for environment conservation should be stricter (Shah et al. 2021). Penalties and reward system should be followed for creating awareness in the society. Solid waste collections are required to be more well practised. More investment and technology is required in solid waste management. Communication between waste management bodies needs to be improved.

4. CONCLUSION

In the urge to improve economic status of the country, we cannot sacrifice the environment. There is need for sustainable development. Development of the country comes with tonnes of solid waste generation, which needs to be effectively managed. The chapter highlights the different solid waste management techniques, its importance and limitations. The different MSW techniques including thermal conversion (incineration, pyrolysis and gasification), biochemical treatments methods (composting and anaerobic digestion), and landfilling are less harmful to the environment but are influenced by various environmental factors like temperature, moisture content, pH, solid content, nutrient availability, etc. All the solid water techniques reduce the total solid waste by significant amount. It is the utmost requirement to study in detail the composition of solid waste and then select the most appropriate treatment technique. Awareness about treatment about 3R's - Reduce, Reuse and Recycle - needs to be spread in primary stages. The objectives have to be set for long-term period. Sensitive and appropriate analysis should be selected and carried out for solid waste treatment. A technique which is less hazardous to environment and generates high-value added products such as pyrolysis is suggested. Further research needs to be done towards finding an ecofriendly, cost effective approach which is acceptable by society and laws. Together, we all can contribute to solid waste management and cleanliness of the surroundings by incorporating small changes, thus leading towards a safer and greener future.

REFERENCES

Abdel-Shafy, H.I. and Mansour, M.S. 2018. Solid waste issue: Sources, composition, disposal, recycling, and valorization. Egyptian Journal of Petroleum 27(4): 1275–1290.

Abu-Dahrieha, J., Orozco, A., Groomb, E. et al. 2011. Batch and continuous biogas production from grass silage liquor. Bioresour. Technol. 102: 10922–8.

Adi, A.J. and Noor, Z.M. 2009. Waste recycling: Utilization of coffee grounds and kitchen waste in vermicomposting. Bioresource Technology 100(2): 1027–1030.

Angima, S.D., Noack, M. and Noack, S. 2011. Composting with worms. https://extension.oregonstate.edu/sites/default/files/catalog/auto/EM9034.pdf.

Arafat, H.A., Jijakli, K. and Ahsan, A. 2015. Environmental performance and energy recovery potential of five processes for municipal solid waste treatment. Journal of Cleaner Production 105: 233–240.

Ayilara, M.S., Olanrewaju, O.S., Babalola, O.O. et al. 2020. Waste management through composting: Challenges and potentials. Sustainability 12(11): 4456.

Bae, Y.J., Ryu, C., Jeon, J.K. et al. 2011. The characteristics of bio-oil produced from the pyrolysis of three marine macroalgae. Bioresource Technology 102(3): 3512–3520.

Bamboriya, O.P., Thakur, L.S., Parmar, H. et al. 2019. A review on mechanism and factors affecting pyrolysis of biomass. Int. J. Res. Advent. Technol. 7(3): 1014–1024.

Bansal, S. and Kapoor, K.K. 2000. Vermicomposting of crop residues and cattle dung with Eisenia foetida. Bioresource Technology 73(2): 95–98.

Bridgwater, A.V. 2007. The production of biofuels and renewable chemicals by fast pyrolysis of biomass. International Journal of Global Energy Issues 27(2): 160–203.

Brunner, P.H. and Rechberger, H. 2015. Waste to energy–key element for sustainable waste management. Waste Management 37: 3–12.

Chen, G., Andries, J. and Spliethoff, H. 2003. Catalytic pyrolysis of biomass for hydrogen rich fuel gas production. Energy Conversion and Management 44(14): 2289–2296.

Chen, L., Yu, Z., Fang, S. et al. 2018. Co-pyrolysis kinetics and behaviors of kitchen waste and chlorella vulgaris using thermogravimetric analyzer and fixed bed reactor. Energy Conversion and Management 165: 45–52.

Chen, J., Ma, X., Yu, Z. et al. 2019. A study on catalytic co-pyrolysis of kitchen waste with tire waste over ZSM-5 using TG-FTIR and Py-GC/MS. Bioresource Technology, 289: 121585.

CPHEEO (Central Public Health & Environmental Engineering Organization). 2016. Manual on Municipal Solid Waste Management. Ministry of Housing and Urban Affairs, Government of India.

Cremiato, R., Mastellone, M.L., Tagliaferri, C. et al. 2018. Environmental impact of municipal solid waste management using Life Cycle Assessment: The effect of anaerobic digestion, materials recovery and secondary fuels production. Renewable Energy 124: 180–188.

CRINIRDPR (Centre for Rural Infrastructure National Institute of Rural Development and Panchayati Raj). 2016. Solid Waste Management in Rural Areas a Step-By-Step Guide for Gram Panchayats. Union Ministry of Rural Development, Government of India, http://nirdpr.org.in/nird_docs/sb/doc5.pdf.

Dahiya, S. and Joseph, J. 2015. High rate biomethanation technology for solid waste management and rapid biogas production: An emphasis on reactor design parameters. Bioresource Technology 188: 73–78.

Dhyani, V. and Bhaskar, T. 2018. A comprehensive review on the pyrolysis of lignocellulosic biomass. Renewable Energy 129: 695–716.

Du, Y., Ju, T., Meng, Y. et al. 2021. A review on municipal solid waste pyrolysis of different composition for gas production. Fuel Processing Technology 224: 107026.

EAI Report, Ministry of new and renewable energy. http://www.eai.in/ref/ae/wte/ concepts.html, Accessed date: 7 April 2019.

Fernández-Nava, Y., Del Rio, J., Rodríguez-Iglesias, J. et al. 2014. Life cycle assessment of different municipal solid waste management options: a case study of Asturias (Spain). Journal of Cleaner Production 81: 178–189.

Golomeova, S., Srebrenkoska, V., Zhezhova, S. et al. 2013. Solid waste treatment technologies. Machines, Technologies, Materials (9): 59–61.

Gunasee, S.D., Danon, B., Görgens, J.F. et al. 2017. Co-pyrolysis of LDPE and cellulose: Synergies during devolatilization and condensation. Journal of Analytical and Applied Pyrolysis 126: 307–314.

He, M., Xiao, B., Liu, S. et al. 2010. Syngas production from pyrolysis of municipal solid waste (MSW) with dolomite as downstream catalysts. Journal of Analytical and Applied Pyrolysis 87(2): 181–187.

Jain, S., Jain, S., Wolf, I.T. et al. 2015. A comprehensive review on operating parameters and different pretreatment methodologies for anaerobic digestion of municipal solid waste. In Renewable and Sustainable Energy Reviews 52: 142–154.

Joseph, L.P. and Prasad, R. 2020. Assessing the sustainable municipal solid waste (MSW) to electricity generation potentials in selected Pacific Small Island Developing States (PSIDS). Journal of Cleaner Production 248: 119222.

Kapetanios, E.G., Loizidou, M. and Valkanas, G. 1993. Compost production from Greek domestic refuse. Bioresour. Technol. 44(1): 13–16.

Kordoghli, S., Khiari, B., Paraschiv, M. et al. 2017. Impact of different catalysis supported by oyster shells on the pyrolysis of tyre wastes in a single and a double fixed bed reactor. Waste Management 67: 288–297.

Kuhlman, L.R. 1990. Windrow composting of agricultural and municipal wastes. Resources, Conservation and Recycling 4(1-2): 151–160.

Kumar, A. and Samadder, S.R. 2017. A review on technological options of waste to energy for effective management of municipal solid waste. Waste Management 69: 407–422.

Kundariya, N., Mohanty, S.S., Varjani, S. et al. 2021. A review on integrated approaches for municipal solid waste for environmental and economical relevance: Monitoring tools, technologies, and strategic innovations. In Bioresource Technology 342: 125982.

Leung, D.Y.C., Yin, X.L., Zhao, Z.L. et al. 2002. Pyrolysis of tire powder: Influence of operation variables on the composition and yields of gaseous product. Fuel Processing Technology 79(2): 141–155.

Liang, C., Gascó, G., Fu, S. et al. 2016. Biochar from pruning residues as a soil amendment: Effects of pyrolysis temperature and particle size. Soil and Tillage Research 164: 3–10.

Malav, L.C., Yadav, K.K., Gupta, N. et al. 2020. A review on municipal solid waste as a renewable source for waste-to-energy project in India: Current practices, challenges, and future opportunities. Journal of Cleaner Production 277: 123227.

Nabegu, A.B. 2010. An analysis of municipal solid waste in Kano metropolis, Nigeria. Journal of Human Ecology 31(2): 111–119.

Nanda, S. and Berruti, F. 2021. Municipal solid waste management and landfilling technologies: A review. In Environmental Chemistry Letters. Springer Science and Business Media Deutschland GmbH 19(2): 1433–1456.

Ozbay, G., Jones, M., Gadde, M. et al. 2021. Design and operation of effective landfills with minimal effects on the environment and human health. Journal of Environmental and Public Health, 1–13.

Pacey, J., Augenstein, D., Morck, R. et al. 1999. The bioreactor landfill-an innovation in solid waste management. MSW Management 1: 53–60.

Panigrahi, S. and Dubey, B.K. 2019. A critical review on operating parameters and strategies to improve the biogas yield from anaerobic digestion of organic fraction of municipal solid waste. In Renewable Energy 143: 779–797.

Patwa, A., Parde, D., Dohare, D. et al. 2020. Solid waste characterization and treatment technologies in rural areas: An Indian and international review. In Environmental Technology and Innovation 20: 101066.

Pujara, Y., Pathak, P., Sharma, A. et al. 2019. Review on Indian Municipal Solid Waste Management practices for reduction of environmental impacts to achieve sustainable development goals. In Journal of Environmental Management 248: 109238.

Ramachandra, T.V., Bharath, H.A., Kulkarni, G. et al. 2018. Municipal solid waste: Generation, composition and GHG emissions in Bangalore, India. Renewable and Sustainable Energy Reviews 82: 1122–1136.

Rapport, J., Zhang, R., Jenkins, B.M. et al. 2008. Current anaerobic digestion technologies used for treatment of municipal organic solid waste. Available at: California Environmental Protection Agency.

Robert Rynk. 1992. On-farm composting handbook. Monogr. Soc. Res. Child. Dev. 77: 132.

Sabbas, T., Polettini, A., Pomi, R. et al. 2003. Management of municipal solid waste incineration residues. Waste Management 23(1): 61–88.

Salman, Z. 2018. Negative Impacts of Incineration-Based Waste-to-Energy Technology.

Scarlat, N., Motola, V., Dallemand, J.F. et al. 2015. Evaluation of energy potential of municipal solid waste from African urban areas. Renewable and Sustainable Energy Reviews 50: 1269–1286.

Shah, A.V., Srivastava, V.K., Mohanty, S.S. et al. 2021. Municipal solid waste as a sustainable resource for energy production: State-of-the-art review. Journal of Environmental Chemical Engineering 9(4): 105717.

Sharholy, M., Ahmad, K., Mahmood, G. et al. 2008. Municipal solid waste management in Indian cities—A review. Waste Management 28(2): 459–467.

Sipra, A.T., Gao, N. and Sarwar, H. 2018. Municipal solid waste (MSW) pyrolysis for bio-fuel production: A review of effects of MSW components and catalysts. Fuel Processing Technology 175: 131–147.

Singh, J.S., Singh, S.P. and Gupta, S.R. 2014. Ecology, Environmental Science & Conservation. S. Chand Publishing.

Sophonrat, N., Sandström, L., Zaini, I.N. et al. 2018. Stepwise pyrolysis of mixed plastics and paper for separation of oxygenated and hydrocarbon condensates. Applied Energy 229: 314–325.

SBMG (Swachh Bharat Mission Gramin), Ministry of Drinking Water and Sanitation, Government of India, 2015. https://www.susana.org/en/knowledge-hub/resources-and publications/library/details/2322.

Thomas, P. and Soren, N. 2020. An overview of municipal solid waste-to-energy application in Indian scenario. Environment, Development and Sustainability 22(2): 575–592.

Tyagi, V.K., Fdez-Güelfo, L.A., Zhou, Y. et al. 2018. Anaerobic co-digestion of organic fraction of municipal solid waste (OFMSW): Progress and challenges. Renewable and Sustainable Energy Reviews 93: 380–399.

Van Fan, Y., Lee, C.T., Klemeš, J.J. et al. 2018. Evaluation of effective microorganisms on home scale organic waste composting. Journal of Environmental Management 216: 41–48.

Vandana Bharti, Jaspal Singh and Singh, A.P. 2017. A review on solid waste management methods and practices in India. Trends in Biosciences 10(21): 4065–4067.

Varjani, S., Rakholiya, P., Shindhal, T. et al. 2021a. Trends in dye industry effluent treatment and recovery of value-added products. Journal of Water Process Engineering 39: 101734.

Varjani, S., Shah, A.V., Vyas, S. et al. 2021b. Processes and prospects on valorizing solid waste for the production of valuable products employing bio-routes: A systematic review. Chemosphere 282: 130954.

Veses, A., Sanahuja-Parejo, O., Callén, M.S. et al. 2020. A combined two-stage process of pyrolysis and catalytic cracking of municipal solid waste for the production of syngas and solid refuse-derived fuels. Waste Management 101: 171–179.

Wang, Y., Zhang, X., Liao, W. et al. 2018. Investigating impact of waste reuse on the sustainability of municipal solid waste (MSW) incineration industry using emergy approach: A case study from Sichuan province. China. Waste Management 77: 252–267.

Wang, C., Jiang, Z., Song, Q. et al. 2021. Investigation on hydrogen-rich syngas production from catalytic co-pyrolysis of polyvinyl chloride (PVC) and waste paper blends. Energy 232: 121005.

Warith, M.A. 2003. Solid waste management: New trends in landfill design. Emirates Journal for Engineering Research 8(1): 61–70.

Wei, Y.S., Fan, Y.B., Wang, M.J. et al. 2000. Composting and compost application in China. Resources, Conservation and Recycling 30(4): 277–300.

Xiang, H., Yin, J., Lin, G. et al. 2019. Photo-crosslinkable, self-healable and reprocessable rubbers. Chemical Engineering Journal 358: 878–890.

Xu, D., Yang, S., Su, Y. et al. 2021. Simultaneous production of aromatics-rich bio-oil and carbon nanomaterials from catalytic co-pyrolysis of biomass/plastic wastes and in-line catalytic upgrading of pyrolysis gas. Waste Management 121: 95–104.

Yaman, S. 2004. Pyrolysis of biomass to produce fuels and chemical feedstocks. Energy Conversion and Management, 45(5): 651–671.

Yang, Q., Yu, S., Zhong, H. et al. 2021. Gas products generation mechanism during co-pyrolysis of styrene-butadiene rubber and natural rubber. Journal of Hazardous Materials 401: 123302.

Zhang, Y., Cui, Y., Liu, S. et al. 2020. Fast microwave-assisted pyrolysis of wastes for biofuels production–A review. Bioresource Technology 297: 122480.

Zhang, Y., Wang, L., Chen, L. et al. 2021. Treatment of municipal solid waste incineration fly ash: State-of-the-art technologies and future perspectives. Journal of Hazardous Materials 411: 125132.

Zhou, H., Long, Y., Meng, A. et al. 2013. The pyrolysis simulation of five biomass species by hemi-cellulose, cellulose and lignin based on thermogravimetric curves. Thermochimica Acta 566: 36–43.

Zhou, H. 2017. Combustible solid waste thermochemical conversion: A study of interactions and influence factors, 1–47.

6

BIOLOGICAL TREATMENT OF LEATHER INDUSTRY SOLID WASTE
CURRENT STATUS AND FUTURE PERSPECTIVES

Merlyn Sujatha R.,[1] Saranya R.,[2] Jayapriya J.,[3,]*
Aravindhan R.[4] and Tamilselvi A.[4]

1. INTRODUCTION

Tanneries are one of the oldest industries established in several parts of the world. The growing demand for leather and leather goods has steered large commercial tanneries and leather goods manufacturing industries to thrive (Bayrakdar 2020a). Today, the leather industry ranks 8th in India's export trade in terms of foreign exchange earnings. India produces nearly 10% of the total global supply of rawhide and skin, which are the basic raw materials for the leather industry (Singh 2017). Tanning refers to the conversion of animal hides and skins to leather using chemicals. The term 'hide' denotes the skin of large animals, such as cows and horses, whereas 'skin' represents those of small animals, such as goats and sheep. Generally, the skin is composed of three layers: epidermis, dermis, and subcutaneous layers. The dermis consists of about 30–35% protein, mostly collagen, and the remaining is water and fat. Leather is generated by tanning, the process in which the derma, epidermis and flesh are chemically converted into a durable non-decaying material (leather). The first step in the process involves cleaning, liming, deliming, and washing. In the case of chrome tanning, after deliming, the hides and skins are subjected to bating and pickling before chrome tanning (Saranya et al. 2016). The effluent from each step of the tanning process is commonly composed of high organic and inorganic dissolved and suspended solid content with high oxygen demand, as well as potentially

[1] Department of Biomedical Engineering, J.N.N.Institute of Engineering Kannigaipair – 601102.
[2] Department of Biotechnology, SRM Institute of Science and Technology, Ramapuram, Chennai - 600089.
[3] Department of Applied Science and Technology, AC Tech, Anna University, Chennai–600025.
[4] Leather Process Technology Department, CLRI -Central Leather Research Institute, Adyar, India.
* Corresponding author: jayapriyachem@gmail.com

toxic metal salt residues. The tanning process is often associated with an unpleasant odour due to the decomposition of solid protein waste and the release of hydrogen sulfide, ammonia and other volatile organic compounds (Ahmed 2015). A large proportion of the chemicals used for tanning is not taken up by the material developed but disposed of in the environment. Although the tanning industry has existed in the country for a long time and has undergone tremendous developments, its contribution to environmental pollution has been considered seriously problematic only in recent years (Saranya et al. 2020). It is now recognized as a substantial contributor to pollution and needs to be addressed. Landfilling and incineration have been the only practices used to manage solid tannery waste due to economic reasons; however, these have some inherent negative impacts on the environment, such as air and groundwater pollution, unpleasant odour in the atmosphere, and the emission of greenhouse gases (GHGs) that eventually play a huge role in climate change (in the case of incineration) (Pujara et al. 2019). To address these issues, many leather industries across the globe now focus on biological routes to manage their solid wastes and recover value-added products towards sustainable development. This chapter reviews the characteristics of tanned waste, untanned waste, keratin wastes and tannery sludge generated from leather-processing units and outlines the different assessment of the various biological treatments currently adopted in the leather industry to promote the 3Rs (Reduce, Reuse and Reutilize) of waste management for environmental sustainability and the challenges involved in implementing these strategies.

2. THE LEATHER MANUFACTURING PROCESS

The basic operations common to all kinds of leather processing, such as pre-tanning, tanning and post-tanning, have been described below:

2.1 Pre-tanning operations

Generally, hides/skins consist of 65% water and 30–35% protein and fat. Due to the high moisture content, they may easily undergo bacterial degradation (IL&FS Ecosmart Limited 2009). Hence, the moisture content is brought down to less than 30% by applying common salt (i.e., sodium chloride) up to 30–45% by weight to the hides/skins. In the soaking stage, the dusted hides or skins are soaked in water with a wetting agent to remove the salt in pits or drums with wooden paddles. Most of the salt applied for the preservation of the skins/hides is removed during the soaking operation. Next, the liming process is performed to aid the fibrous structure to open up for osmotic swelling after removing the hair, flesh, and fat (partially) layers, as well as the inter-fibrillary proteins. The process of liming broadly involves two steps, i.e., dehairing and re-liming. In conventional dehairing, the flesh side of the hides and skins is coated with lime and sodium sulphide and allowed to react for 4 to 6 hours (Saranya et al. 2016). The hides and skins are then dehaired and placed in pits containing lime suspension and sodium sulphide. This operation is carried out to swell the hides and skins and loosen the hair. During the dehairing and fleshing processes, the hair from the grain side and extra flesh adhering to the flesh side are removed either manually or using machines. Afterwards, the hides and skins are washed and delimed with ammonium sulphate or ammonium chloride. The delimed hides and skins are further processed by pickling, a process by which an appropriate pH for the tanning operation is achieved, which also prevents the leather from dehydration. This process uses common salt, formic acid and sulphuric acid (IL&FS Ecosmart Limited 2010). In the case of vegetable tanning, the pH is adjusted to around 5, whereas it is adjusted to around 3.0 for chrome tanning (Seda Badessa et al. 2022). The pickling process helps maintain uniform chemical and physical conditions through the entire collagen matrix to prevent the skin substrate from quickly recombining with tannin, besides facilitating its even penetration into the collagen matrix without forming aggregates. The total period for chrome tanning and vegetable tanning is 10 to 15 days.

2.2 Tanning process

Majorly, there are two types of tanning processes, i.e., chrome tanning and vegetable tanning. In India, more than 80% of leather production is based on chrome tanning and the rest practise vegetable tanning (Suresh et al. 2001). Vegetable tanning is carried out by infusing the delimed skins and hides with plant extracts. Though this process is free of heavy metals, leather obtained from this process is comparatively weaker in terms of heat resistance and dye holding. In chrome tanning, the skins and hides are added into a basic chromium sulphate and drummed for 4 to 6 hours. As chrome tanning is carried out in an acidic pH range, excess acidity is neutralized using sodium formate and sodium bicarbonate. The pH is increased to 4.0 in the final phase of the chrome tanning process in the 'basification' step. Chrome tanning produces semi-finished leather, which is referred to as 'wet blue' (Krishnamoorthy et al. 2013).

2.3 Post-tanning processes

Post-tanning operations comprise of re-chroming of the semi-finished wet blue leather, followed by neutralization, dyeing, fat-liquoring and finishing. However, the operations vary depending on the desired final product. For dyeing, leather is drummed with sufficient float in hot water containing a requisite amount of acid, direct, basic or pre-metalized dyes for about an hour. The primary function of fat-liquoring is to prevent the fibre structure from re-sticking together during drying. Under dry conditions, the interfibrillar water in the leather evaporates, reducing the gap between the constituents of the fine structure and consequently leading to their interaction, which affects leather quality. To circumvent this issue, surface-active softening agents called fat liquors are employed for chrome-tanned leather fibers (Affiang et al. 2018). Different types of fat liquors are used in tanneries, including vegetable fats, animal fats and waxes, while the synthetic ones include paraffin waxes, mineral oils, olefins, processed hydrocarbons, synthetic fatty-acid esters and waxes, fatty alcohols, and alkylbenzenes (Sun et al. 2008). Leather processing involves heavy consumption of chemicals, such as dehairing agents, tanning agents, retanning agents, fat liquors, and dyes to ensure good quality of leather. Such large volumes of chemicals and water employed in leather processing leave the facility with solid and liquid residues as waste.

3. GENERAL CHARACTERISTICS OF SOLID WASTE FROM THE LEATHER INDUSTRY

A typical tannery generates a huge amount of solid waste, which can be classified based on the source as (i) tanned waste (ii) untanned waste and (iii) keratin waste and (iv) tannery sludge. Out of 1000 kg of rawhide processed, nearly 850 kg is generated as solid waste in leather processing. Only 150 kg of the raw material is converted into leather. Solid wastes are generally chrome splits, chrome shaving, and buffing dust (35–40%), flesh (50–60%), skin trimmings (5–7%) and hair (2–5%) (Kanagaraj et al. 2006). The typical composition of tannery solid waste generated per tonne of raw hides/skins (Thanikaivelan et al. 2005) is shown in Fig. 1.

These are contributed by the leather units that handle different processes, including the beamhouse (80%), tanning (19%) and finishing, (1%). The major sources of solid waste produced from tanneries and their characteristics are shown in Table 1. The rich amount of dissolved and suspended solids of organic and inorganic compounds, such as chromium, ammonia and sulfides, elevate the COD (chemical oxygen demand), BOD (biological oxygen demand), hardness, chromium toxicity and salinity increment in the soil and groundwater (Song et al. 2000). However, the solid waste generated at various stages of leather processing contains economically valuable chemicals, such as lime, acids, salts and other metals (Chojnacka et al. 2021). These factors fail to manage the tanning industry waste

Waste composition (in Percent)

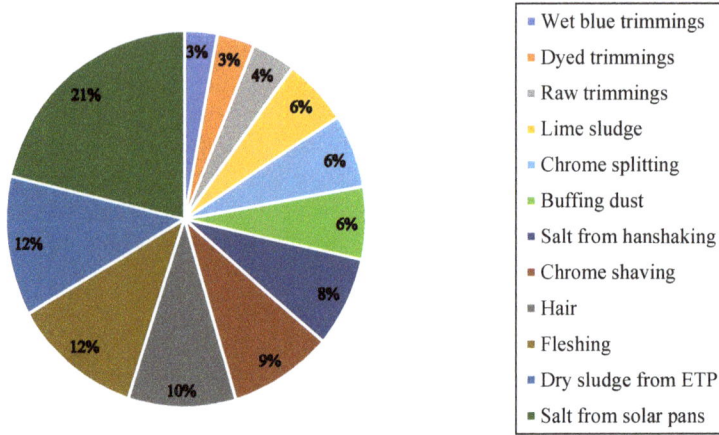

Figure legend:
- Wet blue trimmings
- Dyed trimmings
- Raw trimmings
- Lime sludge
- Chrome splitting
- Buffing dust
- Salt from hanshaking
- Chrome shaving
- Hair
- Fleshing
- Dry sludge from ETP
- Salt from solar pans

Figure 1. Quantity of solid waste generated in a tannery per tonne of raw hides/skins (Thanikaivelan et al. 2005).

Table 1. The attributes of different types of waste generated in tanneries (Ahammed Bin Azam and Saha 2021).

S. no.	Types of waste	Process	Pollutant
1.	**Untanned waste** Green fleshing, lime fleshing and limed splits	Soaking and liming	Sodium chloride, Sodium sulphide and Lime
2.	**Keratin waste** Hair and bristle	Dehairing	-
3.	**Tanned waste** a) Solid waste containing chrome (90% collagen; 1–3% Cr) b) Unusable tanned splits and offcuts c) Shaving dust d) Buffing dust	Leather trimming and shaping Leather trimming and shaping Thickness adjustment Leather buffing	Chromium and other post tanning agents Chromium and other post tanning agents Chromium and other post tanning agents Chromium, tanning agents, dye, synthetic fat, etc.
4.	**Tannery Sludge**	Solid waste generated from the effluent of treated tannery wastewater	Chromium, organic compounds, dyes, etc.

treatment in many cases (Famielec 2020). Improper disposal of tannery solid waste may be hazardous to human health and affect biodiversity (Joyia et al. 2021). In addition to this, untreated tannery solid waste is discharged into the land and water and emits greenhouse gases, polluting natural resources. In the past three centuries, the atmospheric concentration of methane has doubled, with the level increasing by 1% per year in recent times (Vasudevan and Ravindran 2007).

4. BIOLOGICAL TREATMENT OF SOLID LEATHER INDUSTRY WASTE

Leather waste generated from each stage of leather processing have different characteristics. The biological methods can be compelling alternatives to other environmentally unfavourable modes of disposal, such as landfilling and incineration. The negative impact on the environment and energy consumption for incineration can be reduced by the efficient recovery of chromium, methane, enzymes and composite materials via biological routes, which ensure soil amendment, sustainability and a cleaner leather waste management plan. Different biological treatment methods integrated with or without pre-treatment processes that are currently employed by the tanning industry to recover high value-added products (Chojnacka et al. 2021), as shown in Fig. 2, are described below:

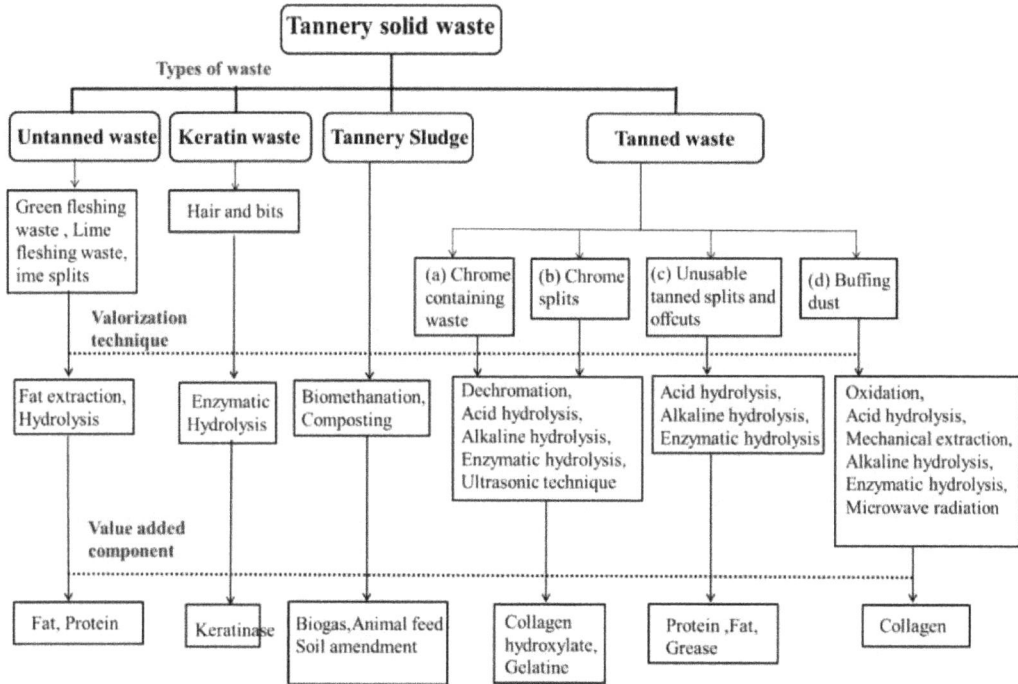

Figure 2. Valorization of tannery solid wastes by biological or integrated solid waste management (Vanitha and Kumar 2016).

4.1 Composting of tannery waste

As per the notification issued by the Government of India (January 2000), if the content of trivalent chromium in tannery sludge is less than 5000 mg/kg (dry weight basis) and that of hexavalent chromium is less than 50 mg/kg, then it is considered low-chrome sludge (UNIDO 2000). For sludge containing prohibited levels of chromium, a properly designed landfill is the best option. Composting and bio methanation are good options for low-chrome sludge disposal since 35–60% of the total solids in the tannery sludge are organic matter (Wahid Murad et al. 2018). Composting relies on the natural decomposition of organic matter by microorganisms (*Bacillus megatherium, Pseudomonas* sp., *Actinoplanes* sp., *Rhizopus nigricans, Saccharomyces* sp., etc.) in the presence of air and moisture, as it occurs naturally on the forest floor and in open fields. However, composting is distinguished from this kind of natural decomposition in that certain parameter, such as temperature and moisture, are controlled to optimize the decomposition process and produce a final product that is sufficiently stable for storage and land application without adverse environmental impacts. Composting encompasses two major stages. In the first stage, the feedstock is decomposed into simpler compounds by microbes, and heat is generated as a product of their metabolic activity. The composting pile reduces in size during this stage. The second stage is called *curing* or finishing of the composted product. As the readily available nutrients in the pile are used up by the microorganisms, microbial activity slows down, which leads to a gradual diminution of heat generation and a dry and crumbly texture of the compost. At the end of curing, the compost is supposedly stabilized or mature (Rawat and Johri

2013). Microbial decomposition beyond this stage, if any, is extremely slow. The key determinants of microbial growth during composting include oxygen, temperature, moisture, C: N content (nutrient levels and balance), particle size, and pH (Jayaprakash and Jagadeesan 2019). The optimization of these parameters has a great influence on the rate of composting and the quality of the composts. Composting may be classified into (i) aerobic, i.e., using the air to compost nitrogen-rich waste with bacteria that create high temperatures to break down organic material; (ii) anaerobic - composting takes place without air and takes long years for organic material to break down, emanating a foul odour and harmful compounds, such as ammonia and methane; (iii) vermicomposting – using various species of worms, usually red wigglers, white worms, and other earthworms, to decompose organic matter in the presence of oxygen and moisture. The advantages and disadvantages of composting are shown in Fig. 3.

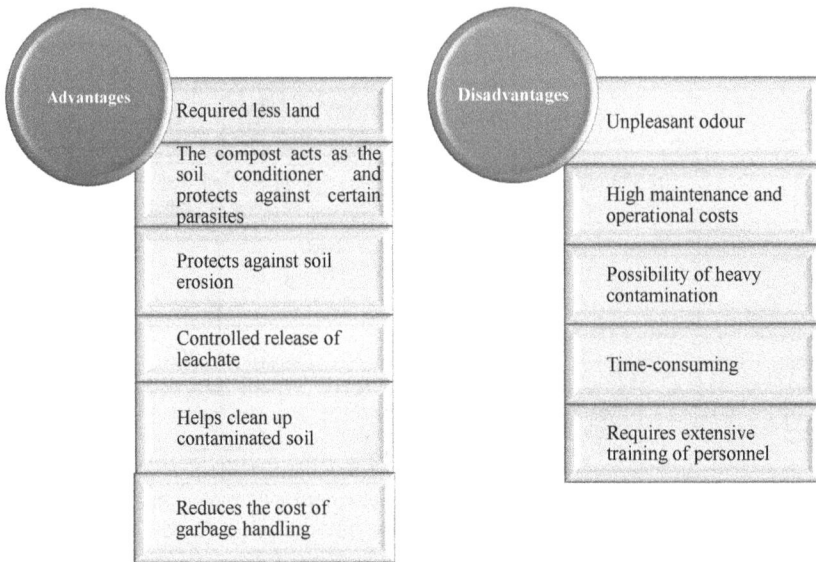

Advantages
- Required less land
- The compost acts as the soil conditioner and protects against certain parasites
- Protects against soil erosion
- Controlled release of leachate
- Helps clean up contaminated soil
- Reduces the cost of garbage handling

Disadvantages
- Unpleasant odour
- High maintenance and operational costs
- Possibility of heavy contamination
- Time-consuming
- Requires extensive training of personnel

Figure 3. Advantages and disadvantages of composting.

4.1.1 Role of microorganisms in composting

Microorganisms play a major role in the composting process; different types of microorganisms are active at different times in the composting pile. Nevertheless, bacteria contribute the most by readily decomposing the macronutrients (primarily proteins and carbohydrates) at the fastest rate. Fungi contribute in the later stages as they can endure low-moisture conditions better than bacteria. Fungi that can survive lower nitrogen conditions than bacteria enable the decomposition of cellulose materials that cannot be digested by bacteria. Foraging animals, such as rotifers, nematodes, mites, springtails, sowbugs, beetles, and earthworms reduce the composting feedstock in size by moving through the pile or chewing its constituents. Besides physically breaking down the materials, these actions increase the surface area and sites for microbial action. Overall, composting is a four-phase temperature-dependent process that is directly proportional to microbial activity. The temperature of the composting matter rises with the microbial growth rate. Therefore, the resultant temperature is a function of the accumulated heat produced by microbial metabolism, and in turn, the temperature determines their metabolic activity. Therefore, temperature control is crucial in composting. The four main composting phases are: (a) mesophilic phase, (b) thermophilic phase, (c) cooling phase, and (d) curing phase (Insam and de Bertoldi 2007).

a) Mesophilic phase (25-40°C)

This initial phase happens at ambient temperature, and the mesophilic microorganisms multiply by feeding on organic matter and generate organic acids at 25–40°C. The activity of primary decomposers, including fungi, bacteria and actinobacteria, at this stage, leads to an increase in the temperature, as well as a reduction in pH due to the acidic by-products.

b) Thermophilic phase (35–65°C)

This stage is also referred to as the sanitation phase. On reaching 40°C, the thermophilic microorganisms convert organic nitrogen into ammonia, thus superseding the mesophilic flora and rendering the mixture alkaline. The decomposition of more complex substances continues and even speeds up until 60°C. The maximum activity of thermophilic fungi happens between 35 and 55°C, while fungal growth is not supported beyond this limit, whereas thermotolerant and thermophilic bacteria and actinobacteria remain active at very high temperatures.

c) Cooling phase (Second mesophilic phase)

In this stage, the temperature dips down to 40–45°C due to substrate exhaustion and the resultant inactivity of thermophilic organisms. The mesophilic bacterial activity begins again, resulting in recolonization and a slow pH drop.

d) Maturation phase

A series of secondary reactions, such as condensation and polymerization of humus, takes place during the maturation phase at ambient temperature The composition of the microbial community is completely modified in the final product compost.

4.1.2 Assessment of tannery waste composting

The composting of tannery sludge along with different organic waste, such as green or limed fleshing, green biomass, tertiary coir pith, and paddy straw, was carried out and the following points were discussed (Organization 2000). The amount of sludge and the different admixtures to be mixed in each heap was initially 1:1 (wet weight basis), while the C/N ratio was maintained > 14. Other quantitative ratios 3:2 and 2:1 between the sludge and admixtures were also tested while ensuring that the C/N ratio of the composting heap was > 14. The sludge and admixtures were arranged layer by layer to build a heap, with cow dung slurry sprinkled on each layer. The following precautions were taken/parameters were maintained during the composting of tannery sludge and can be considered as guidelines for effective composting:

- ✓ The moisture level in the heap was kept at 60–65%.
- ✓ The gravel platform on which the heaps were built obviated the need for intermittent turning over of the composting material.
- ✓ Uniform particle size of the ingredients was achieved by crushing/chopping before they were mixed and piled up. Unlike piling the ingredients layer by layer, this method offered uniform density.
- ✓ Wooden chips and twigs were added to aid natural aeration.
- ✓ The density of the heaps was maintained between 0.4 and 0.6 t/m^3.
- ✓ A C/N ratio between 15 to 35 is generally appropriate.
- ✓ Approximately, 1 tonne of sludge can be combined with the following quantities of tannery waste to accomplish the desired density and C: N ratio: fleshing (250 to 350 kg), green biomass (200 to 300 kg), paddy straw (200 to 300 kg) and coir pith (200 to 300 kg).

The duration of composting varied between 80 and 110 days depending on the admixture. The temperature of the heap was found to rise to above 70°C within the first 10 to 20 days. Then, the temperature remained above 45°C for up to 35 to 60 days (thermophilic phase). After 40 to 60 days, the temperature dropped from 45 to 15°C (mesophilic phase). Mechanical aeration using blowing and suction techniques to turn the composting heap for good microbial action did not shorten the duration for complete composting; both naturally and mechanically aerated processes needed 60–70 days for completion. After this process, the NPK values (on a dry weight basis) of the harvested compost were: nitrogen as TKN –1.1 to 2%, potassium as K_2O –0.7 to 2.6% and phosphorous as P_2O_5 –0.7 to 0.9%. The chromium content in the compost was reported as 1315–3286 mg/kg (Wilson 2022).

Bou Serra (2018) studied the prospects of composting non-chrome waste (hair, trimmings and fleshings). Of the eight samples prepared by mixing different proportions of non-chrome waste (3 kg) with other organic waste (6 kg), such as browns, greens, cattle manure, ash, soil and old compost, crushed wood with tannery sludge (ratio 3:1) could be dried and reduced in weight by 70%. This composting method employed huge turned windrows measuring up to $80 \times 18 \times 3$ m. After 8 weeks, the final compost was screened (Bou Serra 2018). The environmental impact of the operational-state composting plant was also evaluated. The results showed that the total amount of carbon dioxide released to the environment every year is 41.23 tonnes. Compost prepared from tannery sludge mixed with agricultural waste, when applied as fertilizer to cultivate capsicum, significantly increased the number of leaves, fruits, and the amount of chlorophyll in the leaves (Silva et al. 2010). Berilli et al. (2019) also reported that the potential of tannery sludge compost as the substrate for the quality improvement of sweet pepper seedlings was 46%, while that associated with urban compost was 54% (Berilli et al. 2019). Vermicomposting has been shown to stabilize the tannery sludge and yield a nutrient-rich fertilizer; its benefits have been reported by several researchers (Nunes et al. 2018, Ravindran et al. 2019, Vig et al. 2011). The concentration of chromium ions in the sludge feedstock is taken into consideration in the vermicomposting process since chromium is toxic to earthworms as well. A ten-year study has demonstrated that composted tannery sludge can bring about beneficial changes in the soil properties since organic content and macro elements were high, which is desirable. At the same time, the content of chromium in the soil and its pH were elevated, along with changes in the soil microflora and a decline in enzyme activity. The maximum increase in the soil chromium content occurred in the first 5 years of the experiment (Araujo et al. 2020).

4.1.3 *Environmental concerns during composting*

Indian soils sadly are poor in organic content, which can retain moisture and make crops drought-proof, as well as restore soil vitality and fertility. It is estimated that each year India is short of 6 million tonnes of organic manure. The implementation of an Integrated Plant Nutrient Management, that is, achieving optimal soil enrichment using a mixture of organic manure or city compost and chemical fertilizers, can greatly contribute to national savings. Since urea is a cheap and easy-to-use option, farmers prefer it albeit knowing that its indiscriminate use can render the soil infertile in a few years. It was reported that the 11.6 million hectares of agricultural soil in India has turned saline, leading to a loss in annual productivity worth Rs 100–300 crore due to the overuse of fertilizers, overwatering and gradual nutrient depletion (Kumar and Sharma 2020). Compost improves soil structure, restores the organic carbon and fertility of the soils, retains moisture in the soil and thereby drought-proofing crops, improves crop yield and quality, reduces the cost of farm inputs, reduces the need for pesticides and moderates the soil temperature (Sangamithirai et al. 2015). Thus, composts can be considered the best soil amendment, as well as soil conditioner, as it neutralises soil toxicity, especially that caused by heavy metals, in the contaminated site. Nevertheless, certain features of the composting process can be potentially detrimental to the environment and health, or raise safety concerns, which should be evaluated before proceeding with the planning and implementation of a composting facility. Unregulated composting of tannery waste has the potential to foster several environmental issues, including air and water pollution, foul odour, fires, and debris (Ferronato and Torretta 2019). Yet, many of these concerns can be alleviated by adopting appropriate designs based on the raw material and diligently following the applicable operational protocols in the facility.

4.2 Anaerobic digestion (bio methanation) of tannery solid waste

In general, there are two methods to derive energy from the organic waste segment (both biodegradable and non-biodegradable), namely thermo-chemical conversion and biochemical conversion. The thermochemical conversion process involves the thermal decomposition of organic matter to generate heat energy, gas or fuel oil by incineration and pyrolysis/gasification. Biochemical conversion using microbes generates methane gas or alcohol as by-products during the enzymatic breakdown of organic matter. Therefore, it is suitable for wet waste with a high biodegradable organic content. Clean technology refers to those that implement the reuse of waste for energy recovery, value-added product generation and/or clean fuel production.

Anaerobic digestion (AD) is a natural process facilitated by bacteria, in which microorganisms are employed to break down biodegradable matter without oxygen. The end products of this process include clean energy (biogas) and the digested residue, which can be used as nutrient-rich organic fertilizer. Biogas consists of methane (55–70%), carbon dioxide (30–45%) and other trace gases; this is an environmentally and energetically attractive power/heat generation technique, with low environmental impact (Priebe and Gutterres 2012). AD offers the possibility of recovering valuable nutrient-rich manure and energy from different kinds of waste, such as municipal solid waste, animal waste and industrial waste. Among the various biological treatment methods experimented till now, the process of anaerobic digestion has the most advantages, such as low energy needs, low nutrient requirements, minimal sludge production and the possibility for handling a huge amount of organic load at a nearly low hydraulic retention time (Oliveira et al. 2004).

The biological process of AD can be separated into the following four stages/steps: hydrolysis, acidogenesis/fermentation, acetogenesis and methanogenesis (Fig. 4). In the hydrolysis step, bacteria

Anaerobic digestion process

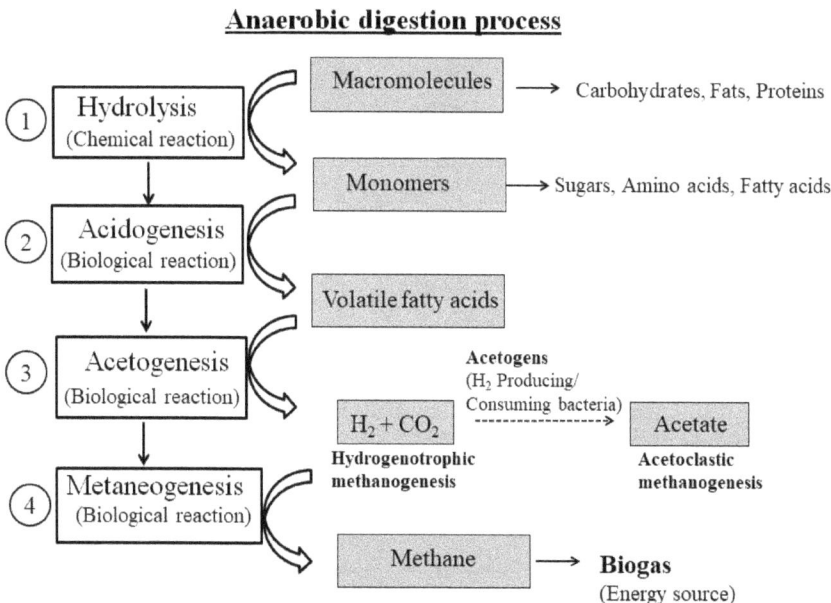

Figure 4. The anaerobic digestion processes (Rajameena and Velayutham 2018).

excrete extracellular enzymes that convert macromolecules, such as carbohydrates, protein and fats, into their respective monomers, namely sugars, amino acids and fatty acids. Acidogenesis (Phase 2) is the biological reaction that converts monomers to short-chain fatty acids or volatile fatty acids (VFA), such as propionate, butyrate, etc. Acetogenesis is the third phase, in which the broken-down monomers or oligomers from phase 2 are converted into acetic acid, or carbon dioxide and hydrogen ($H_2 + CO_2$) depending on the acetogenic organisms present (hydrogen-consuming or producing). Methanogenesis is the biological conversion of acetate to methane and carbon dioxide by acetate-consuming methanogens and that of $H_2 + CO_2$ to $CH_4 + CO_2$ in the presence of CO_2-reducing methanogens (Priebe et al. 2016, Rajameena and Velayutham 2018). The schematic representation of the anaerobic digestion process is shown in Fig. 4.

Generally, 35–60% of total solids in the tannery sludge are organic matter. As the organic matter is amenable for bio methanation, the sludge can be digested along with the wet limed fleshing to generate biogas. The earliest statement on the anaerobic digestion of tannery solid waste is from the early 1980s, followed by several others, validating this technique as an attractive solution for managing tannery solid waste (Bayrakdar 2020b, Cenni et al. 1982, Dhayalan et al. 2007, Priebe et al. 2016) (Table 2). Priebe et al. (2016) described the feasibility of applying microorganisms (inocula) from different sources, such as slaughterhouses and sewage sludge, for the anaerobic degradation of chrome shavings to produce biogas and methane (Priebe et al. 2016). Hydrolyzed collagen with

Table 2. Biogas production from tannery solid waste anaerobic digestion.

S. no.	Type of substrate	Species/Type	Biogas yield	Reference
1.	Tannery solid waste and primary sludge	Bacterial consortium	Methane: 77%	(Thangamani et al. 2010)
2.	Tannery solid waste	Waste-activated sludge inoculum	1087–2933 mL	(Sri Bala Kameswari et al. 2012)
3.	Tannery solid waste	Bacterial consortium	Methane: 73%	(Basak et al. 2015)
4.	Chrome-tanned leather waste	Slaughterhouse sludge	162.2 mL/g	(Priebe et al. 2016)
5.	Leather solid waste	Methanogens and chromium-resistant microorganisms	17 mL/g	(Agustini et al. 2017)
6.	Tannery solid waste with tanning agent	Sludge inoculum	Sludge with chromium: 27.9 mL/g Shavings with chromium: 10.7 mL/g	(Agustini et al. 2018a)
7.	Tannery fleshing	Sludge inoculum	0.301 (Nl_CH_4/gVS$_{add}$)	(Polizzi et al. 2018)
8.	Tannery solid waste and wastewater	Anaerobic microorganism	Methane: 60.5%	(Berhe and Leta 2018)
9.	Tannery solid waste	Wastewater sludge inoculum	21 and 30 mL/kg	(Agustini et al. 2018a)
10.	Tannery waste activated sludge and slaughterhouse sludge	Microbial consortium	215 mL/g	(Mpofu et al. 2019)
11.	Tannery solid waste (leather shavings and the sludge)	Sludge inoculum	30.14 mL/g	(Simioni et al. 2020)
12.	Leather fleshing's and the leather industry wastewater treatment sludge	Methanogenic inoculum	Single-phase: 0.46 m³/kg Two-phase: 0.40 m³/kg	(Bayrakdar 2020b)
13.	Tannery solid waste	*Methanosaetaceae* and *Bacteroidales*	26.1 mL/g	(Agustini et al. 2020)

low level of chromium (lesser than 0.33 wt.%) produced high yields of biogas compared with hide powder and chrome shavings. Agustini et al. (2017) investigated the anaerobic digestion of sludge (beamhouse and beamhouse + chromium tanning) and shavings (pickled hide, chromium-tanned and vegetable-tanned waste), which provided 4.1–11.3 mL methane/g of volatile suspended solids (VSS) (Agustini et al. 2017). Their results also indicated that a low concentration of chromium in the sludge and shavings was in fact more advantageous for biogas and methane production than its complete absence; chromium-containing samples produced about thrice the quantity of biogas, with 55% CH_4 on average, compared with those without chromium. Therefore, chromium in the sludge and shavings improves the reduction of organic matter and enhances the inorganic content (Agustini et al. 2018b). Biogas production from chromium-tanned leather is worthwhile because of the greater efficiency compared with processes using other types of waste. The rationale is that chromium (III) partially removes the organic load in the leather, which enhances anaerobic digestion. Moreover, other compounds (salts, antiseptics, and antibiotics) present in tannery sludge also prove to be potential inhibitors of biogas production (Toniciolli Rigueto et al. 2020).

4.3 Other applications

Furthermore, solid waste from the tannery industry has also been widely explored for the recovery of value-added products, such as the extraction of compounds of industrial interest, isolation of microbial species and enzyme production, besides the applications related to composting and biogas production.

4.3.1 Animal feed

Hydrolyzed protein from vegetable-tanned wastes tested as an animal diet for broilers and cattle (Alam et al. 2010, 2002). Chaudhary and Pati (2016) have tested the protein derived from chromium-tanned wastes to partially substitute soybean meal fed to broiler chicken (Chaudhary and Pati 2016). Undeniably, one of the greatest challenges that restrict the application of chrome tannery wastes as animal feed is the harmful effects of chromium on animal health and the environment despite the rich protein and lipid content of the protein hydroxylates (Chaudhary and Pati 2016, Hossain and Hasan 2014).

4.3.2 Enzymes

Fleshing waste is a major waste component composed of adipose tissues rich in carbohydrates, protein, polypeptides, fat and minerals. These go unexplored despite the rich nutritional value and can create disposal issues. Animal fleshing waste usually possesses proteins (50–70%), fats (4–18%), water content (62–80%), sodium chloride (2–4%), sodium sulfide (2–4%), calcium hydroxide (2–6%) and nitrogenous substances (1–2%), which can be utilized for the extraction of various types of proteases (Kanagaraj et al. 2006, Vasudevan and Ravindran 2007).

Animal fleshing discharged from the tannery industries are processed by a chemical or thermal process, which is energy-intensive and time-consuming (Bajza and Marković 1999). An alternative eco-friendly method is to extract enzymes from solid waste through anaerobic process. This process can be carried out with low chemical and energy input. Aspartate proteases have been produced by solid-state fermentation using Synergistes sp., with a yield of 400–420 U/mL (Kumar et al. 2009), and *Clostridium limosum*, with a yield of 433 U/mL (Ravindran et al. 2016). Similarly, the proteinaceous solid waste of ANFL has been hydrolysed in an anaerobic process by the enzymes from *Selenomonas ruminantium* (Ravindran et al. 2012). Instead of the direct fermentation of animal fleshing waste, the fleshing hydrolysate could be a better inducer for increased protease production when used along with agro-wastes. Shakilanishi et al. (2017) utilized the collagen hydrolysate derived from chrome shavings and fermented it using *Bacillus cereus* VITSN04, which yielded a maximum enzyme output of 203 ± 0.07 U/mL (Shakilanishi et al. 2017). To improve the yield of protease, Rao et al. (2018) generated fleshing hydroxylate by the thermal hydrolysis of delimed fleshing waste. When the fleshing hydrolysate powder (FHP) was employed in protease production together with wheat bran and rice bran, a 1.7-fold increase in protease activity was observed compared with that obtained by the solid-state fermentation of agro residues using *Brevibacterium luteolum* (MTCC 5982) without FHP (Bayramoglu et al. 2014, Rao et al. 2018). These results demonstrate that the FHP obtained by hydrolysing the animal fleshing waste is a better carbon and nitrogen source in fermentation media at the industrial scale. Nevertheless, a high chromium concentration in the protein hydroxylate can inhibit microbial growth, affecting protease yield. Therefore, an integrated process of dechromation and hydrolysis, followed by fermentation would be a sustainable solution to circumvent this issue and recover high -end products from tannery solid waste.

The keratin-rich waste (hair) from tanneries is a cause of environmental concern since this kind of waste is discarded directly in the soil or landfills. Keratin is insoluble in water since it possesses disulfide bonds and other rigid cross-links, such as salt linkages, hydrogen bonds and hydrophobic bonds, which render it resistant to diluted acids and alkalis (Korniłłowicz-Kowalska and Bohacz 2011). However, the amino acids can be denatured by physico-chemical hydrolysis; hence, keratin biodegradation would be an effective method. *Brevibacterium luteolum* MTCC 5982 has been shown to have effective keratinolytic activity towards goat hair waste, with a maximum hydrolysis efficiency of Hydrolysis efficiency of 80% and keratinase activity of 102 U/mL after 72 hours (Thankaswamy et al. 2018).

5. SUSTAINABLE CHEMICALS FROM TANNERY WASTE

Green chemistry refers to the concept of inventing chemical products and processes that minimize or avoid the formation of hazardous substances. Limed fleshing wastes generated in the leather processing industries are rich in fat content and can be used as a promising feedstock for biodiesel and bio lubricant synthesis. However, only a few studies are available on producing biodiesel from fleshing oil with Free fatty acids (FFA) since the pH of limed fleshing's is high, which might dampen the extraction efficiency (Bhatti et al. 2008, Işler et al. 2010). Sandhya et al. (2016) attempted the use of neutralizing agents, such as ammonium chloride (NH_4Cl) and hydrochloric acid (HCl), followed by solvent extraction (Sandhya et al. 2016), which yielded maximum oil from limed fleshing's. Using the alkaline hydrolysis process, 70% of the fleshing fat was recovered by Kubendran et al. (Kubendran et al. 2017). The optimal operational parameters were: 1:12 oil to methanol molar ratio, 80°C temperature, and 1.5 wt% acid concentration, with 120 min of agitation at 350 rpm to achieve the conversion of 94% FFA into esters. Swaminathan et al. (2021) tested the biodiesel extracted from tannery fleshing waste in a diesel engine and found that the blended diesel (D90WLB10) produced lower smoke, hydrocarbons, and CO emission, substantiating that it can be a good sustainable eco-friendly substitute for fossil fuels (Esakki and Mangudi Rangaswamy 2021).

On the other hand, micro-fined collagenous buffing dust is one of the airborne pollutants and is generally discharged into the environment by landfilling, thereby contaminating soil. Thermal incineration, pyrolysis, and bioremediation of buffing dust have proven unsuccessful (Sethuraman et al. 2013). Sekaran et al. (2007) reported that processing a ton of skin/hide generates about 2–6 kg buffing dust (Sekaran et al. 2007). A composite sheet from the buffing dust was fabricated by incorporating various chemicals, such as wetting agent, aluminium sulphate, latex and glycerine. The tensile strength and elongation at break of the finished composite were 6.03 MPa and 23.3%, respectively. The thermal conductivity of dry chrome chips and polishing dust is extremely low, which minimizes their adverse effect on the strength of construction materials (Lakrafli et al. 2017, 2013). The excellent insulating properties of these waste materials were verified by the thermal simulations of a structure insulated with panels made of a mixture of leather (polishing dust, chrome chips) and carpentry waste. Tudose et al. (2020) also demonstrated that leather waste can be utilized as raw materials for good thermoinsulating materials. In this work, they investigated thermal diffusion in seven such materials; the least values were obtained for untreated wool (6×10^{-8} m^2/s) and powdered leather waste (8.5×10^{-8} m^2/s) (Tudose et al. 2020).

6. CONCLUSIONS

In the present article, various solid wastes generated from the leather industry have been listed and the sustainable disposable techniques of the same has been reviewed and discussed. It is understood that the solid wastes generated during leather processing poses serious threat to the environment. Various methods currently being employed for converting the solid waste to useful products through biological means have been discussed in detail. In the future, the possibility of fabricating novel composites and extracting value-added materials from solid tannery waste may be considered. Biological methods would pave way for producing these value-added materials in a sustained way. The future directions would be to extract and utilize the major constituents of the skins like collagen, keratin, fat and the minor components of the skins like the sulphates of dermatan, chondroitin and heparin in to useful high-end products.

REFERENCES

Affiang, S.D., Ggamde, G., Okolo, V.N. et al. 2018. Synthesis of sulphated-fatliquor from neem (Azadirachta Indica) seed oil for leather Tannage, 215–221.

Agustini, C., da Costa, M. and Gutterres, M. 2018. Biogas production from tannery solid wastes – Scale-up and cost saving analysis. Journal of Cleaner Production 187: 158–164. https://doi.org/10.1016/j.jclepro.2018.03.185.

Agustini, C.B., da Costa, M. and Gutterres, M. 2020. Biogas from tannery solid waste anaerobic digestion is driven by the association of the bacterial order bacteroidales and archaeal family methanosaetaceae. Applied Biochemistry and Biotechnology 192: 482–493. https://doi.org/10.1007/s12010-020-03326-6.

Agustini, C.B., Neto, W.L., Priebe, G. et al. 2017. Biodegradation of leather solid waste and manipulation of methanogens and chromium-resistant microorganisms. Journal of the American Leather Chemists Association 112: 7–14.

Agustini, C.B., Spier, F., da Costa, M. et al. 2018. Biogas production for anaerobic co-digestion of tannery solid wastes under presence and absence of the tanning agent. Resources, Conservation and Recycling 130: 51–59. https://doi.org/10.1016/j.resconrec.2017.11.018.

Ahammed Bin Azam, F. and Saha, B. 2021. Probable ways of Tannery's solid and liquid waste management in Bangladesh—An overview. Textile & Leather Review 4. https://doi.org/10.31881/TLR.2020.25.

Ahmed, S.A.E.K. 2015. An Assessment of Environmental Concerns in Leather Industry Case study Khartoum North Tanneries. https://doi.org/10.13140/RG.2.1.1232.1440.

Alam, J., Hossain, M., Beg, A.H. et al. 2010. Effects of tannery wastes on the fattening of growing cattle, carcass, and meat quality. Korean Journal for Food Science of Animal Resources 30: 190–197. https://doi.org/10.5851/kosfa.2010.30.2.190.

Alam, M.J., Amin, M.R., Samad, M.A. et al. 2002. Use of tannery wastes in the diet of broiler. Asian-Australasian Journal of Animal Sciences 15: 1773–1775. https://doi.org/10.5713/ajas.2002.1773.

Araujo, A.S.F., de Melo, W.J., Araujo, F.F. et al. 2020. Long-term effect of composted tannery sludge on soil chemical and biological parameters. Environmental Science and Pollution Research 27: 41885–41892. https://doi.org/10.1007/s11356-020-10173-9.

Bajza, Ž. and Marković, I. 1999. Influence of enzyme concentration on leather waste hydrolysis kinetics. Journal of the Society of Leather Technologists and Chemists 83: 172–176.

Basak, S., Rouf, M., Hossain, M. et al. 2015. Anaerobic digestion of tannery solid waste by mixing with different substrates. Bangladesh Journal of Scientific and Industrial Research 49: 119–124. https://doi.org/10.3329/bjsir.v49i2.22006.

Bayrakdar, A. 2020a. Anaerobic co-digestion of tannery solid waste: Optimum leather fleshings waste loading. Pamukkale University Journal of Engineering Sciences 26: 1133–1137. https://doi.org/10.5505/pajes.2020.22465.

Bayrakdar, A. 2020b. Anaerobic co-digestion of tannery solid wastes: A comparison of single and two-phase anaerobic digestion. Waste and Biomass Valorization 11: 1727–1735. https://doi.org/10.1007/s12649-019-00902-8.

Bayramoglu, E.E., Yorgancioglu, A., Yeldiyar, G. et al. 2014. Extraction of keratin from unhairing wastes of goatskin and creating new emulsion formulation containing keratin and calendula flower (Calendula offlclnalis L.). Journal of the American Leather Chemists Association 109: 49–55.

Berhe, S. and Leta, S. 2018. Anaerobic co-digestion of tannery waste water and tannery solid waste using two-stage anaerobic sequencing batch reactor: Focus on performances of methanogenic step. Journal of Material Cycles and Waste Management 20: 1468–1482. https://doi.org/10.1007/s10163-018-0706-9.

Berilli, S. da S., Valadares, F.V., Sales, R.A. et al. 2019. Use of tannery sludge and urban compost as a substrate for sweet pepper seedlings. Journal of Experimental Agriculture International, 1–9. https://doi.org/10.9734/jeai/2019/v34i430181.

Bhatti, H.N., Hanif, M.A., Qasim, M. et al. 2008. Biodiesel production from waste tallow. Fuel 87: 2961–2966. https://doi.org/10.1016/j.fuel.2008.04.016.

Bou Serra, J. 2018. Assessment of tannery solid waste management: A case of Sheba Leather Industry in Wukro (Ethiopia).

Cenni, F., Dondo, G. and Tombetti, F. 1982. Anaerobic digestion of tannery wastes. Agricultural Wastes 4: 241–243. https://doi.org/10.1016/0141-4607(82)90016-6.

Chaudhary, R. and Pati, A. 2016. Poultry feed based on protein hydrolysate derived from chrome-tanned leather solid waste: Creating value from waste. Environmental Science and Pollution Research 23: 8120–8124. https://doi.org/10.1007/s11356-016-6302-4.

Chojnacka, K., Skrzypczak, D., Mikula, K. et al. 2021. Progress in sustainable technologies of leather wastes valorization as solutions for the circular economy. Journal of Cleaner Production 313: 127902. https://doi.org/https://doi.org/10.1016/j.jclepro.2021.127902.

Dhayalan, K., Fathima, N.N., Gnanamani, A. et al. 2007. Biodegradability of leathers through anaerobic pathway. Waste Management 27: 760–767. https://doi.org/10.1016/j.wasman.2006.03.019.

Esakki, T. and Mangudi Rangaswamy, S. 2021. Extraction and utilization of biodiesel from tannery fleshing waste in a diesel engine equipped with common rail direct injection system for cleaner emission. Energy Sources, Part A: Recovery, Utilization and Environmental Effects. https://doi.org/10.1080/15567036.2021.1900455.

Famielec, S. 2020. Chromium concentrate recovery from solid tannery waste in a thermal process. Materials 13. https://doi.org/10.3390/ma13071533.

Ferronato, N. and Torretta, V. 2019. Waste mismanagement in developing countries: A review of global issues. International Journal of Environmental Research and Public Health 16. https://doi.org/10.3390/ijerph16061060.

Hossain, M.A. and Hasan, Z. 2014. Excess amount of chromium transport from tannery to human body through poultry feed in bangladesh and its carcinogenic effects.

IL&FS Ecosmart Limited. 2009. Government of India Technical EIA guidance manual for Leather/Skin/Hide processing industry.

IL&FS Ecosmart Limited. 2010. Technical EIA Guidance Manual for Leather/Skin/Hide Processing Industry.

Insam, H. and de Bertoldi, M. 2007. Chapter 3 Microbiology of the composting process. Waste Management Series. https://doi.org/10.1016/S1478-7482(07)80006-6.

Işler, A., Sundu, S., Tüter, M. et al. 2010. Transesterification reaction of the fat originated from solid waste of the leather industry. Waste Management 30: 2631–2635. https://doi.org/10.1016/j.wasman.2010.06.005.

Jayaprakash, J. and Jagadeesan, H. 2019. Sustainablewaste management in higher education institutions—A case study in AC Tech, Anna University, Chennai, India. Green Engineering for Campus Sustainability, 163–172. https://doi.org/10.1007/978-981-13-7260-5_12.

Joyia, F.A., Ashraf, M.Y., Shafiq, F. et al. 2021. Phytotoxic effects of varying concentrations of leather tannery effluents on cotton and brinjal. Agricultural Water Management 246. https://doi.org/10.1016/j.agwat.2020.106707.

Kanagaraj, J., Velappan, K.C., Chandra Babu, N.K. et al. 2006. Solid wastes generation in the leather industry and its utilization for cleaner environment—A review. Journal of Scientific and Industrial Research 65: 541–548. https://doi.org/10.1002/chin.200649273.

Kandasamy, R., Venkatesan, S.K., Uddin, M.I. et al. 2020. Solid waste and production of high value-added. Biovalorisation of Wastes to Renewable Chemicals and Biofuels 3–25.

Korniłowicz-Kowalska, T. and Bohacz, J. 2011. Biodegradation of keratin waste: Theory and practical aspects. Waste Management 31: 1689–1701. https://doi.org/10.1016/j.wasman.2011.03.024.

Krishnamoorthy, G., Sadulla, S., Sehgal, P.K. et al. 2013. Greener approach to leather tanning process: D-Lysine aldehyde as novel tanning agent for chrome-free tanning. Journal of Cleaner Production 42: 277–286. https://doi.org/10.1016/j.jclepro.2012.11.004.

Kubendran, D., Salma Aathika, A.R., Amudha, T. et al. 2017. Utilization of leather fleshing waste as a feedstock for sustainable biodiesel production. Energy Sources, Part A: Recovery, Utilization and Environmental Effects 39: 1587–1593. https://doi.org/10.1080/15567036.2017.1349218.

Kumar, P. and Sharma, P.K. 2020. Soil salinity and food security in India. Frontiers in Sustainable Food Systems 4. https://doi.org/10.3389/fsufs.2020.533781.

Kumar, Venkatesan, Rao, B.P., Swarnalatha, S. and Sekaran, G. 2009. Utilization of tannery solid waste for protease production by Synergistes sp. in solid-state fermentation and partial protease characterization. Engineering in Life Sciences 9: 66–73. https://doi.org/10.1002/elsc.200700040.

Lakrafli, Tahiri, Albizane, Bouhria and El Otmani. 2013. Experimental study of thermal conductivity of leather and carpentry wastes. Construction and Building Materials 48: 566–574. https://doi.org/10.1016/j.conbuildmat.2013.07.048.

Lakrafli, Tahiri, Albizane, Houssaini, E. and Bouhria. 2017. Effect of thermal insulation using leather and carpentry wastes on thermal comfort and energy consumption in a residential building. Energy Efficiency 10: 1189–1199. https://doi.org/10.1007/s12053-017-9513-8.

Mpofu, A.B., Oyekola, O.O. and Welz, P.J. 2019. Co-digestion of tannery waste activated sludge with slaughterhouse sludge to improve organic biodegradability and biomethane generation. Process Safety and Environmental Protection 131: 235–245. https://doi.org/10.1016/j.psep.2019.09.018.

Nunes, R.R., Pigatin, L.B.F., Oliveira, T.S. et al. 2018. Vermicomposted tannery wastes in the organic cultivation of sweet pepper: Growth, nutritive value and production. International Journal of Recycling of Organic Waste in Agriculture 7: 313–324. https://doi.org/10.1007/s40093-018-0217-7.

Oliveira, S.V.W.B., Moraes, E.M., Adorno, M.A.T. et al. 2004. Formaldehyde degradation in an anaerobic packed-bed bioreactor. Water Research 38: 1685–1694. https://doi.org/10.1016/j.watres.2004.01.013.

Organization, U.N.I.D. 2000. Regional programme for pollution control in the tanning industry in South-East Asia. *In*: United Nations Industrial Development Organization, p. 27.

Polizzi, C., Alatriste-Mondragón, F. and Munz, G. 2018. The role of organic load and ammonia inhibition in anaerobic digestion of tannery fleshing. Water Resources and Industry 19: 25–34. https://doi.org/10.1016/j.wri.2017.12.001.

Priebe, G.P.S. and Gutterres, M. 2012. Olein production from pre-fleshing residues of hides in tanneries. Latin American Applied Research 42: 71–76.

Priebe, G.P.S., Kipper, E., Gusmão, A.L. et al. 2016. Anaerobic digestion of chrome-tanned leather waste for biogas production. Journal of Cleaner Production 129: 410–416. https://doi.org/10.1016/j.jclepro.2016.04.038.

Pujara, Y., Pathak, P., Sharma, D.A. et al. 2019. Review on Indian Municipal Solid Waste Management practices for reduction of environmental impacts to achieve sustainable development goals. Journal of Environmental Management 248: 1–14. https://doi.org/10.1016/j.jenvman.2019.07.009.

Rajameena, S. and Velayutham, D.T. 2018. A critical review on biochemical process of anaerobic digestion. International Journal of Science Technology & Engineering 4: 1–4.

Rao, R., Muralidharan, V. and Saravanan, P. 2018. Preparation and application of unhairing enzyme using solid wastes from the leather industry—An attempt toward internalization of solid wastes within the leather industry. Environmental Science and Pollution Research 25. https://doi.org/10.1007/s11356-017-0550-9.

Ravindran, B., Nguyen, D.D., Chaudhary, D.K. et al. 2019. Influence of biochar on physico-chemical and microbial community during swine manure composting process. Journal of Environmental Management 232: 592–599. https://doi.org/10.1016/j.jenvman.2018.11.119.

Ravindran, B., Wong, J.W.C., Selvam, A. et al. 2016. Microbial biodegradation of proteinaceous tannery solid waste and production of a novel value added product – Metalloprotease. Bioresource Technology 217: 150–156. https://doi.org/10.1016/j.biortech.2016.03.033.

Ravindran, Gayathri, B., Mohana Priya, D., Kanimozhi, S. et al. 2012. Hydrolysis of proteinaceous tannery solid waste for the production of extracellular acidic protease by Selenomonas ruminantium. African Journal of Biotechnology 11: 11978–11990. https://doi.org/10.5897/ajb11.2484.

Rawat, S. and Johri, B.N. 2013. Role of thermophilic micro flora in composting. Thermophilic Microbes in Environmental and Industrial Biotechnology: Biotechnology of Thermophiles 137–169. https://doi.org/10.1007/978-94-007-5899-5_5.

Sandhya, K.V., Abinandan, S., Vedaraman, N. et al. 2016. Extraction of fleshing oil from waste limed fleshings and biodiesel production. Waste Management 48: 638–643. https://doi.org/10.1016/j.wasman.2015.09.033.

Sangamithirai, K.M., Jayapriya, J., Hema, J. et al. 2015. Evaluation of in-vessel co-composting of yard waste and development of kinetic models for co-composting. International Journal of Recycling of Organic Waste in Agriculture 4: 157–165. https://doi.org/10.1007/s40093-015-0095-1.

Saranya, R., Prasanna, R., Jayapriya, J. et al. 2016. Value addition of fish waste in the leather industry for dehairing. Journal of Cleaner Production 118: 179–186. https://doi.org/10.1016/j.jclepro.2015.12.103.

Saranya, R., Tamil Selvi, A., Jayapriya, J. et al. 2020. Synthesis of fat liquor through fish waste valorization, characterization and applications in tannery industry. Waste and Biomass Valorization 11: 6637–6647. https://doi.org/10.1007/s12649-020-00944-3.

Seda Badessa, T., Hailemariam, M.T. and Ahmed, S.M. 2022. Greener approach for goat skin tanning. Cogent Engineering 9: 2018959. https://doi.org/10.1080/23311916.2021.2018959.

Sekaran, Swarnalatha, Somasundaram and Dandaiah, Srinivasulu. 2007. Solid waste management in leather sector. J. Des. Manuf. Technol. 1: 47–52. https://doi.org/10.18000/ijodam.70008.

Sethuraman, Srinivas, Kota and Sekaran, Ganesan. 2013. Double Pyrolysis of chrome tanned leather solid waste for safe disposal and products recovery. International Journal of Scientific & Engineering Research 4: 61.

Shakilanishi, S., Chandra Babu, N.K. and Shanthi, C. 2017. Exploration of chrome shaving hydrolysate as substrate for production of dehairing protease by Bacillus cereus VITSN04 for use in cleaner leather production. Journal of Cleaner Production 149: 797–804. https://doi.org/https://doi.org/10.1016/j.jclepro.2017.02.139.

Silva, J.D.C., Leal, T.T.B., Araújo, A.S.F. et al. 2010. Effect of different tannery sludge compost amendment rates on growth, biomass accumulation and yield responses of Capsicum plants. Waste Management 30: 1976–1980. https://doi.org/https://doi.org/10.1016/j.wasman.2010.03.011.

Simioni, T., Agustini, C.B., Dettmer, A. et al. 2020. Nutrient balance for anaerobic co-digestion of tannery wastes: Energy efficiency, waste treatment and cost-saving. Bioresource Technology 308. https://doi.org/10.1016/j.biortech.2020.123255.

Singh, S.P. 2017. Health hazards among workers of leather industries in unnao district—A statistical review. International Journal of Current Engineering and Scientific Research 4: 50–61.

Song, Z., Williams, C.J. and Edyvean, R.G.J. 2000. Sedimentation of tannery wastewater. Water Research 34: 2171–2176. https://doi.org/10.1016/S0043-1354(99)00358-9.

Sri Bala Kameswari, K., Kalyanaraman, C., Porselvam, S. et al. 2012. Optimization of inoculum to substrate ratio for bio-energy generation in co-digestion of tannery solid wastes. Clean Technologies and Environmental Policy 14: 241–250. https://doi.org/10.1007/s10098-011-0391-z.

Sun, D., He, Q., Zhang, W. et al. 2008. Evaluation of environmental impact of typical leather chemicals. Part I: Biodegradability of fatliquors in activated sludge treatment. Journal of the Society of Leather Technologies and Chemists 92: 14–18.

Suresh, V., Kanthimathi, M., Thanikaivelan, P. et al. 2001. An improved product-process for cleaner chrome tanning in leather processing. Journal of Cleaner Production 9: 483–491. https://doi.org/10.1016/S0959-6526(01)00007-5.

Thangamani, A., Rajakumar, S. and Ramanujam, R.A. 2010. Anaerobic co-digestion of hazardous tannery solid waste and primary sludge: Biodegradation kinetics and metabolite analysis. Clean Technologies and Environmental Policy 12: 517–524. https://doi.org/10.1007/s10098-009-0256-x.

Thanikaivelan, P., Rao, J.R., Nair, B.U. et al. 2005. Recent trends in leather making: Processes, problems, and pathways. Critical Reviews in Environmental Science and Technology 35: 37–79. https://doi.org/10.1080/10643380590521436.

Thankaswamy, S.R., Sundaramoorthy, S., Palanivel, S. et al. 2018. Improved microbial degradation of animal hair waste from leather industry using Brevibacterium luteolum (MTCC 5982). Journal of Cleaner Production 189: 701–708. https://doi.org/10.1016/j.jclepro.2018.04.095.

Toniciolli Rigueto, C., Rosseto, M., Krein, D. et al. 2020. Alternative uses for tannery wastes: A review of environmental, sustainability, and science. Journal of Leather Science and Engineering 2: 1–20. https://doi.org/10.1186/s42825-020-00034-z.

Tudose, E.T.I., Mindru, T.B. and Mamaliga, I. 2020. Wool and leather waste materials with thermo-insulating properties. Revista de Chimie 71: 70–78. https://doi.org/10.37358/RC.20.7.8226.

UNIDO. 2000. Regional Programme for Pollution Control in the Tanning Industry, pp. 1–38.

Vanitha, S. and Kumar, M. 2016. Valorization of solid waste from the tannery industry: Preparation of adsorbent by cost effective method. International Journal of Pharmacy and Technology 8: 23377–23386.

Vasudevan, N. and Ravindran, A.D. 2007. Biotechnological process for the treatment of fleshing from tannery industries for methane generation. Current Science 93: 1492–1494.

Vig, A.P., Singh, J., Wani, S.H. et al. 2011. Vermicomposting of tannery sludge mixed with cattle dung into valuable manure using earthworm Eisenia fetida (Savigny). Bioresource Technology 102: 7941–7945. https://doi.org/10.1016/j.biortech.2011.05.056.

Wahid Murad, A.B.M., Mia Abusayid Md. and Rahman, M.A. 2018. Studies on the Waste Management System of a Tannery: An overview. International Journal of Science, Engineering and Technology Research (IJSETR) 7: 253–267.

Wilson, M. 2022. Recommended Methods of Manure Analysis. Recommended Methods of Manure Analysis. https://doi.org/10.24926/9781946135858.

7

BIOREFINING OF ORGANIC SOLID
CURRENT TECHNOLOGY AND FUTURE PERSPECTIVES

Mridusmita Barooah,[1] *Ramya Chandrasekaran*[2]
and Rajmohan Soundararajan[1,*]

1. INTRODUCTION

Biorefining of organic waste to produce bioproducts has immense potential in terms of environmental and economic aspects. Organic solid waste contains large amounts of useful organic material that can be efficiently converted to useful nutrients and energy and applied for other potential applications. Proper management and utilization of solid waste by incorporating efficient waste management framework is the key to achieve the required solution. The physico-chemical complexity in the properties of the various organic solid waste requires identification of varied technological schemes. The conversion of non-recyclable waste material to useful heat and energy can lead to value recovery of waste (Shahid et al. 2022, Subhash et al. 2022, Rodionova et al. 2022, Jones et al. 2022).

For efficient waste-to-energy conversion, various solid waste management frameworks have been identified by mandating government policies and programs. The key strategy is to encourage a circular economy, keeping the environmental aspect into consideration. Biorefining involves the conversion of biomass feedstock to valuable bioenergy and bioproducts. It can be considered as an equivalent to petroleum refinery with the raw materials being bio-based ones. Various biorefining processes are available that effectively convert the biomass feedstock to bioenergy and bioproducts. Some of the most commonly utilized biorefining processes include thermo-chemical, physical or biological strategies. In most of the cases, pretreatment steps in biorefining processes plays a major role for the concept to be completely realized. Pretreatment steps may include physical methods such as grinding, milling, sonication, use of chemicals like alkali, acids, organic solvents and physico-chemical methods such as steam treatment, freeze explosion or fractionation solvents and hydrothermolysis (Chen et al. 2018, Ahmed El-Imam et al. 2019). The table given below (Table 1) depicts the energy content of organic solid waste.

---
[1] National Institute of Technology, Warangal, Telangana 506 004, India.
[2] Lee Kong Chian School of Medicine, Nanyang Technological University Singapore, Singapore.
* Corresponding author: rajmohan@nitw.ac.in

Table 1. Typical values of energy content of different organic solid waste.

Component	Energy content (Btu/lb)	
	Range	Typical
Wood	6500–8500	8000
Food waste	1500–3000	2400
Mixed yard waste	1000–8000	2800
Textile	6500–8300	8000
Leather	7000–9000	8000
Mixed paper	5000–8000	7000
Plastic	12000–17000	14000
Rubber	9000–12000	11000

Biofuels are fuels derived from biological matter such as biomass. The different forms of biomass include food waste, wood scrap, microfiber product, animal residue and algae. Unlike fossil fuels such as crude oil obtained from geological matter which faces dire consequences of extinction over time, biofuels serve as an efficient replacement due to its desirable characteristics such as cleaner burning capacity, control of greenhouse gas pollutant emission, and no dependency on supply from foreign entities. In a broader aspect, biofuel depicts the existence of fuel in all three forms such as solid, liquid and gas. The usage of biofuel diminishes the consumption of fossil fuel thus reducing carbon dioxide emission. The source of biofuel can vary from agricultural crops to waste from municipal, agricultural and food sectors. Biofuels can be generally divided into primary biofuels and secondary biofuels. For primary fuels, the raw organic materials including fuel wood and wood chips are utilized in the unprocessed form (Kuisma et al. 2013).

1.1 Using waste as a useful resource

The way to a sustainable future depends on the efficient disposal of tons of waste produced from varied activities related to the household, agricultural, and industrial sectors on a day-to-day basis. Also, the usage of alternative, and renewable fuel sources to tend to the ever-increasing power demand could provide the ample solution needed to tackle the challenges faced by the scientific community. Also, in this aspect, the waste to energy technology could provide the much-needed solution by converting waste products to energy resources. Several under-utilized waste streams from various feedstocks like agricultural, food, industrial, and municipal solid waste feedstocks can be explored for renewable fuel and product generation. The conversion of waste matter to useful biofuels can be achieved by utilizing varied processes such as combustion, pyrolysis, gasification, or biological process (Omran et al. 2022, Ashrafi et al. 2022, Kee et al. 2021).

Solid waste management remains an environmental challenge owing to unprecedented population growth, rapid industrialization, and proper disposal strategies. Among them, biomass solid waste streams provide the potential alternative energy feedstock due to their enormous availability and access (Tripathi et al. 2019).

1.2 Circular economy and Industry 4.0

The industry wheel of production works by two routes, conventional linear economy, and circular economy. As illustrated in Fig. 1, the fundamental difference between these two economies is their modus operandi. The economic pattern of the linear economy made by bulk production and selling it essentially follows the "take it- make it- use it- dispose of it" approach. However, the challenges

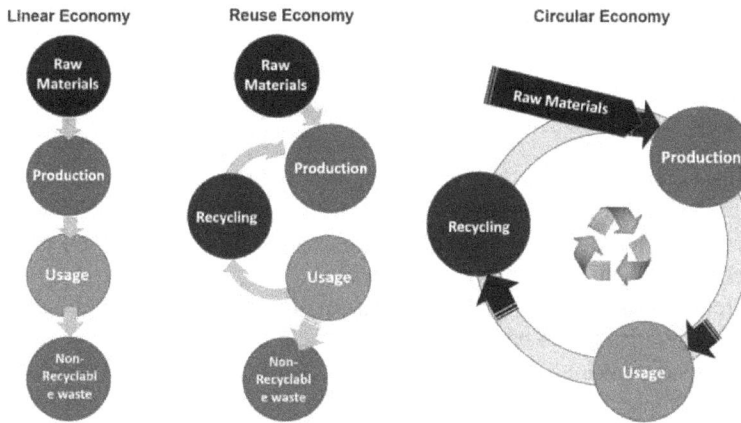

Figure 1. Circular economy in comparison with linear economy.

of linear model economy include irrational usage and exhaustion of raw material and, its negative impact on the environment and climate changes.

Although this theory originated in the early 1960s, it was named as "circular economy" by Pierce and Turner in 1989 (Ekins et al. 2019). They insisted on a conscious and enhanced way of production of any product.

In the business model of the linear economy, the emphasis is on the product whereas the focus is on services in the circular economy. The life of the product ends after the use by the customer in a linear economy, while the life of the product is used as a basic resource in the production cycle.

The focus is on diverting more solid wastes from being disposed of in landfills or incineration and transforming them into useful biofuels, decorative paints, solvents and stick glues.

Non-recyclable waste such as plastic waste pose a great threat to the environment. Technical advancement in the thermochemical conversion such as the addition of catalyst and its processing can play a sustainable role in conversion of the non-recyclable waste into syngas and subsequently translation it into liquid fuel (Yang et al. 2022). The demand for bioethanol which is biodegradable and is blended with gasoline is imminent, as a minimum of 65 countries have renewables targets or mandates. Bioethanol may be used as oxygenate in fuel cars because of its high-octane level.

Due to the greater ecological benefits, concepts on circular economy (CE) practices and Industry 4.0 has become prominently important. These measures aim to create sustainable and greener environment by reducing carbon emissions. The decentralization technique to reduce the waste is the main objective of Industry 4.0. Figure 2 depicts the vision of Industry 4.0. In addition to waste

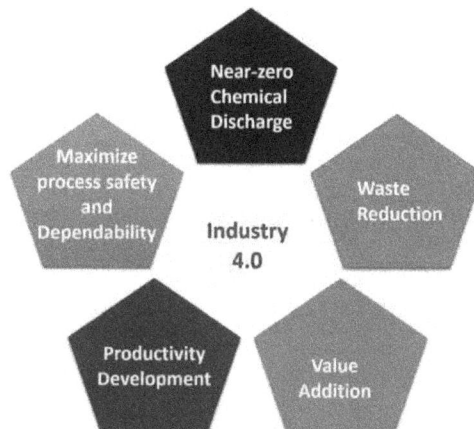

Figure 2. Vision of Industry 4.0.

mitigation, a circular economy contributes to environmental sociability and is a source of revenue for the industry (Khan et al. 2021, Mastos et al. 2021).

1.3 Biorefinery

Rapid industrialization has led to the ever-increasing demand for fuel reserves. Conventional fossil fuel reserves face the future of exhaustion which may lead to a high imbalance in the demand-supply chain. This thereby leads to actively looking for alternative and sustainable energy resources (Shah et al. 2022, Cheirsilp et al. 2022, Igbokwe et al. 2022).

As illustrated in Fig. 3, bio-based circular economy has several common features with circular economy. Biorefinery can be broadly defined as a facility that involves the conversion of biomass feedstock to a wide range of value-added products such as food, materials, chemical, energy, fuel, and power. To achieve this, it involves the cascading of various biorefining processes. Some of the commonly applied biorefineries include the paper and pulp industry, biofuel, and food industry. The operation of biorefinery involves the usage and amalgamation of several processing steps such as mechanical pressing, chemical, biochemical, and thermochemical processes in a sustainable way.

Biorefineries can be classified as simple or advanced biorefineries based on their process. Simple biorefinery involves the application of traditional biomass whereas advanced biorefinery involves the use of advanced biorefining processes such as membrane separation, advanced methodologies

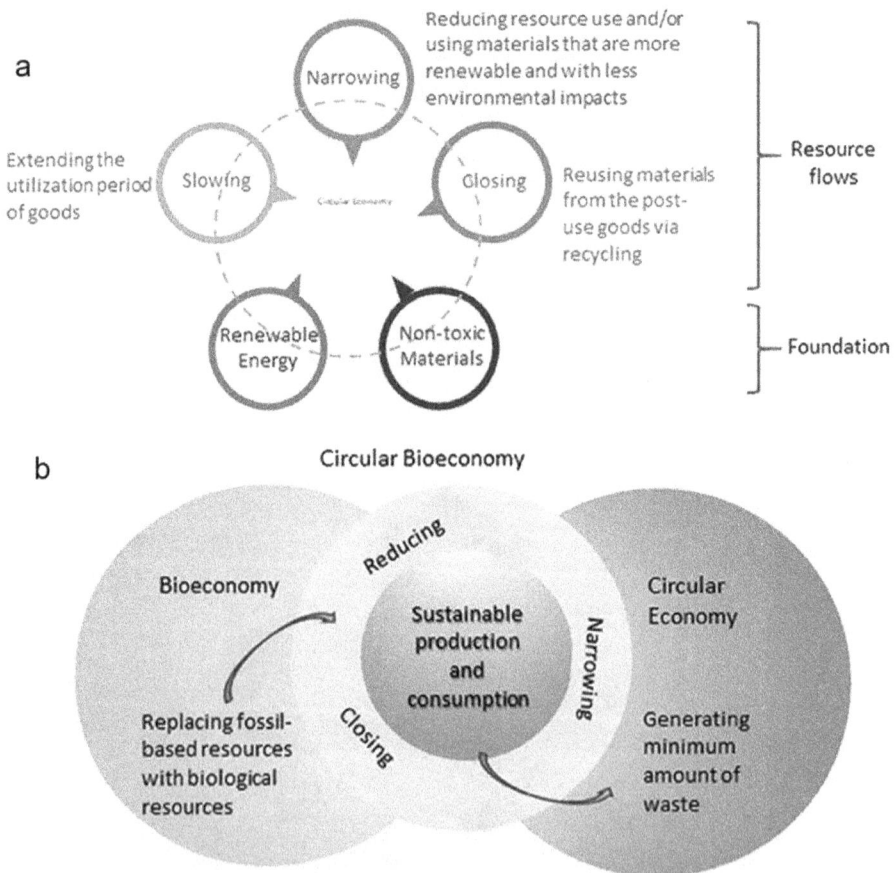

Figure 3. (a) Methodological structure of working circular economy (b) Flowchart of a circular bioeconomy showing the overlap between circular economy and bioeconomy (CCL, Copyright © 2021 Tan and Lamers).

and techniques and high flexibility in its operation. Also, based on the nature of the feedstock used, it can be broadly classified into different categories. Lignocellulosic biorefinery involves the use of lignocellulosic biomass whereas green, algal, and whole crop biorefinery involve the application of green biomass, aquatic biomass, and cereal feedstock, respectively.

A plethora of value added products consisting of biofuels and chemicals can be obtained from biorefinery. These include bioethanol, biodiesel, biogas, biomethanol, bio-oils, solvents, plastics, paints and coatings, chemical intermediates, dyes and detergents. Bio oil is obtained from biomass by applying two approaches. In the pyrolysis process, the biomass feedstock is heated at elevated temperatures (between 200 and 750°C) under anaerobic conditions. The resulting product is bio-oil and biochar, which is a promising alternative to the more conventional fuel resources such as petroleum fuel due to the advantages it showcases. Apart from being biodegradable, it is highly neutral to greenhouse gas, emits a very minimal amount of SOx and Nox, thus proving to be energy efficient compared to its petroleum fuels. However, for bio-oil to be accepted as an efficient fuel in transportation and industrial uses, it should be able to surpass the restrictions such as high viscosity, high corrosiveness, and thermal instability. This can be provided by opting for the right biomass feedstock. In this aspect, algal biomass can serve as the potential and energy-rich feedstock (Rajendran et al. 2021, Okoligwe et al. 2022).

1.3.1 Concept of bioleaching

Bioleaching is also known as the biomining process. It finds high application as sediment heavy metal pollution remediation technology and also to tend to the increased demand of metals in industrialization. It is defined as the extracting of high-value metals from ores by utilizing microorganisms. In this process, the microbes interact with the minerals. This application has seen huge use since prehistoric times. It is considered to be more effective than the traditional mining processes. Commonly used metals extracted from bioleaching include gold, silver, uranium, copper, cobalt, zinc, etc. In the case of the bioleaching process, the *thiobacillus* bacteria oxidize the reduced sulfur content thereby producing hydrogen ions (Bosecker et al. 1997). This results in the decrease in pH of the system and changing the state of heavy metal from bound to free state. Since bioremediation technology plays a pivotal role in the field of environmental protection, bioleaching is an important subject in this aspect. In bioleaching, bacteria act as the catalyst, and the mixed culture species have been found to give good results. The future study needs to check on the strategies for improved leaching efficiency and microbial mechanisms (Shen et al. 2022, Chu et al. 2022).

1.4 Production of biofuels

Given the advantages of a green solution utilizing biofuels, it can be a viable source of renewable energy and an alternative to the exhaustive energy from non-renewable sources like fossil fuel or petroleum. Biofuel can be described as the energy derived through a biological process. It can be produced from a wide variety of feedstock ranging from plant or animal products (Joshi et al. 2022). The biofuel produced can be broadly classified into two types: primary and secondary biofuels. For primary biofuel, the burning of the woody or cellulosic biomass directly produces the biofuel. In the case of secondary biofuel, the biofuel is generated from plant and animal matter. As shown in Fig. 4, the production of biofuel can be conducted by various technologies which are broadly classified into 1st, 2nd and 3rd generation based on the type of feedstock used. First generation basically utilizes the edible feedstock such as starch, sugar, rapeseed to form bio oil, bioethanol, and bio-diesel. Non-edible, cellulosic biomass feedstock such as waste wood, sewage sludge, municipal solid waste (MSW) falls into the 2nd generation conversion technology. All these conversion technologies require a higher degree of pretreatment to access and process the organic material. 3rd generation utilizes algae for energy production. As illustrated in Fig. 5, the major conversion technologies for solid waste management have been discussed in detail in this section (Ambaye et al. 2021).

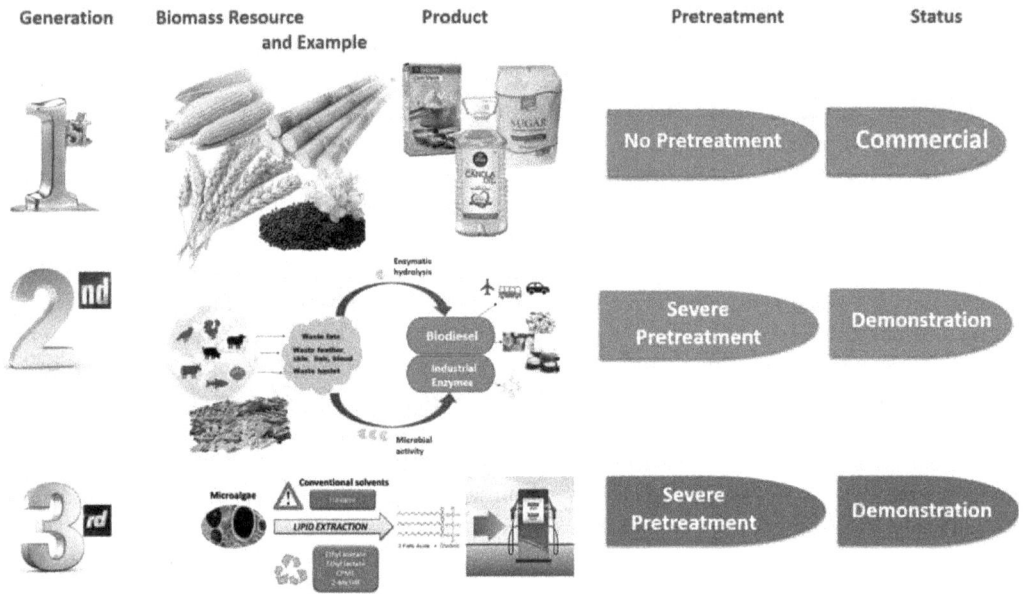

Figure 4. Diverse sources of biomass feedstocks and resulting products.

Figure 5. Technologies available for solid waste management.

1.4.1 Overview of biochemical pathways

Biochemical conversion is a slower process, that does not need any energy input externally (Gouveia et al. 2017, Vigot et al. 2018). As compared to thermochemical method, this technique involves bacteria, enzymes and microorganisms to break down biomass molecules to biofuels.

Thermochemical and biochemical routes can be used to convert biomass into useful products. As shown in Fig. 5, gasification, liquefaction, pyrolysis and direct combustion are examples of thermochemical routes and anaerobic digestion, fermentation and bio algae processes are examples of biochemical routes. Products such as syngas, bio-oil, dimethyl ether, biodiesel, biogas, ethanol, and methanol are obtained.

1.4.1.1 Anaerobic digestion (AD)

In this technique, the breakdown of organic matter occurs in the absence of oxygen to produce biogas. The organic residue produced from this process is called digestate. Methane, the major component of biogas, has vast application for processes involving the breakage of wet organic solids; anaerobic digestion (AD) plays a pivotal role. The solid waste which has high moisture content is suitable for the AD process. In most cases, the digestate obtained after AD contains effective nutrients which cannot be directly discharged. Recycling of the residue is then considered. In this technology, the pre-treatment steps are crucial for the substantial production of biomethanol. The AD process involves the hydrolysis of complex molecules into simpler monomeric units. Thereafter, the simpler monomers are converted to compounds like volatile fatty acids, acetate, ammonia, hydrogen sulfide and carbon dioxide and forms methane gas. In hydrothermal processing, the wet organic waste such as AD digestate, manure, food and municipal waste at high pressure is converted to high carbon content called hydrochar (Kruczek et al. 2020). The water content in the biomass acts as the chemical reactant in the decomposition process.

1.4.1.2 Fermentation

Fermentation is a biochemical process which utilizes microorganisms to convert biofeedstock such as sugar and starch to biofuels (bioethanol, biobutanol). Unlike anaerobic digestion which primarily produces biogas and digestate in an oxygen-free environment, fermentation can produce a broader range of products, including organic acids, alcohols, gases, and other metabolites, in both aerobic and anaerobic conditions. The fermentation of starch and sugar-based feedstock to produce ethanol has been commercialized in many countries including the USA and Brazil.

1.4.1.3 Transesterification

In the transesterification process, the oil crops, fat, waste oil, algae, and cyanobacteria are converted to biodiesel. It is a process in which alcohol reacts with oil/fat leading to the formation of gylcerol and ester. To improve the product yield, catalyst is utilized. The biodiesel produced can be an efficient replacement for petroleum diesel owing to the lower viscosity.

1.4.2 Pathways of thermochemical conversion

Thermochemical conversion encompasses methods such as direct combustion, pyrolysis, and gasification, wherein controlled high temperatures, heating rates, and specified time durations facilitate the oxidative transformation of biomass, ultimately yielding power, heat, and energy within the conversion process. In this process, the bonding between the molecules (carbon, oxygen, and hydrogen) is broken down to release stored chemical energy (Weir et al. 2022).

1.4.2.1 Direct combustion

Combustion is an exothermic process in which the thermochemical conversion occurs in the presence of excess oxygen. The biomass is converted to heat, as well as water and carbon dioxide. It is extremely simple in operation and involves burning. However, the release of toxic gas and contaminants such as carbon dioxide, sulfur dioxide, SO_x, and NO_x favors the utilization of other thermochemical processes.

1.4.2.2 Gasification

Gasification involves the partial combustion of biomass feedstock conversion at a high temperature range ($> 700°C$) in gasifiers. The gasification process produces synthetic gas and producer gas which can be applied for heat generation and electricity production. The syngas is a mixture of carbon monoxide (CO), hydrogen gas (H2), CO_2 and CH_4. It can also be converted to useful chemicals by using the Fischer-Tropsch process. However, the removal of contaminants from the syngas (tar, HCl) restricts the application of this technology. The conventional gasification of biomass occurs in four stages:

1. **Heating and Drying:** High moisture content affects the composition of gas leading to higher carbon dioxide presence in the final mixture. Thus, the feedstock is heated and dried as the first step.

2. **Pyrolysis:** The biomass feedstock is thermally decomposed in the complete absence of O_2 at 300–700°C. This converts around 90% of biomass to vapors and gas.

3. **Gas solid reaction:** The char produced from pyrolysis and the surrounding gas converts the gas to CO, H2O, and CH4.

4. **Gas phase reaction:** Water gas shift reaction and methanation are the gas phase reaction for the production of synthetic natural gas.

1.4.2.3 Pyrolysis

In this process, the conversion of biomass feedstock to bio-oil as well as biochar and pyrolytic gas occurs at high temperature (300–700 C) without the oxygen. The liquid fuels produced as a product makes the pyrolysis technology desirable as compared to other technology. Based on the varied temperature, pressure and reaction time condition, pyrolysis can be classified as torrefaction, slow pyrolysis, and fast pyrolysis. Slow pyrolysis produces gas and solid bio charcoal whereas fast pyrolysis produces bio-oils. Bio-oil consists of a mixture of oxygenated organic compounds including carbolic acids, saccharides, alcohols, esters, aldehydes, and other compounds.

1.4.2.4 Hydrothermal liquefaction

The hydrothermal liquefaction process is a promising thermochemical process that can be operated at relatively low-temperature (300–400°C) and effectively process both wet and dry biomass. In this method, thermochemical conversion of biomass into liquid fuels occurs by processing in a hot, pressurized water environment. As the result, solid biopolymeric structure was broken down into the liquid components. Biomass with high initial water content may be straight away converted into a bio-oil and platform chemicals without any energy-sensitive pretreatments.

1.5 Microwave-assisted conversion techniques

Microwaves find high applications in the synthesis of various chemicals. The microwave assisted biomass conversion technique has gained interest over the years. Microwave applied biofuel production finds major potential as compared to conventional techniques. Microwave finds application in the preparation of the biomass feedstock extraction process. The extraction of oils from biomass feedstock can be done using thermal pretreatment or process enhancement techniques. Microwave when compared to conventional extraction requires low energy consumption. Microwave energy plays a pivotal role in determining extraction efficiency. Also, the use of solvents can be avoided in the case of the microwave-assisted extraction process. The penetration of the microwaves determines the extraction capability of oils from the plant seeds. This type of extraction process is fast and simple. As an example, the Soxhlet extraction was modified by integrating microwave energy into the process. This advanced form called the microwave integrated Soxhlet (MIS) was tested for the extraction of oil and fats. It was observed that MIS extraction showed similar results when compared to the conventional Soxhlet extraction (Vignesh et al. 2022, Hoang et al. 2021). Some advanced thermochemical conversion methods utilize internal heating such as microwave heating. The heat and mass flowing are concurrent. It offers advantages such as ability to handle inhomogeneity in biomass feedstock. Microwave-assisted pyrolysis and microwave-assisted gasification have been largely utilized.

2. BIOMASS POTENTIAL – CHALLENGES AND OPPORTUNITIES

Biomass as a feedstock has varied applications ranging from facility heating to power and energy generation. The source of biomass can range from plant stock, waste from agricultural land, industrial activity, human and animal activity. As a major source of renewable energy, biomass has enormous potential to address the issues of climatic changes, acid rain, air pollution, and other environmental hazards.

2.1 Factors affecting cultivation of algae and energy extraction

The cultivation and production of biomass were found to be influenced by various factors. This included temperature, pH (Rai et al. 2014), light, nutrient stress and photobioreactor cultivation system (Chinnasamy et al. 2009). The temperature has been shown to have a different influence on algal growth depending on the strain. The composition of the algae thrives at the temperature of 4 to 35°C, with some species of *Chlorella Vulgaris* were found to have a maximum growth rate of 15°C as compared to *Scenedesmus subspicatus*, which can grow at 4°C (Kim et al. 2014). The pH of the algae influences not only its biomass composition but also its lifespan as many of the cellular functions and cell wall-associated enzymes were related by its pH. Some marine algae, including *diatoms* and *cyanobacteria*, thrive in a neutral to alkaline pH range, while some prefer a slightly acidic to neutral pH range (Rai et al. 2014). The intensity of the light is related to the photosynthesis process of algae, which results in light saturation or inhibition, resulting in irreversible damage and a decline in biomass production. At lower light intensities, algae-like *Botryococcus braunii KMITL 2* and *Chlorella vulgaris ESP-31* tend to produce higher biomass. The nutrient and carbon supply in the form of CO_2 is necessary for photoautotrophic cultivation. In algae *Chlorella vulgaris*, limiting the nitrogen input for saccharification increased carbohydrate content (Kim et al. 2014).

The high amount of carbohydrate and lipid in the algae plays an important role in alternative energy generation. Algal biomass finds its application in various sectors. Algae generates a vast range of bioactive chemicals, including vitamins, protein, carbohydrates, steroids, polysaccharides, and fatty acids that can be useful in the pharmaceutical, cosmetics, and food sectors (Wang et al. 2015). Biofuels such as bioethanol and biobutanol can be extracted from algal polysaccharides (Gao et al. 2016). The substrate for the fermentation of ethanol and butanol is derived from algae starch/cellulose. Due to its low lignin concentration, the carbohydrate generated by algae is highly fermentable, making it ideal for the generation of bioethanol. The higher biomass growth rate of algae, shorter life cycle, and lipid and oil content would indirectly help to mitigate the greenhouse gas (GHG) emissions, resulting in more eco-friendly solutions (Yoo et al. 2010).

2.2 Systems for productions/cultivation

Over the last few decades, an increasing interest in algae-to-energy conversion techniques is observed. In particular, in the mid-1990s, main focus was on discovering strains with high lipid content to exploit lipids derived from algae to generate liquid fuels. Algal cultivation systems are categorized into two, namely, closed and open methods. The common closed algae production system encompassed photobioreactors, flat-panel photobioreactors, tubular horizontal photobioreactors, and plastic bags. On the other hand, the open production system consists of aquaculture tanks and raceway ponds.

There are few established methods available for the cultivation of algae. These included open ponds and photobioreactor (PBR) systems. PBR technique is a highly preferred technique to obtain ultra-high algal density cells; however, the demand for high energy for the cultivation system remains a hindrance to a large-scale commercial application.

2.3 Post-harvest treatments, preservation and storage

As shown in Fig. 6, microalgae species can be grown using open type or closed type PBR. The algae biomass may be separated from the sludge using flocculation, centrifugation, or air flotation, and water can be removed from the biomass by natural drying, oven drying, or using thermal rollers. Then, the dry microalgae biomass is collected. Subsequently, lipid extraction using chemical or mechanical extractors is performed. Later, algae biodiesel is produced using a transesterification process. Thus, prepared biodiesel is tested for performance, combustion, and emissions characteristics in a calibrated internal combustion engine (ICE) engine.

Figure 6. Microalgae-based biofuel production method.

3. SUSTAINABLE TECHNOLOGIES FOR MUNICIPAL SOLID WASTE TREATMENT

Urban municipal solid waste management is one of the major challenges in the recent times. A major chunk of biomass belongs to the urban municipal solid wastes (MSW). MSW refers to the waste generated from community activity such as residential, commercial, or business activity. The rapid industrialization has led to an upsurge in the production of large portions of urban solid waste worldwide. Waste collection, storage, transportation, handling, recycling, and disposal of waste are the commonly employed waste management techniques. The disposal of MSW needs to surpass several restrictions. There is a pressing need to integrate the informal waste management using regulated management system, like the Municipal Solid Wastes (Management and Handling) Rules

MSWM Rules 2016 in India. The unclear price estimates of informal waste pickers to local scrap dealers to junkyard owners also add more complications. Lack of domestic reserves and supply of rare earth elements required for making electronic gadgets is also one of the restrictions observed in the disposal of the organic waste. Along with it, effective urban waste management technologies, stakeholder and economic intervention, legal and administrative perspectives need to be considered.

The MSW collected from various sources may be initially treated to reduce the weight of the material and recover some materials, heat, and energy for recycling and reuse purposes. The use of microorganisms and heat process technology may be applied to break down the disposal part into heat, power, and energy (Rajmohan et al. 2019, Sun et al 2021).

3.1 Incineration

Incineration involves the burning of usually the raw/residual part of waste inside a furnace under carefully controlled conditions. Upon combining with oxygen, the waste with combustible parts burns into carbon dioxide, water vapor, and heat. Incineration can reduce the percentage of inert residue (ash, glass, and other solid materials) by less than 10 percent (Joseph et al. 2018). The solid inert residue is known as bottom ash. The gaseous residue which includes soot and dust is known as fly ash. These inert materials are then disposed of in a landfill. If the inert material contains toxic material, it is segregated as hazardous waste. Incineration with energy recovery is a widely used technique for municipal waste treatment.

3.2 Landfill

Landfilling is the most common and cost-effective disposal technique for MSW management. It involves the utilization of land area as a disposal site that is carefully selected, designed, and developed to prevent any interaction between a landfill and the adjacent atmosphere. The commonly utilized types of landfill include open dumps and semi-controlled and sanitary landfills. The main objective while constructing a landfill is to ensure minimum contact with surface or groundwater and avoid leakage.

3.3 Vermicomposting

Vermicomposting, an economically viable technology, is widely used for the disposal and treatment of solid waste. In this process, the microbes play a pivotal role in the bond breaking of complex organic biodegradable waste. Heavy metals are known to hinder the process of vermicomposting. The composting of organic biowaste by utilizing earthworms and microorganisms under humidifying conditions can lead to the formation of dark-colored, high nutrient storing capability and superior soil structure.

3.4 Recycling

MSW contains a complex mixture of various materials that can be recycled or reused. Recycling aids in decreasing the quantity of landfill storage. The rag pickers contribute a major part in recycling of solid waste. The steps involved in recycling include collection, subsequent separation and selling in the market, processing and reuse of materials. The scrap brokers hold the intermediate market for recyclable material. The materials that can be effectively recycled from solid waste include paper, food scrap, vegetative waste, and other recyclable materials.

3.5 Solid food waste technologies

About 1.6 billion tons of food is wasted globally every year. This large number showcases the adverse environmental impact food waste can have. Food waste can be solved if the advanced technology available is applied correctly. It contains high content of moisture, oil, and organic matter. The commonly utilized technology for handling food waste include digestion, landfill, incineration, composting, vermicomposting, and anaerobic digestion. The landfill has been used highly. However, it is sometimes restricted by the emission of large amounts of greenhouse gas (GHG). In a few places, incineration of food waste is restricted due to the release of large quantities of dioxins which have serious health issues.

3.6 Slurry waste technologies

Urban waste generates sludge and slurry-type waste in good quantities. The concentration of slurry depends on the size of the pollutant and its physico-chemical properties. For the efficient conversion of slurry waste to product, the moisture needs to be removed. The technologies utilized are drying, pyrolysis, anaerobic digestion, and aerobic digestion. For direct thermal drying, high operating cost is entailed. The combination of peat with a solid fraction of slurry can be effectively utilized to produce compost for soil landscaping.

4. GLOBAL SCENARIO ON HANDLING BIOFUELS SUSTAINABILITY

Bioenergy as a sustainable source of energy can be an efficient replacement owing to the rising energy prices, its ability to reduce GHG emissions, and increased energy security. Despite such requirements, fossil fuel consumption still plays a pivotal role in fulfilling the energy requirement. In terms of the utilization of biofuels, there are few players in the global market. Countries like Brazil and the USA produced over 70% of total worldwide biofuel production followed by European Union countries like Germany and France producing around 9% of biofuel (Cadillo-Benalcazar et al. 2021).

However, in most cases, the active participation of more countries is not seen due to the lack of understanding by the policymakers and the lack of firm regulatory laws. In the near future, biofuels would likely replace the transport fuels.

Energy-rich biomass can be grown in the tropical countries. Second generation technologies would enable using grasses and trees that would grow in drought-prone and less fertile lands. Rapid expansion of bioenergy can potentially reduce the accessibility of staple food to the under-privileged people. Continuous harvesting of energy-rich crops may affect the quality of the soil and water, and affect the flora and fauna. The impact of global biofuel development and growth has its own pros and cons (Liu et al. 2019).

5. FUTURE PERSPECTIVES

Several bottlenecks need to be addressed to implement the biorefineries concept in the energy sector. However, the recent advancements in science and technology targeted at energy recovery from biomass represent significant progress in meeting these challenges. With innovations being developed and current technologies improved, a strong political will and investors play a major role to implement biomass usage at large-scale immediate commissioning which makes an impact. In the Indian context, the Union government's flagship projects such as 'Make in India', 'Atma Nirbhar Bharat' promotes using renewable resources and smart waste management techniques. It paves way for innovative startups and enterprises to promote circular economy and new job opportunities. Also in terms of e-waste management, urban mining will provide an opportunity for producers to ensure resources are affordable and plentiful such that production is not hampered. The successful deployment of a biorefinery and circular economy strategy hinges on factors such as public perception, political acceptance, and financial viability.

The implementation of strict laws and increased awareness among the masses may propagate the efficient collaboration among the stakeholders from different sectors such as government, industry, and academia. Regular monitoring of the strategies from time to time can lead to taking prompt actions and incorporating changes to the framework. This can thus lead to effective handling of the waste management structures.

REFERENCES

Ahmed El-Imam, A.M., Greetham, D., Du, C. et al. 2019. The development of a biorefining strategy for the production of biofuel from sorghum milling waste. Biochem. Eng. J. 150: 107288.

Ambaye, T.G., Vaccari, M., Petriciolet, A.B. et al. 2021. Emerging technologies for biofuel production: A critical review on recent progress, challenges and perspectives. J. Environ. Manag. 290: 112627.

Ashrafi, G., Nasrollahzadeh, M., Jaleh, B. et al. 2022. Biowaste- and nature-derived (nano) materials: Biosynthesis, stability and environmental applications. Advan. Coll. Interfac. Sc. 301: 102599.

Bosecker, K. 1997. Bioleaching: metal solubilization by microorganisms. FEMS Microbiology Review 20: 591–604.

Cadillo-Benalcazar, J.J., Bukkens, S.G.F., Ripa, M. et al. 2021. Why does the European Union produce biofuels? Examining consistency and plausibility in prevailing narratives with quantitative storytelling. Energ. Res. & Soc. Sci. 71: 101810.

Cheirsilp, B. and Maneechote, W. 2022. Insight on zero waste approach for sustainable microalgae biorefinery: Sequential fractionation, conversion and applications for high-to-low value-added products. Biores. Technol. Rep. 18: 101003.

Chen, P., Anderson, E., Addy, M. et al. 2018. Breakthrough technologies for the biorefining of organic solid and liquid wastes. Eng. 4(4): 574–580.

Chinnasamy, S., Ramakrishnan, B., Bhatnagar, A. et al. 2009. Biomass production potential of a wastewater alga Chlorella vulgaris ARC 1 under elevated levels of CO_2 and temperature. Int. J. Mol. Sci. 10(2): 518–32.

Chu, H., Qian, C., Tian, B. et al. 2022. Pyrometallurgy coupling bioleaching for recycling of waste printed circuit boards. Res. Conser. Recyc. 178: 106018.

Ekins, P., Domenech, T., Drummond, P. et al. 2019. The Circular Economy: What, Why, How and Where, Background paper for an OECD/EC Workshop within the workshop series. Managing environmental and energy transitions for regions and cities, Paris.

Gao, K., Orr, V. and Rehmann, L. 2016. Butanol fermentation from microalgae-derived carbohydrates after ionic liquid extraction. Bioresour. Technol. 206: 77–85.

Gouveia, L. and Passarinho, P.C. 2017. Biomass conversion technologies: Biological/biochemical conversion of biomass. In: Rabaçal, M., Ferreira, A., Silva, C. et al. (eds.). Biorefineries. Lecture Notes in Energy, vol 57. Springer, Cham.

Hoang, A.T., Niżetić, S., Ong, H.C. et al. 2021. Insight into the recent advances of microwave pretreatment technologies for the conversion of lignocellulosic biomass into sustainable biofuel. Chemosphere 281: 130878.

Igbokwe, V.C., Ezugworie, F.N., Onwosi, C.O. et al. 2022. Biochemical biorefinery: A low-cost and non-waste concept for promoting sustainable circular bioeconomy. J. Environ. Manag. 305: 114333.

Joshi, S. and Mishra, S.D. 2022. Recent advances in biofuel production through metabolic engineering. Biores. Technol. 127037.

Jones, R.E., Speight, R.E., Blinco, J.L. et al. 2022. Biorefining within food loss and waste frameworks: A review. Renew. Sustain. Energ. Rev. 154: 111781.

Joseph, A.M., Snellings, R., Van den Heede, P. et al. 2018. The use of municipal solid waste incineration ash in various building materials: a belgian point of view. Materials (Basel). 16, 11(1): 141.

Kee, S.H., Chiongson, J.B.V., Saludes, J.P. et al. 2021. Bioconversion of agro-industry sourced biowaste into biomaterials via microbial factories—A viable domain of circular economy. Environ. Poll. 271: 116311.

Khan, I.S., Ahmad, M.O. and Majava, J. 2021. Industry 4.0 and sustainable development: A systematic mapping of the triple bottom line, circular economy and sustainable business models perspectives. J. Clean. Product. 297: 126655.

Kim, K.H., Choi, I.S., Kim, H.M. et al. 2014. Bioethanol production from the nutrient stress-induced microalga Chlorella vulgaris by enzymatic hydrolysis and immobilized yeast fermentation. Bioresour. Technol. 153: 47–54.

Kruczek, H.P., Niedzwiecki, L., Sieradzka, M. et al. 2020. Hydrothermal carbonization of agricultural and municipal solid waste digestates—Structure and energetic properties of the solid products. Fuel 275: 117837.

Kuisma, M., Kahiluoto, H., Havukainen, J. et al. 2013. Understanding biorefining efficiency—The case of agrifood waste. Biores. Technol. 135: 588–597.

Liu, X., Lendormi, T. and Lanoisellé, J.-L. 2019. Overview of hygienization pretreatment for pasteurization and methane potential enhancement of biowaste: Challenges, state of the art and alternative technologies. J. Clean. Product. 236: 117525.

Mastos, T.D., Nizamis, A., Terzi, S. et al. 2021. Introducing an application of an industry 4.0 solution for circular supply chain management. J. Clean. Product. 300: 126886.

Okoligwe, O., Radu, T., Leaper, M.C. et al. 2022. Characterization of municipal solid waste residues for hydrothermal liquefaction into liquid transportation fuels. Waste Manag. 140: 133–142.

Omran, B.A. and Baek, K.-H. 2022. Valorization of agro-industrial biowaste to green nanomaterials for wastewater treatment: Approaching green chemistry and circular economy principles. J. Environ. Manag. 311: 114806.

Rajendran, N., Gurunathan, B. and Selvakumari, A.E. 2021. Optimization and techno economic analysis of bio oil extraction from Calophyllum inophyllum L. seeds by ultrasonic assisted solvent oil extraction. Ind. Crops. Product. 162: 113273.

Rajmohan, K.S., Ramya, C. and Varjani, S. 2019. Trends and advances in bioenergy production and sustainable solid waste management. Energ. & Env. 1–27.

Rai, S.V. and Rajashekhar, M. 2014. Effect of pH, salinity and temperature on the growth of six species of marine phytoplankton. J. Algal Biomass Utln. 5(4): 55–59.

Rodionova, M.V., Bozieva, A.M., Zharmukhamedov, S.K. et al. 2022. A comprehensive review on lignocellulosic biomass biorefinery for sustainable biofuel production. Int. J. Hyd. Energ. 47(3): 1481–1498.

Shah, A.V., Singh, A., Mohanty, S.S. et al. 2022. Organic solid waste: Biorefinery approach as a sustainable strategy in circular bioeconomy. Biores. Technol. 3491: 26835.

Shahid, M.K., Batool, A., Kashif, A. et al. 2022. Biofuels and biorefineries: Development, application and future perspectives emphasizing the environmental and economic aspects. J. Environ. Manag. 297: 113268.

Shen, C., Zhang, G., Li, K. et al. 2022. A pathway of the generation of acid mine drainage and release of arsenic in the bioleaching of orpiment. Chemos. 298: 134287.

Subhash, G.V., Rajvanshi, M., Kumar, G.R.K. et al. 2022. Challenges in microalgal biofuel production: A perspective on techno economic feasibility under biorefinery stratagem. Biores. Technol. 343: 126155.

Sun, Y., Qin, Z., Tang, Y. et al. 2021. Techno-environmental-economic evaluation on municipal solid waste (MSW) to power/fuel by gasification-based and incineration-based routes. J. Env. Chem. Eng. 9(5): 106108.

Tripathi, N., Hills, C.D., Singh, R.S. et al. 2019. Biomass waste utilisation in low carbon products: Harnessing a major potential resource. NPJ. Clim. Almos. Sci. 2: 35.

Vignesh, N.S., Soosai, M.R., Chia, W.Y. et al. 2022. Microwave-assisted pyrolysis for carbon catalyst, nanomaterials and biofuel production. Fuel 313: 123023.

Vigot, M.A., Damartzis, T. and Maréchal, F. 2018. Thermoeconomic design of biomass biochemical conversion technologies for advanced fuel, heat and power production. pp. 1801–1806. *In*: Mario R. Eden, Marianthi G. Ierapetritou and Gavin P. Towler (eds.). Computer Aided Chemical Engineering. Elsevier, 44.

Wang, H.D., Chen, C.C., Huynh, P. et al. 2015. Exploring the potential of using algae in cosmetics. Bioresour. Technol. 184: 355–362.

Weir, A., Jiménez del Barco Carrión, A., Queffélec, C. et al. 2022. Renewable binders from waste biomass for road construction: A review on thermochemical conversion technologies and current developments. Const. Build. Mater. 330: 127076,

Yang, R.-X., Jan, K., Chen, C.-T. et al. 2022. Thermochemical conversion of plastic waste into fuels, chemicals, and value-added materials: A critical review and outlooks. ChemSusChem. 15: 11 bio.

Yoo, C., Jun, S.Y., Lee, J.Y. et al. 2010. Selection of microalgae for lipid production under high levels of carbon dioxide. Bioresour. Technol. 101 Suppl 1: S71–4.

8

SUSTAINABLE MANAGEMENT OF MUNICIPAL SOLID WASTE BY A COMBINATION OF BIOLOGICAL AND CHEMICAL PROCESS APPROACHES

Yalakala Praneetha and *Lalit M. Pandey**

1. INTRODUCTION

The rapid acceleration of global advances has resulted in an increase in energy consumption throughout the world. As a result, fossil fuels are extensively burnt to meet the increased demand of energy. On the other hand, as the population is radially growing, so is the usage of everyday products, resulting in a huge rise in waste generation. MSW is mostly made up of papers, plastics, metal, organic waste, rubber, leather, glass, ceramics, soil materials and other miscellaneous materials. The worldwide annual generation of MSW is expected to increase by 70% to reach 3.40 billion tons as compared to 2.01 billion tons in 2016 (Kaza et al. 2018).

In addition to that, the rapid increase in the volume of garbage across the world results in the scarcity of locations for waste disposal, which in turn causes non-sustainable waste management. These wastes are rich in carbon contents. The organic fractions of MSW are very high. The reported value of organic fractions ranges from 80 to 97.5% of volatile solid/total solid contents (Van Fan et al. 2018). In this regard, various technological inventions have been explored for the utilization of these wastes for the generation of energy. MSW can be converted into energy and useful chemicals using the following processes, namely biological, chemical and hybrid/composite (Rabaey and Verstraete 2005, Van Fan et al. 2018). Figure 1 shows the various biological, chemical and bio-chemical waste-

Bio-interface & Environmental Engineering Lab, Department of Biosciences and Bioengineering, Indian Institute of Technology Guwahati, Assam, 781039, India.
* Corresponding author: lalitpandey@iitg.ac.in

Figure 1. Depicting Biological, Chemical and Bio-chemical WTE technologies for the management of MSW.

to-energy (WTE) technologies for the management of solid wastes. Chemical methods utilize thermal energy for the conversion of solids into gas, liquid and ash (Das and Tiwari 2018). These methods are energy-intensive and quick processes. Biological methods exploit microorganisms and enzymes to utilize organic contents as substrate/feed to metabolize them into gases and solvents through a series of biochemical reactions (Datta et al. 2018). These methods are eco-friendly but slow. Bio-chemical methods directly convert the chemical energy of biodegradable substrates to electricity through oxidation with the aid of exo-electrogenic microorganisms (Sevda et al. 2019). Bio-electrochemical systems are evolving technologies and require further interventions to be applied at a large scale (Bhattacharjee and Pandey 2020, Sevda et al. 2018).

In order to overcome the current energy demand and to face the challenges of MSW management, alternative sustainable methods have to be used for the effective utilization of solid wastes coming from municipalities and to produce useful energy and chemicals from it. This chapter mainly focuses on the recycling of MSW and then converting them into energy and valuable chemicals. The next section provides a detailed idea of biological, chemical and bio-chemical methods followed for the sustainable management of MSW.

2. BIOLOGICAL PROCESSES

The major biological processes for the management of MSW include anaerobic digestion (AD), composting and ethanol fermentation. Apart from treating the MSW, the biological processes also produce value-added by-products like biogas, biofuel and compost. These processes are discussed in this section.

2.1 Anaerobic digestion

AD is the process for the conversion of organic matter of MSW into digestate and biogas by the action of bacteria in the absence of oxygen as shown in Fig. 2. It can be considered as a major alternative for the treatment of different wastes. It is also a potential solution to improve current energy supply security (Van Fan et al. 2018).

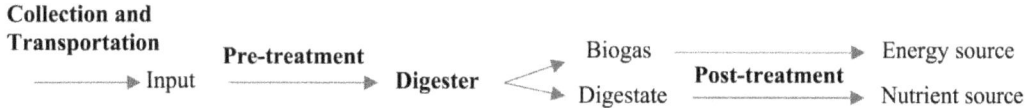

Collection and Transportation ——————▶ Input   **Pre-treatment** ——————▶ **Digester** ⟨ Biogas ——————▶ **Post-treatment** Energy source / Digestate ——————▶ Nutrient source

Figure 2. Overview of AD process including pre-treatment, digestion and post-treatment. Adapted with permission from (Van Fan et al. 2018).

The stages of AD are depicted in Fig. 2 and are discussed as follows:

2.1.1 Pre-treatment

Pre-treatment of MSW prior to AD is mainly for sorting, separating, sterilization and reduction of size. Pre-treatment is an important step for removing the impurities and in turn improving the biogas yield. It is mainly categorized into the five major groups: chemical, biological, mechanical, thermal, and combined methods. But, mechanical and thermal methods like thermal hydrolysis, microwave, grinding and sonication are comparatively more favorable due to being relatively significant in energy intake.

2.1.2 Digestion process

The digestion efficiency varies depending on the operation mode and kind of AD system. For the metabolic activity of the anaerobe, waste properties like pH (6.3–7.8), C:N ratio (25–30), volatile fatty acid (2000–3000 mg/L) and other properties like nutritional contents, organic loading rate must be fine-tuned.

Reports on MSW studies mainly show that a two-phase AD is more favorable when compared to a single-phase AD (Ganesh et al. 2014). In the case of a two-phase AD, individual bacterial species execute hydrolysis and methanogenesis separately, whereas in a single phase AD, all the processes, namely hydrolysis, acidogenesis, acetogenesis and methanogenesis occur in the same reactor.

Hydrolysis results in the transformation of MSW into simple sugars, fatty acids and amino acids, which are converted to H_2, CO_2, volatile fatty acids (VFA), acetate and alcohols during acidogenesis. VFA and alcohols are converted to H_2 and acetic acid during acetogenesis. Finally, CH_4 is produced during methanogenesis from acetic acid, and CO_2 and H_2. The common genera used in AD include *Ruminococcus*, *Acetobacterium*, *Acetoanaerobium*, *Butyribacterium*, *Clostridium*, *Eubacterium*, *Aminobacterium* and *Methanobacterium*.

2.1.3 Post-treatment

Biogas and digestate are the two main products of the AD process. The quality of the energy produced has a big impact on the further applications of biogas. The calorific value of biogas varies in the range of 16 to 28 MJ/m^3 (Salunkhe et al. 2012). High content of CO_2 in the biogas reduces its

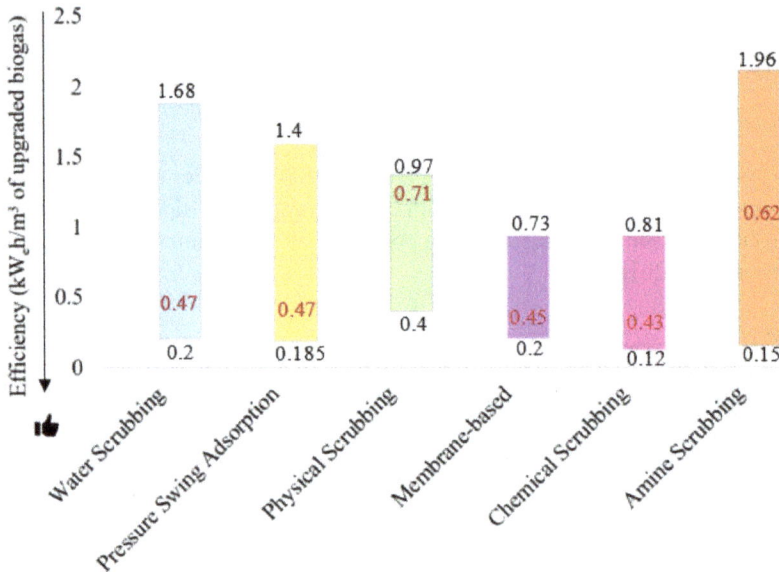

Figure 3. Energy consumption efficiency of the different post-treatment processes. The texts in red font refer to the average energy efficiency of post-treatment processes. Adapted with permission from (Van Fan et al. 2018).

heating value, which results in low combustion efficiency, corrosion and then low economic value. Temperature swing adsorption, amine scrubbing, cooling effect dehydration, and pressure water scrubbing are some of the methods which can be used for CO_2 and water removal. For reducing the concentration of H_2S, methods like absorption, biological system and the use of regenerable liquid or solid media are used (Leme and Seabra 2017). Figure 3 depicts the energy consumption efficiencies of the different post-treatments, which are measured as necessary input energy per m^3 of enhanced biogas. On the basis of energy consumption, chemical scrubbing is reported as the most efficient post-treatment for MSW (0.43 $kW_e h/m^3$), followed by membrane-based, pressure swing adsorption, water scrubbing, amine scrubbing and physical scrubbing (Fig. 3).

The appropriate selections at the individual stages should be made to improve the energy efficiency of the whole AD process. A few observations are summarized as follows:

(a) Waste segregation at the source can significantly contribute to the viability of AD of MSW. Segregation mainly helps to reduce the energy requirements of all the stages.

(b) Microwave pre-treatment is found to be the best among the other pre-treatment processes because it enhances the biogas outputs but with a limitation of high energy input. The energy input range for microwave pre-treatment is around 114.24 to 8040 $kW_e h/t$ with a percentage output enhancement of +4 to 39.28% (Pecorini et al. 2016, Savoo and Mudhoo 2018, Zhang et al. 2016).

(c) Two-phase AD is considered to be more suitable than single-phase AD as mentioned above. Concerning the conditions for AD, thermophilic is suggested to be more suitable than mesophilic conditions for AD of MSW. The specific energy consumption for the AD of MSW under thermophilic conditions (8–14 MJ/Nm^3 CH_4) is much lesser than in mesophilic conditions (12–22 MJ/Nm^3 CH_4) (Van Fan et al. 2018, Wu et al. 2015).

(d) The best post-treatment method can be decided based on the amount of carbon emission footprint. Among the post-treatment methods given above, amine scrubbing offers a relatively carbon emission footprint (1.028 kg CO_2/Nm^3 of upgraded biogas) while chemical scrubbing contributes the least (0.063 kg CO_2/Nm^3 of upgraded biogas) carbon emission footprints. So, chemical scrubbing is considered to be the best among the other post-treatment methods.

The merits and limitations of AD are outlined as follows. The major outputs are biogas and digestate. Biogas is a major energy source which can be used for the combined generation of power and heat. It can be used as a fuel for generating electricity. It can be further processed for the production of bio-methane, which can be used as a fuel for cars. The byproduct digestate is used as a fertilizer.

On the other hand, pre-treatment of complex materials is necessary for efficient AD. Also, post-treatment of the generated waste by AD is necessary before it is discharged into the environment. In addition, key parameters like pH, temperature and feed rate must be monitored continuously. The variations in substrate characteristics, different accessible digestion and gas upgrading processes, as well as product consumption, affect the overall energy efficiency of this sophisticated biological system.

2.2 Composting

Composting of MSW offers a method for processing sewage sludge and garbage in one operation. With the increasing environmental rules, composting can be used as an alternative to other MSW management methods like incineration, and landfill with gas capture. Composting is a biological process in which the organic element of solid waste is allowed to decay under carefully controlled conditions (Hassen et al. 2001). It is another technique for handling MSW. Diverse microbial community, namely yeasts, fungi and bacteria are reported to be present in a compost of MSW (*Escherichia coli*, *Bacillus*, *Streptococcus*, *Staphylococcus*, *Salmonella*) (Hassen et al. 2001). The organic waste material is metabolized by microbes, which reduces its volume by up to 50%. Compost or humus is the stabilized product and is rich in nitrogen, phosphorous and potassium contents.

The steps involved in the process of composting are as follows (Kumar 2011):

2.2.1 Sorting and shredding

Sorting and separating procedures separate decomposable materials from glass, metal, and other inorganic components of trash (Kumar 2011). Mechanically, using distinctions in physical properties, these separations are carried out. The size of the waste pieces is reduced by shredding or crushing for producing a homogeneous mass of material.

2.2.2 Digesting and processing

Composting of pulverized garbage can be done in an open windrow or in an enclosed mechanical facility. Windrows are long and low rubbish heaps. After every few days, windrows are turned or stirred to allow air to reach the microbes that are digesting the organics. Before it can be utilized as a mulch or soil conditioner, dug compost must be treated. Drying, screening, and granulating or pelletizing are all steps in the processing process (Kumar 2011).

The key merits and limitations of composting are outlined as follows. Composting is an economical process and saves resources likes water, fertilizers and pesticides. It is a low-cost alternative to regular landfill cover and artificial soil additions and acts as a marketable product. Composting also extends the life of municipal landfills by diverting organic debris from them, and it is a less expensive alternative to traditional methods of soil remediation (cleaning). However, composting in windrows and aerated static piles takes up a lot of space, and odor control is a regular issue. Windrow and aerated static pile composting are affected by ambient temperatures and meteorological conditions.

2.3 Ethanol Fermentation

Apart from commonly used waste disposal methods such as AD, composting, refuse derived fuel, and land filling, ethanol production from the organic fraction of MSW (OMSW) could be seen as a potential source of energy generation and trash management (Dornau et al. 2020). Typically, OMSW contains about 50% of lignocellulose-rich materials, which can be explored as feedstock. The complex hydrocarbons are first converted to fermentable sugars. Fermentation is a method of converting glucose and fructose into ethanol, a type of alternative biofuel for motors. Principally, the following three experiments were carried out in series.

2.3.1 Hydrolysis

In hydrolysis, the fibers from the samples of MSW are milled till the particle size is consistent and then loaded into a hydrolysis vessel. The sample is now diluted into a slurry mass with water and hydrolysis is performed using physical and chemical methods like thermal treatment, alkali treatment and acid treatment. Hydrolysis results in the formation of fermentable sugars. Adjustments can be made at this stage and the resulting slurry is centrifuged in order to separate the desired hydrolysate from un-hydrolyzed solids.

2.3.2 Fermentation

In this, microorganisms which are intrinsically suited for growing on OMSW hydrolysates are screened through substrate utilization and tolerance. Microbes are incubated in a fermentation medium that includes filter-sterilized MSW fiber hydrolysate as carbon source. *S. cerevisiae* and *Z. mobilis* are commonly utilized for the anaerobic fermentation of ethanol (Dornau et al. 2020). The key factors impacting bioethanol fermentation include sugar concentration, inoculum size, temperature, pH, fermentation time and agitation rate (Zabed et al. 2014). The optimum pH and temperature for bioethanol fermentation are 4.0–5.0 and 20–35°C, respectively (Zabed et al. 2014).

2.3.3 Distillation

Ethanol and water mixture forms a minimum boiling azeotrope, which inhibits fractional distillation to produce pure ethanol from the fermented broth. Pressure swing distillation with an entrainer (acetone) is utilized to avoid azeotropic point and facilitate the separation. A schematic representation of ethanol fermentation utilizing MSW (Meng et al. 2021) is shown in Fig. 4.

Autoclaved treatment of MSW is considered to be the best method for the production of bio-ethanol from MSW. Developing an integrated bio-refinery utilizing MSW has the potential to considerably improve the environmental burden of existing waste management practices. The main advantages and limitations of this route are as follows: (a) Bioethanol burns more cleanly, so the exhaust gases from this are much cleaner. (b) Bioethanol reduces greenhouse gases. (c) The fuel spills for this are more biodegradable compared to other fuels. However, the energy content of ethanol is much lesser than gasoline. Also, bioethanol as fuel can adversely affect electric fuel pumps due to corrosive effects.

3. CHEMICAL PROCESSES

The key chemical processes for the management of MSW, namely incineration, pyrolysis and gasification, have been described in this section.

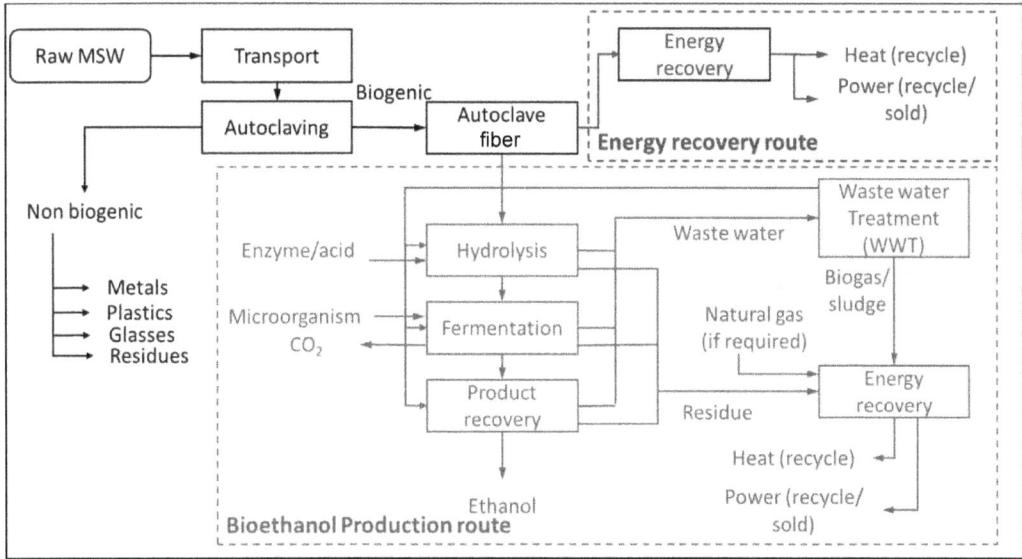

Figure 4. Different processes involved in the ethanol/energy conversion from MSW. Adapted with permission from (Meng et al. 2021).

3.1 Incineration

Waste management is done by the combustion of substances in the incineration process (Brunner and Rechberger 2015). Energy recovery was not an objective for the early MSW incinerators (Brunner and Rechberger 2015). Initially, incinerators were developed for sanitary and volume reduction purposes. Emissions were excessive due to the lack of flue gas purification and the poor conditions for incineration as shown in Fig. 5. Due to a lack of understanding in the realms of air pollution and its control, improvements in furnace technology and developments of efficient filters took several decades (Brunner and Mönch 1986).

It required a new generation of research, technology and regulation to make the quantum leap that is required to make improvements in incineration so that it could be used to meet waste management goals. Modern multi-stage filter systems have supplanted older filter technologies like cyclones and other times one-chamber electrostatic precipitators. These wet and dry systems are capable of extracting tiny particulates as well as a bulk of fly ash that the previous filters had removed. The most recent development is the extraction of secondary resources from these leftovers, for example, metals like iron, copper, aluminium, zinc and others. Figure 6 explains

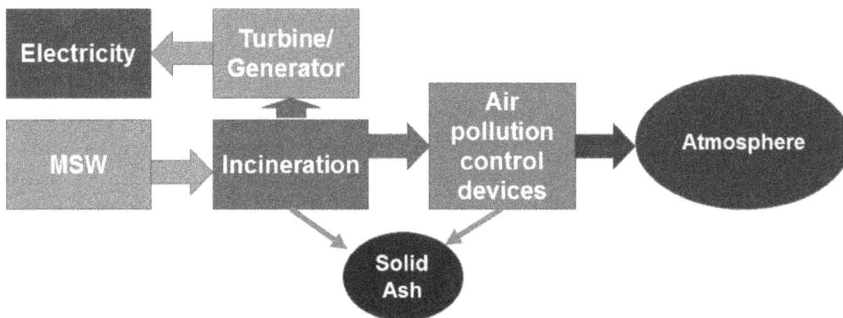

Figure 5. Illustration of the incineration process of MSW.

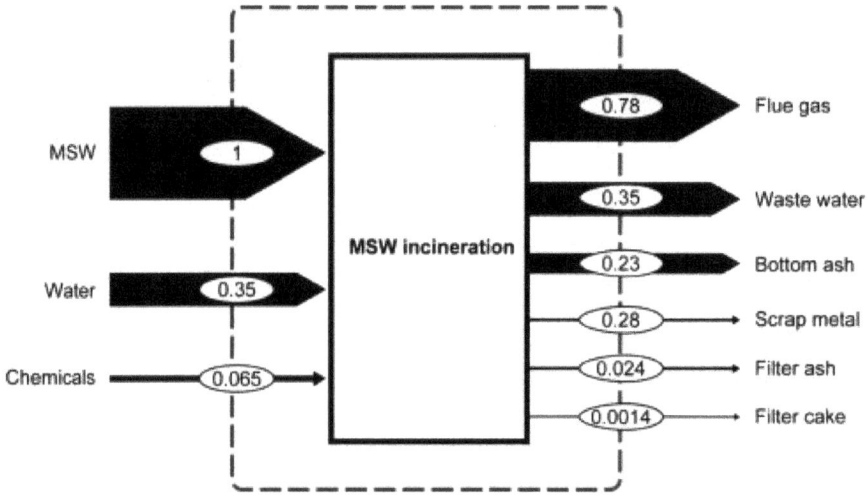

Figure 6. Mass flow of different streams (flue gas, ash and cake) in kg/kg of MSW through a WTE facility, which is equipped with both dry and wet air pollution control devices. Adapted with permission from (Brunner and Rechberger 2015).

the mass flow of different streams (in kg/kg of MSW) through a WTE facility, which is equipped with both dry (Electrostatic Precipitators) and wet air pollution control devices (Brunner and Rechberger 2015).

All the possible energy conversion systems used for the incineration process have an exergetic efficiency of about 20% and efforts are being made to increase the exergetic efficiency up to 30% (Brunner and Rechberger 2015). The optimal energy conversion system for a specific site needs to be chosen considering economic and demand constraints. The modern incinerators have gone a long way to develop into an economically favorable method of waste treatment and can be used as a better alternative to landfills.

The main advantage of incineration as compared to other treatment processes is the mineralization of organic substances (*CxHyOz*) to CO_2 and H_2O. Incineration is an effective mechanism for the sustainable management of hazardous organic materials which cannot be recycled. For such wastes, thermal destruction by WTE appears to be the best solution. It is an effective volume reduction method for MSW.

However, the installation of an incineration plant is expensive and the plant needs regular maintenance which adds to the already existing expenses. The smoke produced from burning waste during incineration pollutes the environment. The ash produced from the process needs to be disposed of appropriately, or else it can cause serious harm to the environment. The fly ash residues can be utilized for co-processing in the cement industry (Quina et al. 2018).

3.2 Pyrolysis

Pyrolysis is the process of thermal breakdown of MSW at high temperatures (300–900°C) in an inert atmosphere. Pyrolysis is gaining popularity these days because of its ability to produce a variety of gaseous (syngas), liquid (bio-oil) and solid (bio-char) products in different proportions just by changing the operating parameters like temperature or heating rate (Shukla 2022). At low temperatures < 450°C and a slow heating rate, MSW pyrolysis yields mostly solid residue. At intermediate temperatures between 450–800°C, the main product is liquid products (bio-oil). At high temperatures > 800°C and a rapid heating rate, syngas is preferably obtained. The gas yields vary from different feedstocks and are positively related to the pyrolysis temperature.

The main advantages and limitations of pyrolysis are as follows. Pyrolysis can be done on a small scale, which helps in reducing transportation and waste. The bio-oil produced from pyrolysis is a dark brown liquid which can be upgraded to engine fuel or further gasification of this gives syn gas and then bio-diesel. Syngas is made up of a mixture of energy-rich gases, so it can be combusted to generate electricity. Bio-char is practically pure carbon and can be used to boost soil fertility and also helps soil get rid of heavy metals and other pollutants. However, the release of volatiles into the atmosphere during the process is a major disadvantage of this method. Also, pyrolysis is a complex process and needs high operational and investment costs.

3.3 Gasification

Gasification is a process of conversion of the organic compounds of MSW into a mixture of gaseous species comprising of CO_2, CO, H_2 and CH_4. The main advantage of gasification is its ability to convert MSW into energy carrier gases (Arafat and Jijakli 2013). Gasification products are produced through a complex reaction mechanism that includes a phase change from solid to gas and sequence reactions in the gas phase. In the first phase, any remaining moisture evaporates. The second phase is pyrolysis. During pyrolysis, volatiles are liberated from the solid phase, which react with each other and the solid char leading to the formation of new species. Lastly, the complex char-gas and gas-gas interactions provide a way to the controlling reactions and the final products of gasification are generated. The following equation explains the reaction between the reactants (feedstock ($C_nH_yO_xN_z$) and air) and products in the gasification process. The air is considered a mixture of 1 mole of O_2 and 3.76 moles of N_2.

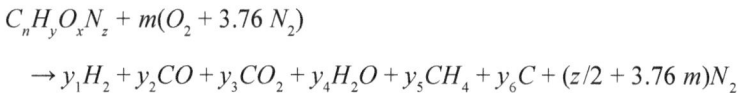

$$C_nH_yO_xN_z + m(O_2 + 3.76\ N_2)$$

$$\rightarrow y_1H_2 + y_2CO + y_3CO_2 + y_4H_2O + y_5CH_4 + y_6C + (z/2 + 3.76\ m)N_2$$

The generation of gasification products is also affected by the reaction temperature. The combustion reaction becomes more complete with an increase in the gasification temperature. Thus, the quantities of combustible gas products (H_2, CH_4) decrease, while the extents of O_2, H_2O and CO_2 increase. Also, the amount of energy produced per kg of mixed MSW depends on the relative composition. Gasification products have a wide range of energy and industrial applications. Gasification has the potential to be (i) more efficient, (ii) operate at a lower temperature, and (iii) release fewer pollutants when compared to incineration.

The combustible gases are the energy outputs of the gasification of MSW, which can be utilized for thermal energy generation. So, the total energy output increases with an increase in the extent of combustible gases. Another important point to remember is that the efficiency of gasification relates to the efficiency of the end usage of its products. For example, the efficiencies of a steam cycle energy recovery system, an ignition engine and a gas turbine are about 23%, 25% and 40%, respectively (Belgiorno et al. 2003). The performance of a gasification process can be evaluated in terms of electrical efficiency. The net electrical efficiency of a gasification process is expected to be about 20% (Arena 2012).

The merits and limitations of the gasification process are as follows. It produces comparatively lower quantities of air pollutants when compared to incineration. Because it allows waste products to be used as feedstock, gasification also eases the environmental influence of waste disposal. In addition, gasification plants require less amount of process water than coal-based traditional power plants. Gasification byproducts are harmless. However, the presence of halogens, heavy metals, particulates, various tars and alkaline compounds can result in accumulation in the gasification reactor.

4. BIO-CHEMICAL PROCESSES

Bio-chemical processes directly convert the chemical energy of biodegradable substrates to electricity through oxidation with the aid of microbes (Sevda et al. 2019). In this section, landfill with gas capture and bio-electrochemical processes are discussed. Microbial fuel cells (MFC), microbial electrolysis cells (MEC) and the generation of biological hydrogen have been described as bio-electrochemical processes.

4.1 Landfill with gas capture

Landfill with gas capture is a technique in which waste is buried in pits and then letting it to naturally deteriorate/degrade for a period of time. There are particular standards for operating a landfill, like eliminating seepage, minimizing odor and controlling greenhouse gas emissions.

Advancements have been made in this technology to improve its performance, like the application of microwave irradiation for the activation of carbon in MSW, Fenton and persulfate oxidation for leachate treatment (Usman et al. 2020) and leachate removal via advanced oxidation processes followed by adsorption (Poblete et al. 2019).

The key merits and limitations of this process are as follows. Landfills are an excellent source of energy. Methane can be used as a source for the generation of electricity and heat. A landfill is generally more cost-effective than the burning of MSW in incineration. However, this method requires large and remote lands. Methane is comparatively more effective as compared to other greenhouse gases. It contaminates underground water with leachates.

4.2 Bio-electrochemical process

4.2.1 Microbial fuel cells

MFCs are devices that use bio-electrogenic microorganisms to generate energy utilizing a range of substrates. The bacteria act as a catalyst for the generation of bio-hydrogen in both aerobic and anaerobic conditions. A wide range of wastes like sludge, animal waste and household waste can be utilized as feedstock for MFCs. This process is based on the notion of electron allocation, which occurs within the bacteria via a redox reaction. Bacteria receive electrons that are present in proteins, lipids, and other compounds as a result of catabolism and transport them through the metabolic route as shown in Fig. 7 (Logroño et al. 2015, Rabaey and Verstraete 2005). The main advantages of MFCs include direct generation of electricity, a small carbon footprint, low operating cost and less generation of sludge. However, the major limitation for MFC is the expensive membrane and other materials.

4.2.2 Microbial electrolysis cells

MEC is built similar to a MFC, with the exception that the cathode isn't exposed to air. When electrons reach the cathode and the reduction process begins, hydrogen is generated. A single chamber MEC is shown in Fig. 8. MECs have the ability to produce a wide range of products including hydrogen, methane, hydrogen peroxide, acetate, formic acid, ethanol and others (Beyene et al. 2018).

The main advantages include higher hydrogen recovery. Other merits include the utilization of a wide variety of substrates as compared to MFC and an energy-positive process. However, MEC must operate at optimal conditions for both microbial processes and electrolysers to function. High current densities during the industrial application disrupt the optimal conditions of the system.

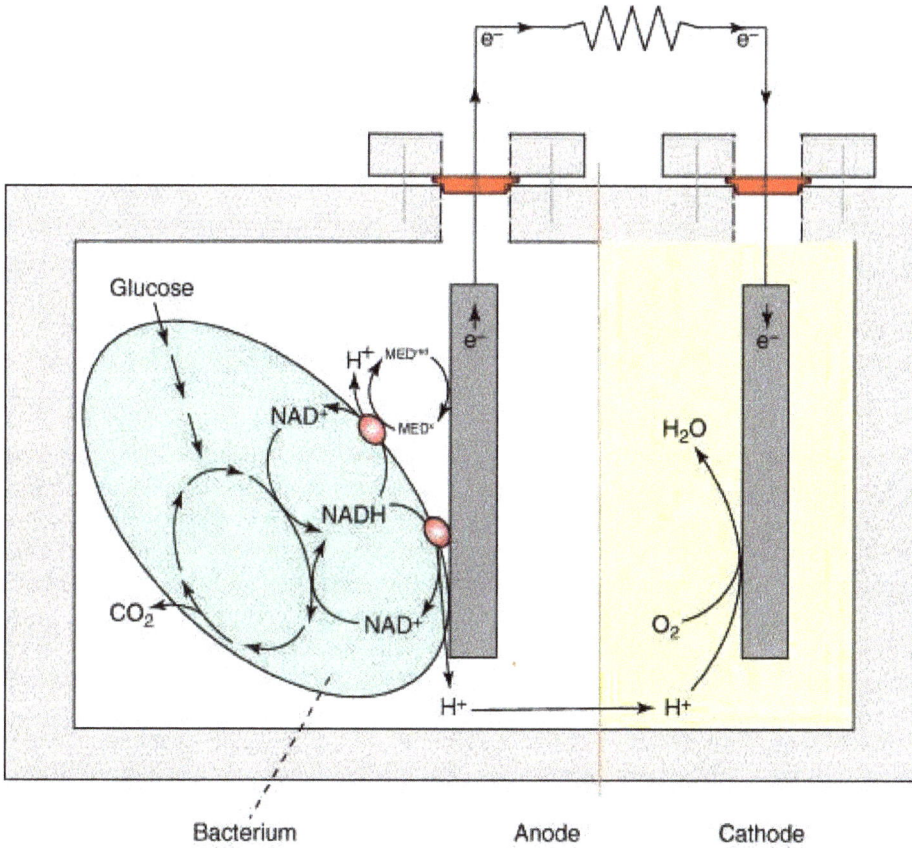

Figure 7. Schematic diagram of MFC. Substrates (bio-macromolecules) are catabolized by bacteria, which then transfer the gained electrons to the anode through the membrane or mobile redox shuttles. MED refers redox mediator and the red oval represents a terminal electron shuttle. Adapted with permission from (Rabaey and Verstraete 2005).

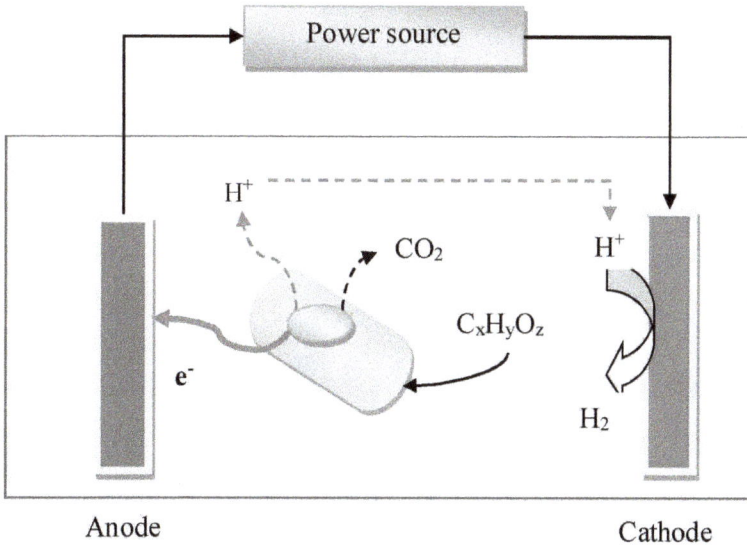

Figure 8. A schematic diagram of a single chamber MEC representing the generation of H_2. Adapted with permission from (Beyene et al. 2018).

4.2.3 Generation of biological hydrogen

Waste also has been discovered to be a long-term and renewable hydrogen source. Microorganisms are responsible for turning waste into hydrogen. Two main hydrogen production methods include biological and physical-chemical. These methods include thermal conversion, photo/dark fermentation and photo-biological route. Bio-hydrogen energy can be obtained from MSW by the action of microorganisms having different and adaptable digestive machineries. Due to the low energy need, biological hydrogen generation is more favorable and economical (Goud et al. 2014). The energy yield from hydrogen is about 142 kJ/g, which is three folds greater than other hydrocarbon fuels (Lai et al. 2014, Rasheed et al. 2021).

The merits and limitations of this process are as follows. Bio-hydrogen is a clean energy source and reduces CO_2 (greenhouse gas) emissions. The utilization of renewable energy sources makes the process sustainable and eco-friendly. Bio-hydrogen has high calorific values and can be used as a renewable fuel. However, the process of bio-hydrogen production is expensive. It has storage complications due to its low density. Bio-hydrogen is a highly inflammable and volatile matter.

Summarily, energy potentials vary for the different streams of MSW depending on the WTE technologies applied. Arafat et al. (2015) compared the average recoverable energy contents from various streams of MSW in terms of electrical energy efficiency through chemical and biological technologies of WTE as shown in Fig. 9 (Arafat et al. 2015, Ganesh et al. 2014). AD and gasification are best suited for food wastes and plastics, respectively, while incineration can be applied for various streams of MSW (Fig. 9). Further, the approach of bio-electrochemical processes seems to be more sustainable when compared to chemical and biological methods. Chemical processes like incineration, pyrolysis and gasification need a large amount of energy for operation and cost efficiency also comes as an obstacle. Biological processes are more efficient than chemical methods. Furthermore, the bio-electrochemical processes already include biological methods like anaerobic digestion in the first chamber of the cell. Additionally, bio-electrochemical processes can produce a wide range of products apart from methane. These hybrid methods can also operate on a wide range of feedstocks, which are cost-efficient. Lastly, MFCs, MECs and the generation of bio-hydrogen are the most advanced and recent techniques for the sustainable management of MSW.

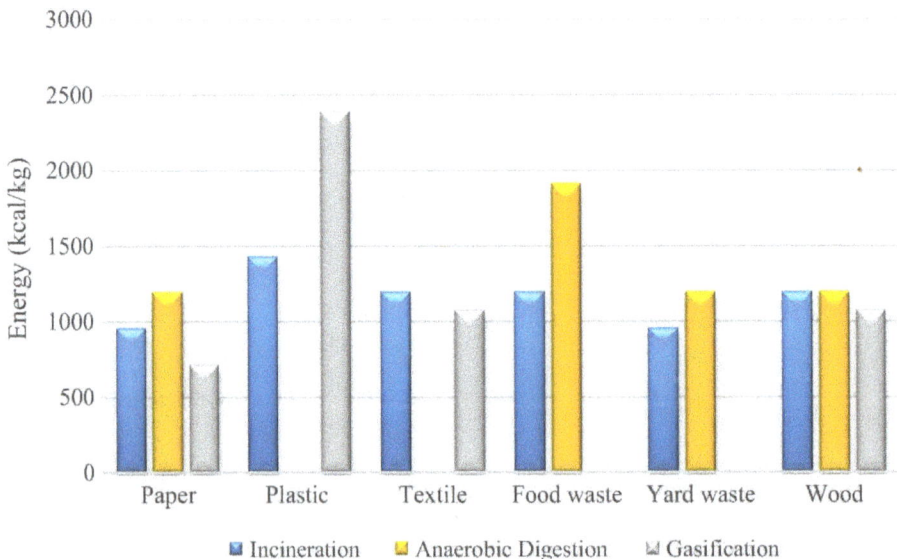

Figure 9. A comparison of energy recovery potential of different WTE technologies utilizing various streams of MSW. Adapted with permission from (Kumar and Samadder 2017).

5. CONCLUSIONS

Different WTE technologies acquainted with the recovery of energy from MSW have been reviewed. MSW is considerably a potential source of energy when WTE technologies are adopted. The best possible methods for sustainable management of MSW have been discussed in this chapter and in the end, the best-suited method has also been suggested. The adaption of the WTE technologies not only solves the traditional problem of energy crises but also provides a solution for the effective management of a bulk quantity of MSW. Due to the current situation across the world, developing countries have been using WTE technologies meticulously for MSW management. But these infrastructures/set-ups still lack in areas like maintenance, infrastructure and pollution control which are also to be kept in mind. Though on a smaller scale, a few developed countries have already installed WTE plants. Developing countries can work on strengthening the WTE technologies by providing them with the necessary financial support and making government strategies. A sincere effort will certainly aid the policymakers and researchers in coming up with the development of the best WTE technologies.

REFERENCES

Arafat, H.A. and Jijakli, K. 2013. Modeling and comparative assessment of municipal solid waste gasification for energy production. Waste Management 33(8): 1704–1713.

Arafat, H.A., Jijakli, K. and Ahsan, A. 2015. Environmental performance and energy recovery potential of five processes for municipal solid waste treatment. Journal of Cleaner Production 105: 233–240.

Arena, U. 2012. Process and technological aspects of municipal solid waste gasification. A review. Waste Management 32(4): 625–639.

Belgiorno, V., De Feo, G., Della Rocca, C. et al. 2003. Energy from gasification of solid wastes. Waste Management 23(1): 1–15.

Beyene, H.D., Werkneh, A.A. and Ambaye, T.G. 2018. Current updates on waste to energy (WtE) technologies: A review. Renewable Energy Focus 24: 1–11.

Bhattacharjee, U. and Pandey, L.M. 2020. Novel nanoengineered materials-based catalysts for various bioelectrochemical systems. pp. 45–71. In: Lakhveer Singh, D.M.M. and Hong Liu (eds.). Novel Catalyst Materials for Bioelectrochemical Systems: Fundamentals and Applications. American Chemical Society, Washington.

Brunner, P.H. and Mönch, H. 1986. The flux of metals through municipal solid waste incinerators. Waste Management & Research 4(1): 105–119.

Brunner, P.H. and Rechberger, H. 2015. Waste to energy–key element for sustainable waste management. Waste Management 37: 3–12.

Das, P. and Tiwari, P. 2018. The effect of slow pyrolysis on the conversion of packaging waste plastics (PE and PP) into fuel. Waste Management 79: 615–624.

Datta, P., Tiwari, S. and Pandey, L.M. 2018. Bioethanol production from waste breads using Saccharomyces cerevisiae. pp. 125–134. In: Ghosh, S.K. (ed.). Utilization and Management of Bioresources. Springer, Singapore.

Dornau, A., Robson, J.F., Thomas, G.H. et al. 2020. Robust microorganisms for biofuel and chemical production from municipal solid waste. Microb. Cell Fact. 19(68): 1–18.

Ganesh, R., Torrijos, M., Sousbie, P. et al. 2014. Single-phase and two-phase anaerobic digestion of fruit and vegetable waste: Comparison of start-up, reactor stability and process performance. Waste Management 34(5): 875–885.

Goud, R.K., Sarkar, O., Chiranjeevi, P. et al. 2014. Bioaugmentation of potent acidogenic isolates: A strategy for enhancing biohydrogen production at elevated organic load. Bioresource Technology 165: 223–232.

Hassen, A., Belguith, K., Jedidi, N. et al. 2001. Microbial characterization during composting of municipal solid waste. Bioresource Technology 80(3): 217–225.

Kaza, S., Yao, L., Bhada-Tata, P. et al. 2018. What a waste 2.0: A global snapshot of solid waste management to 2050. World Bank Publications, Washington, pp. 3.

Kumar, A. and Samadder, S.R. 2017. A review on technological options of waste to energy for effective management of municipal solid waste. Waste Management 69: 407–422.

Kumar, S. 2011. Composting of municipal solid waste. Critical Reviews in Biotechnology 31(2): 112–136.

Lai, Z., Zhu, M., Yang, X. et al. 2014. Optimization of key factors affecting hydrogen production from sugarcane bagasse by a thermophilic anaerobic pure culture. Biotechnology for Biofuels 7(1): 1–11.

Leme, R.M. and Seabra, J.E. 2017. Technical-economic assessment of different biogas upgrading routes from vinasse anaerobic digestion in the Brazilian bioethanol industry. Energy 119: 754–766.

Logroño, W., Ramírez, G., Recalde, C. et al. 2015. Bioelectricity generation from vegetables and fruits wastes by using single chamber microbial fuel cells with high Andean soils. Energy Procedia 75: 2009–2014.

Meng, F., Dornau, A., McQueen Mason, S.J. et al. 2021. Bioethanol from autoclaved municipal solid waste: Assessment of environmental and financial viability under policy contexts. Applied Energy 298: 117118.

Pecorini, I., Baldi, F., Carnevale, E.A. et al. 2016. Biochemical methane potential tests of different autoclaved and microwaved lignocellulosic organic fractions of municipal solid waste. Waste Management 56: 143–150.

Poblete, R., Cortes, E., Bakit, J. et al. 2019. Landfill leachate treatment using combined fish scales based activated carbon and solar advanced oxidation processes. Process Safety and Environmental Protection 123: 253–262.

Quina, M.J., Bontempi, E., Bogush, A. et al. 2018. Technologies for the management of MSW incineration ashes from gas cleaning: New perspectives on recovery of secondary raw materials and circular economy. Science of the Total Environment 635: 526–542.

Rabaey, K. and Verstraete, W. 2005. Microbial fuel cells: Novel biotechnology for energy generation. Trends in Biotechnology 23(6): 291–298.

Rasheed, T., Anwar, M.T., Ahmad, N. et al. 2021. Valorisation and emerging perspective of biomass based waste-to-energy technologies and their socio-environmental impact: A review. Journal of Environmental Management 287: 112257.

Salunkhe, D., Rai, R. and Borkar, R. 2012. Biogas technology. International Journal of Engineering Science Technology 4(12): 4934–4940.

Savoo, S. and Mudhoo, A. 2018. Biomethanation macrodynamics of vegetable residues pretreated by low-frequency microwave irradiation. Bioresource Technology 248: 280–286.

Sevda, S., Sharma, S., Joshi, C. et al. 2018. Biofilm formation and electron transfer in bioelectrochemical systems. Environmental Technology Reviews 7(1): 220–234.

Sevda, S., Singh, S., Garlapati, V.K. et al. 2019. Sustainability assessment of microbial fuel cells. pp. 313–330. *In*: Lakhveer Singh, D.M.M. (ed.). Waste to Sustainable Energy. CRC Press, Boca Raton.

Shukla, I. 2022. Potential of renewable agricultural wastes in the smart and sustainable steelmaking process. Journal of Cleaner Production 370: 133422.

Usman, M., Cheema, S.A. and Farooq, M. 2020. Heterogeneous Fenton and persulfate oxidation for treatment of landfill leachate: A review supplement. Journal of Cleaner Production 256: 120448.

Van Fan, Y., Klemeš, J.J., Lee, C.T. et al. 2018. Anaerobic digestion of municipal solid waste: Energy and carbon emission footprint. Journal of Environmental Management 223: 888–897.

Wu, L.-J., Kobayashi, T., Li, Y.-Y. et al. 2015. Comparison of single-stage and temperature-phased two-stage anaerobic digestion of oily food waste. Energy Conversion and Management 106: 1174–1182.

Zabed, H., Faruq, G., Sahu, J.N. et al. 2014. Bioethanol production from fermentable sugar juice. The Scientific World Journal 2014: 957102.

Zhang, J., Lv, C., Tong, J. et al. 2016. Optimization and microbial community analysis of anaerobic co-digestion of food waste and sewage sludge based on microwave pretreatment. Bioresource Technology 200: 253–261.

9

THE RECENT DEVELOPMENT OF WASTE TO ENERGY SYSTEM FOR MUNICIPAL SOLID MANAGEMENT SYSTEM IN DEVELOPING COUNTRIES

Shital Potdar, Malkapuram Surya Teja and *Shirish H. Sonawane**

1. INTRODUCTION

1.1 Urbanization and municipal waste

In present time, urbanization is taking place rapidly. The main reason for this is the high rate of population growth and less amount of earning opportunities in the rural area. Lack of hospital and other facilities also motivates the people to migrate from rural to urban area. However, this transformation has put the urban area in many problems. Out of several issues, the lack of living space, load on the water and electricity, heavy transportation required and higher living cost can be listed as a few of them. Another serious concern that needs to be focused is the management of waste generated, as urbanization directly contributes to waste formation, and leads to environmental pollution (Simatele et al. 2017). Thus, in the urban area the increased population generates a lot of waste that is difficult to be collected and treated by municipal authorities. This chapter describes the solid waste management system in developing countries as against the developed countries, the current practices of solid waste collection and disposal, classification and characterization of solid waste and the need to convert this waste into energy as an alternative to depleting fossil fuel resources.

Chemical Engineering Department, National Institute of Technology, Warangal, Telangana State, India 506 004.
* Corresponding author: shirish@nitw.ac.in

1.2 Municipal solid waste management

The collection and proper disposal of the huge volume of solid waste generated is a challenging task in many developing countries like India, China, and Pakistan. Moreover, the quantity of solid waste produced will continue to increase with an increase in population and migration from rural to urban area. Economical limitations, industrial poor waste management systems and treatment, inappropriate selection of technology and public lack of interest towards the solid waste management have exacerbated the situation (Vij 2012). The current routines of the unconstrained disposal of waste on the urban fringe have led to critical environmental pollution and health issues. Municipal solid waste encompasses the domestic, industrial, institutional wastes and the garbage collected during the street sweeping (Jain 2007). Moreover, the term municipal solid waste management (MSWM) refers to the process of collection, transport, recovery, recycling, and processing of solid waste (Mehta et al. 2017). Hence, it is essential to have knowledge of the solid waste collection transfer, recovery, recycling, and processing of solid waste process in detail. Before any treatment of solid waste, segregation of solid waste is a very essential step for the further treatment and selection of technology for conversion of waste to energy.

1.2.1 Primary collection of municipal solid waste

In the primary collection process, the sweepers take the roadside dumped garbage to the immediate depots. Also, the cleaning programme adopted by local municipal authority, where the garbage trucks pass through residential areas to collect the domestic waste and it is transferred to the collection point, is termed as a primary collection. A number of primary collection points are situated across the city. The depots are large space for collection closed from three sides and a masonry wall of about 1.35 m height. The storage capacity of the depots varies from 15 to 40 m³ (Sharholy et al. 2007).

1.2.2 Secondary collection of municipal solid waste

The primary collection points are generally within 15 km; hence, the solid waste from primary collection depots is transferred to landfill. Since in this collection process, the solid waste is collected from primary sites and carried to secondary collection site, it is called secondary collection of municipal solid waste. For the secondary collection, transportation and disposal of Municipal solid waste (MSW), the hauled container, stationary container, manually loaded dumper and mechanically loaded dumper are in practise. The use of hauled container for waste collection allows the collection of waste at source and then the containers are carried to the disposal sites; these containers are emptied at disposal sites and are placed at their original locations, whereas stationary containers are waste collectors which remains stationary. The waste collected in stationary containers is adopted in the areas where dumper trucks are unable to reach (Sharholy et al. 2007, Pattnaik and Reddy 2010).

1.3 Classification and characteristics of municipal solid waste

Once the solid waste is collected, its classification and characterization are very much important to differentiate hazardous and non-hazardous waste. In general, the domestic and other non-industrial sector is referred as non-hazardous waste. The industrial waste sector contains hazardous waste which needs separate collection and treatment process. Non-hazardous waste typically includes garbage, package covers, and leaf litter (Rajput et al. 2009). The classification of MSW is depicted in Fig. 1 below (Zhou et al. 2014).

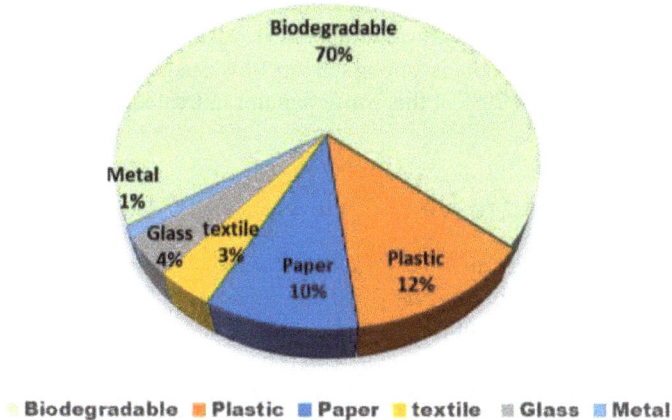

Figure 1. Classification of municipal solid waste (Zhou et al. 2014).

The waste characteristics and composition differ in great extent for developed and developing countries and even in the same country the characteristics and composition of solid state may differ in different cities. The physical constituents of solid waste is influenced by different parameters like socioeconomic profile, weather climatic conditions, extent of recycling, collection frequency and demography.

From the numerous survey carried out till date, it is observed that the developed countries like US has less moisture content (20% to 30%) as compared to developing countries like China and India (50% to 70%) (Mohee and Mudhoo 2012, Cheng et al. 2007). The solid waste collected from developed countries largely contains paper waste, textiles waste and other dry organic recyclable wastes and thus, it has high calorific values (2000–4000 kcal/kg) in comparison with the developing countries (700–1600 kcal/kg). The composition of waste is also a function of income status of the society. The high-income societies have higher percentage of recyclable wastes in their solid waste stream. In addition, the organic content of solid waste stream is less than 30% in developed countries like Japan, USA, South Korea and Singapore whereas in developing countries the percentage is more than 50% (Aleluia and Ferrao 2016).

Particularly India, where the population is more than 1.4 billion, is at high risk of disposal of all the waste it produces per day. Annual waste generation in top 10 cities (lakh tons) of India is represented in Fig. 2 (World Bank 2018). The total solid waste produced in India per year is

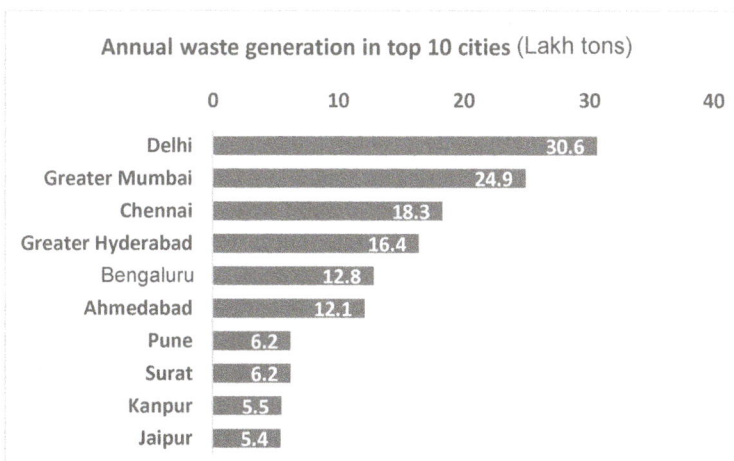

Figure 2. Annual waste generation in the major cities of India (World Bank 2018).

62 million tons. The detailed classification of this total waste includes 7.90 million tons of hazardous waste, 5.5–5.7 million tons of plastic waste, biomedical waste contributes 0.17 million tons and e-waste contributes 15 lakh tons. India is among the top 10 waste generation countries in the world. The major concern is that almost 70% of this waste remains untreated and most of this unprocessed waste is discarded into landfills.

1.4 Source reduction and reuse

The term MSW management encapsulates control on the source of waste generation, reuse, recycling, composting, land filling, and energy recovery. The main objective of MSWM is to maintain the health of the population, protect environment from pollution, and establish energy source from waste as an alternative to fossil fuels. To achieve these goals, a well-planned waste management system must be executed not only by municipal authorities but also by an individual. Even though the solid waste generated from lower economic group and developing countries have less calorific value, its proper management is crucial. The reuse of material will reduce the cost of raw material in the production process.

Source reduction: The generation of solid waste can be achieved by proper product design. The storage containers of food packaging should be designed in such a way that they can be repeatedly used. The use of steel glass, and plates instead of use and throw plastic glasses and plates will reduce the solid waste. Vendors should encourage the customers to use existing packaging materials instead of using new every time. This will not only reduce the solid waste but will also help to reduce the cost of product (Chaaban 2001).

Recycling: The recycling of waste is reprocessing of used materials to obtain same material or other possible material of consumer benefit. Recycling will help to minimize the overall cost of product but will also reduce the harm to an environment (Troschinetz and Mihelcic 2009).

Composting: Another method of recycling is by composting. Controlled biological decay of organic wastes will form a compost, beneficial for soil health, as the compost will provide the minerals necessary for plant growth (Gajalakshmi and Abbasi 2008).

1.5 Waste to energy conversion: Technological aspects

To meet the present demand of fuel, there is high thrust to convert waste to energy. The increased population has put load on the use of fossil fuel. With the current rate of use of fossil fuel, they are depleting at a very fast rate. By the end of this century, the world's energy requirement will be six folds than the current requirement (Kothari 2010). In the present scenario, the energy provided is far less from the actual energy requirement in most of the developing countries. The primary source of energy in the present time is fossil fuels which contributes approximately 80–85% of the total energy supplied (Ouda et al. 2016). The large volume of solid waste generated causes a serious problem of waste disposal and leads to environmental pollution. Thus, there is a wide scope to generate wealth from waste. In this scope, the conversion of solid waste into energy will not only help to reduce the greenhouse gas emission from solid waste, but it will also contribute to meet the fuel requirement. To convert solid waste into energy at present, anaerobic digestion, incineration, gasification, landfill gas-fired power generation are the potential ways to convert waste to wealth. However, the energy production from solid waste is mainly a function of the characteristic properties of solid waste, namely moisture content, density, particle size and calorific value of source (Aleluia and Ferrao 2016).

The developed countries are aware of the need to convert waste to energy and also the need to develop technology for the same for efficient solid waste management. Developing countries have a common concern of appropriate solid waste management and also the reliable source of energy. The only solution to this concern is through technological breakthroughs, and government policies.

2. TECHNOLOGIES FOR WASTE TO ENERGY CONVERSION

2.1 Conventional technologies

London's '*dust-yard system*' of 18th century is the first organized waste management system in modern world. The dust (coal ash) produced during early industrialization was recovered and successfully used for brick making with public-private intervention. Since then, waste valorisation and waste to energy (WtE) conversion methods have gained huge attention. Providing value to waste ultimately reduced the quantity of residual waste and resulted in replacement of fossil fuels. However, increase in consumption levels, since industrialization, has generated huge quantity of waste that ultimately led to creation of "*destrcuctors*", which are also called as "*incinerators*", "*gasifiers*", and "*pyrolysis plants*" (Malinauskaite et al. 2017).

Incineration, gasification, and pyrolysis are thermal decomposition methods where the feedstock is reacted at high temperatures to convert them into an energy resource. Henceforth, it eventually shrinks the consumption levels of fossil fuels. Various waste management systems adopt these WtE conversion technologies to decrease the quantity of waste as well as to valorise the waste for other applications. In incineration plant, waste is combusted at very high temperatures to convert it into a high temperature heat. Incineration of solid waste is supposed to reduce solid mass and volume at least by 80% and 95%, respectively. Thus, the size of the land required for '*landfilling*' will be greatly subsided. In addition to heat, burning of waste results in the generation of flue gas, and fly ash. The fly ash can be recovered and used for brickmaking and Portland cement production. However, the presence of fly ash, as a particulate matter, in flue gas causes severe health adversities such as cancer (Wienchol et al. 2020).

The major advantage of gasification over incineration is the '*synthesis gas*' production. Synthesis gas, popularly known as '*syngas*', is a mixture of hydrogen and carbon monoxide gases. Syngas coming out from gasifier can be transformed into higher commercial value products such as fertilizers, transportation fuels, and chemicals. It also can be used as an alternative for natural gas as a clean energy resource. Over a period of time, various kinds of gasifiers were developed. '*Biogasification*' exploits bacterial metabolic activity to decompose the waste material, thereby producing energy releasing products. The microbes present in an anaerobic digester break down the organic material whereupon biogas is released. Generally, food waste, organic matter of MSW, crop residue and garden waste is provided as substrate for the biogasification process. The process is highly regulated by the temperature and the pH of the system (Pujara et al. 2021). Recently, '*distributed modular gasification (DMG)*' and '*plasma arc gasification (PAG)*', which will be discussed in upcoming sections, have gained huge attention due to their better conversion efficiencies and environmental security (Loizidou et al. 2021).

Unlike incineration and gasification, pyrolysis is an anaerobic thermal treatment method. In the absence of air, the heat given to the feedstock produces oil, tar, and volatile compounds. In general, pyrolysis is employed for treating the organic content present in materials. Pyrolysis of biomass produces bio-oil, which is considered as a bio-fuel. When sewage sludge is pyrolyzed in the presence of alkaline, known as '*alkaline pyrolysis*', hydrogen rich gas is produced which can be directly exploited for fuel cell applications. Pyrolysis of vehicle tires results in forming asphalt and carbon black. However, the rubber pyrolysis produces high sulphur content oil, which is hazardous to environment as well as human health. However, recent advances in pyrolysis such as '*microwave pyrolysis*', and '*catalytic co-pyrolysis*', have made the system more sustainable and eco-friendly (Mukherjee et al. 2020).

The decomposition of organic waste by bacteria in the absence of oxygen in a bio-reactor is called '*anaerobic digestion*' (AD). It is conventionally used for disintegrating the sludge produced from municipal wastewater treatment plants. The breakdown of bio-waste results in methane production which is why it is also referred as biomethanization. In general, anaerobic digestion comprises of four steps: hydrolysis (liquefaction of larger compounds), acidolysis (disintegration of acids), acetogenesis (formation of acetic acid), and methanogenesis (production of methane). The process is able to recover approximately 90% of energy available from waste in the form of methane. The methane produced from this process can directly be used for household purposes such as cooking, and electricity. Nonetheless, the foul smell and slowness of the process make the system less compatible (Kumar and Samadder 2017).

Even though the thermal WtE systems can process greater volumes of waste, their real practice is still an environmental issue as they produce harm causing materials from their operation. Table 1 shows a comparison among thermal waste to energy treatment methods. The AD of bio-waste is a time consuming process that produces nauseating odor. Hence, researchers have developed new techniques as discussed below to provide sustainable solutions for transforming the waste materials into energy.

Table 1. Comparison among thermal waste to energy treatment methods.

Sr. no.	Parameter	Incineration	Gasification	Pyrolysis	HTC
1.	**Temperature**	up to 1200°C	750–900°C	500–800°C	180–250°C
2.	**Pressure**	~ 1 atm	~ 1 atm	> 1 atm	10–50 atm
3.	**Oxidizing agent**	air	air, oxygen, steam, or combination	-no-	hot compressed air
4.	**Products**	fly ash, flue gas, heat	fly ash, syngas, tar	char, asphalt, oil	hydrochar, syngas, bio-oil
5.	**Physical state**	predominantly gas	predominantly gas	solid, liquid, gas	solid, liquid, gas
6.	**Environmental impact**	particulate matter	particulate matter, tar	particulate matter, sulphur	–
7.	**Recent advancements**	operational improvements	plasma arc gasification (PAG), distributed modular gasification (DMG)	microwave pyrolysis and catalytic co-pyrolysis	microwave-assisted hydrothermal carbonization (MAHC)

2.2 Updated conventional methods

2.2.1 Distributed modular gasification

DMG technology enables production of hydrogen gas (a clean energy resource) from non-recyclable plastic waste. Additionally, the products can be used for electricity generation. The *Power House Energy Group PLC* of London has developed DMG technology to help local communities to fight air and plastic pollution. It is a single thermal unit chamber where all the kinds of waste is being processed into useful hydrogen gas and electricity. Further, syngas produced can be effectively separated from hydrogen gas by using pressure swing absorption technique (Ward et al. 2020).

2.2.2 Plasma arc gasification

PAG uses both the electrical and thermal energy to change biomass into syngas without burning the materials. The inorganics present in the waste, upon PAG, produces glassy substance called slag

(Janajreh et al. 2021). The temperature of plasma arc can reach up to 14,000°C. However, only 2–5% of supplied energy is consumed by plasma torch and about 80% of total energy input can be recovered as synthesis gas. Hence, PAG contributes to solid waste management and WtE developments.

2.2.3 Microwave pyrolysis

Microwave irradiation induces hotspots that improve the yield as well as characteristics of the products. The yields of solid and gaseous products are higher than the conventional pyrolysis and solid products will have greater surface areas and higher heating values. Microwave pyrolysis derives more bioenergy due to higher yields of hydrogen and carbon monoxide. In addition, presence of proper catalyst will improve the selectivity. However, the products formed by using this technique are different from the products obtained from pyrolysis (Ge et al. 2020). For instance, pyrolysis produces biochar with uneven pore shape and size. It implies the possibility of occurrence of heterogeneous reactions, whereas pore size and shapes of biochar produced from microwave pyrolysis is even. Additionally, the pores on biochar are not contaminated in the case of microwave irradiation. It also implies the efficiency of volatile organic compound removal. Like conventional pyrolysis, microwave pyrolysis also produces bio-oil, a condensable vapor, with higher carbon content and insignificant contents of oxygen. Various researchers used different catalysts to improve the microwave pyrolysis of different kinds of biomass. The effect of catalyst, as mentioned earlier, is advantageous. However, the reason for the effect is still unclear (Huang et al. 2016).

2.2.4 Catalytic co-pyrolysis

Catalytic co-pyrolysis (CCP) helps in processing polymer wastes and in providing energy security by harnessing energy from it. The CCP is primarily employed for the production of deoxygenated bio-oil, known as drop-in biofuel, due to proportionate blend of lignocellulosic biomass and plastic waste. This kind of pyrolysis reduces the amount of oxygen present in bio-oil. The catalysts used for CCP are broadly categorized into acid and base catalysts. The presence of acid catalysts promotes certain reactions such as aromatization, cracking and dehydration, whereas base catalysts promote carbon-coupling reactions that help converting low molecular weight materials into fossil fuel range products. Various kinds of catalysts such as zeolites, metal modified zeolites, and metal catalysts such as Zn/lignin-char, MgO, MgO/C, MgO/ZrO$_2$, MgO/Al$_2$O$_3$, and SiO$_2$/Al$_2$O$_3$ are extensively employed for pyrolysis of lignin and cellulose. The major drawback of the process is clogging of catalyst pore by coke produced in the process (Miandad et al. 2017).

2.3 Advanced technologies

2.3.1 Hydrothermal carbonization

Transforming the organic matter into energy resource has always been an active research topic. There are various methods to convert the organic matter into fuels. Most of the second-generation biomasses have larger content of moisture. Thus, pre-treatment such as drying the feedstock makes the processes non-economical. Additionally, the conversion efficiencies of these processes is also still lower than 70%. Hydrothermal carbonization (HTC) is a thermochemical process where waste material is converted into a valuable resource at lower operating temperatures than gasification and pyrolysis. The processing of waste by using HTC, its products and applications are shown in Fig. 3. In brief, the wet feedstock in HTC reactor is converted into more than 60% dry carbon (hydrochar), which avoids further drying treatment methods as such in conventional methods. Hydrochar is

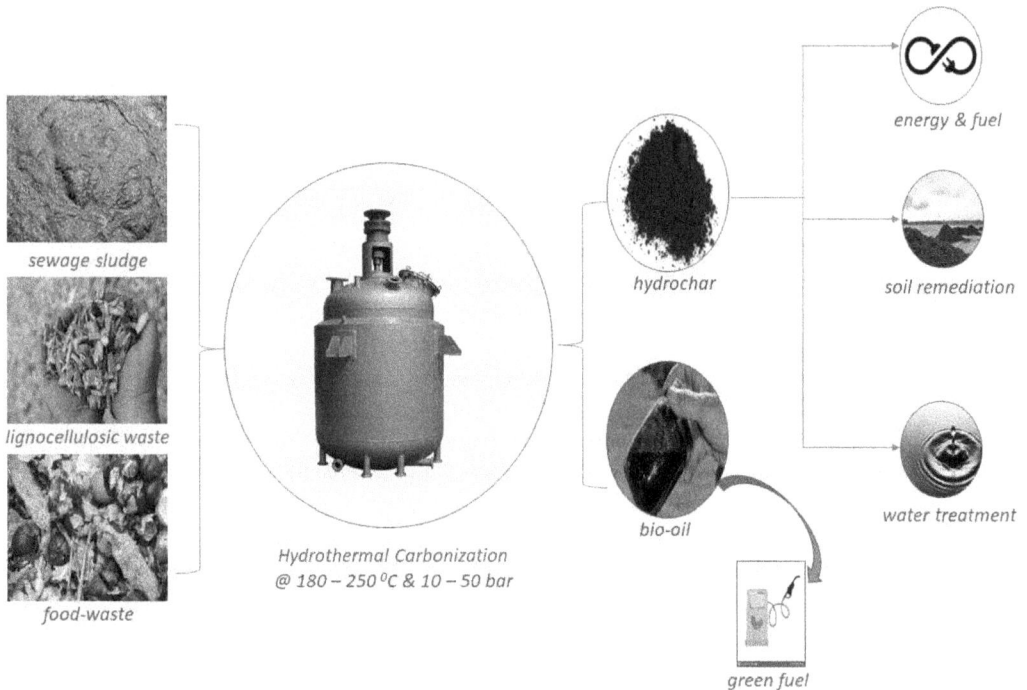

Figure 3. Processing of waste by using hydrothermal carbonization: Its products and applications.

considered as an equivalent to fossil fuels and is also used for wastewater treatment, soil remediation, and carbon sequestration processes (Azzaz et al. 2020). The advantages of HTC is compared with conventional thermal treatment methods and are presented in Table 1.

In a natural pathway, conversion of biomass into brown coal takes at least 50,000 years but HTC produces it along with syngas only in few hours. Further, it prevents carbon dioxide release into environment. Various researches have successfully exploited and commercialized the HTC for producing fuel from discarded biomass. When the wet feedstock is subjected to HRC reactor at 180–250 C and 10–15 bar, various reactions such as polymerization, hydrolysis, and dehydration occur. Water acts as a solvent as well as a reactant and it will decompose to produce hydrochar with other products. HTC is majorly applied to transform sewage sludge, and lignocellulosic and organic biomasses. Various researchers have also investigated on liquid fuel production through HTC of algal biomass. Hydrochar produced from HTC is highly capable of adsorbing fatty acids required for biodiesel production. Indeed, HTC is proposed to be an intermediate step for biodiesel production (Ischia amd Fiori 2021). Further, HTC is also studied by combining with other WtE methods such as AD, gasification, and pyrolysis. The hydrochar produced from HTC+gasification process reduces tar generation and promotes the syngas production and HTC+pyrolysis produces a hydrochar that is better in its physical and chemical properties than the actual hydrochar produced from HTC.

Recently, HTC is further upgraded by inducing microwave irradiation on feedstock. This integration is called '*microwave assisted hydrothermal carbonization (MAHC)*'. Assistance of microwave in a process is due to its selectivity, homogeneity, and quickness in completing the conversion at low costs. Researchers have converted waste materials such as cellulose, lignocellulosic materials and dairy manure into hydrochar. However, there is need of more research in this field to render maximum benefits (Azzaz et al. 2020).

2.3.2 Dendro Liquid Energy

Dendro Liquid Energy (DLE) is the most recent WtE technology that claims to be 400% efficient than AD. It was first developed in Germany where all sort of waste, including plastic discards, is treated in a reactor that results in syngas production (Sharma et al. 2021). The special characteristic of the method is zero waste generation. The peculiar features of DLE include wide variety of feedstock, dry and wet feedstock, low-cost decentralised units, and absence of burning of material. As material is not subjected for burning, the DLE is considered as an emission-free technology, thus reducing the additional cost required for cleaning of flue gases and ensures the environmental safety. The temperatures are moderate (100–250°C), and the energy conversion is higher (up to 80%). This technology has great scope of improvement and industrial application as very limited research is carried out (Sharma et al. 2021).

2.3.3 Microbial fuel cell

The power released from the metabolic activity of bacteria is harnessed and transformed into electrical energy by using a microbial fuel cell (MFC). The replication of bacteria in the cell and redox reactions promoted by them makes the system sustainable for its pollution removal operations. The schematic of working principle for MFC is shown in Fig. 4 below. In MFC, bacteria is grown on anode by oxidizing the organic waste (glucose/acetate) present in the cell. The decomposition of organic matter on anode produces protons (H^+ ions), electrons, and carbon di-oxide. The anaerobic (anode) and aerobic (cathode) chambers are separated by a semi-permeable membrane where only protons can diffuse. The flow of electron from anode to cathode via an external circuit generates electricity whereas the reaction of electrons, dissolved oxygen, and protons at cathode produces pure water molecules (Sevda et al. 2020).

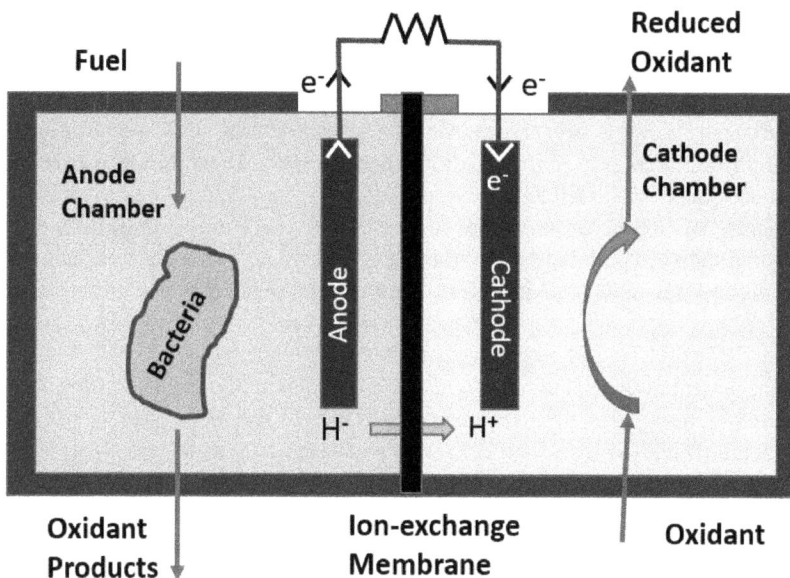

Figure 4. Schematic representation of microbial fuel cell.

Anodic Reaction:

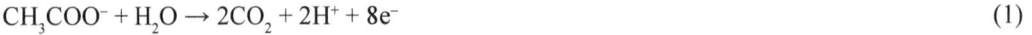

$$CH_3COO^- + H_2O \rightarrow 2CO_2 + 2H^+ + 8e^- \tag{1}$$

Cathodic Reaction:

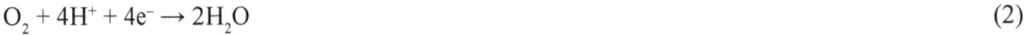

$$O_2 + 4H^+ + 4e^- \rightarrow 2H_2O \tag{2}$$

In general, the technology is primarily employed for sewage, and industrial wastewater treatment processes. However, MFCs are non-economical to operate at industrial scale. Additionally, the energy density achieved is also very low. Hence, lot of research is to be done to enhance the MFC technology.

2.3.4 Microbial electrolysis cell

When both anodic and cathodic chambers are kept under anaerobic conditions and an external energy is provided, an electrolysis cell produces hydrogen at cathode and labels the cell as microbial electrolysis cell (MEC). Biocatalytic oxidation occurs at the anode of the MEC, while cathode promotes reduction processes. It is believed that MECs make the hydrogen economy feasible. These cells, through biochemical reactions, convert organic carbon into value added products (Yang et al. 2021). Even though energy is externally supplied, the production of clean hydrogen by MEC is economical, as conventional hydrogen production methods require ten times more energy for hydrogen production. Instead of hydrogen, in some scenarios, methane is generated depending on the type of bacteria present in the cell (Logan et al. 2008).

3. ECONOMIC ASPECT OF WASTE-TO-ENERGY TECHNOLOGIES

3.1 Waste-to-energy: Its role in the circular economy

The '*linear economy: take-make-dispose*' concept has resulted in indiscriminate exploitation of natural resources by posing severe environmental as well as socio-economic threats. The conceptualization of '*circular economy*' as alternative for liner economy has revolutionized the way resources are being consumed and the way waste is being treated. The '*3R principle: Reduce, Reuse, and Recycle*' of circular economy paved a sustainable path for the utilization of resources by meeting the requirements of population. Environmental enthusiasts, activists, and policy-makers have promoted circular economy since its inception to minimize the dependency on fossil fuels, harness clean energy, reduce waste generation, and valorise no-value/low-value substances (Fernandez-Gonzalez et al. 2017).

Waste management, waste valorisation and WtE principles are interlinked to one another through which assets (money, mass, or energy) are created from useless materials. The technologies discussed so far are phenomenal in their own way. However, the burden to run the process must overcome by the fruits it bear. Otherwise, the concept of circular economy cannot be opted in most of the countries. Thus, the technologies must guarantee the positive output from its employability.

The economy of waste to energy generation largely depends on the calorific value of solid waste generated. In developing countries, namely China, India, and Pakistan, the MSW largely contains the food waste with low calorific value and high moisture content. Thus, the technologies practised in developed countries for waste to energy conversion may not be economical and useful to treat such solid waste. Thus, in China circulating fluidised bed based incineration plants are established (Zhao et al. 2016).

3.2 Economic and environmental review of waste-to-energy systems

The conventional treatment methods employed for the conversion of waste into energy consume lot of energy. They are less skilful and environmentally burdensome. In addition, landfilling method of waste management requires huge footprint area. They are also environmentally unfit. Hence, the advancements in the existing technologies and development of new technologies are highly encouraged. However, the feasibility of the technology depends on several factors such as economy of the community, type of waste produced, geography of the location, and the awareness among the community.

To exercise WtE technologies in reality, economic study is needed. The fixed and variable operating cost including waste disposal must overcome the revenues the system generate such as fees on waste producer, income generated from selling fuels, and other products produced (Khoshnevisan et al. 2021).

Countries with larger areas with less population can opt for landfilling as they produce less waste but have sufficient space for landfilling. However, they must be ready with provisions for handling consequences such as leaching of hazardous materials and foul smell. Countries with greater portions of forests can choose thermochemical treatment methods as trees can absorb carbon di-oxide released from them. However, proper precautions must be practised. HTC and DLE are suitable for European countries where environmental concern is relatively higher and can sustain economic burden tied with modern technologies.

For instance, India's MSW characteristics are different from developed nations: 51% of waste constitutes biodegradable waste, where developed countries have greater fractions of paper waste. Due to higher percentage of moisture content (50%), the caloric value of MSW in India is almost half of the US and UK. Therefore, selection of treatment method plays a major role in waste management. Non-recyclable waste can be directly used for boilers as a fuel. This is known as '*refuse-derived fuel (RDF)*'. The pellets made from RDF will have a calorific value of 1912–3346 kcal/kg. Gasification is always a great option for developing nations for WtE technological practice. It produces a syngas of calorific value 4–10 MJ/Nm³. The largest incineration plant in Delhi, India treats 300 tons/day despite having poor MSW characteristics. AD generates at least 2 times of methane in 4 weeks from 1 ton of MSW than landfilling of MSW for at least 6 years (Chand Malav et al. 2020).

However, the particulate matter, fly ash, tar, metal ions, and other substances coming out from WtE system pose severe threat to human and environment. The particulate matter is labelled under PM 2.5 which is regarded as air pollutant. Gases such as carbon dioxide and methane cause global warming. Hence, advancements in each technology are essential to mitigate the environmental effect of WtE technologies for their implementation across the communities.

A case study on WtE technologies for MSW of Iran reveals that at least 0.5% of annual greenhouse gases reduced with additional benefit of generating over 5000 GWh electricity per year (Rajaeifar et al. 2017). The calorific values of products of WtE technologies, and other advantages such as reduction of land usage, and environmental safety further maintained the trade-off between economy, energy, and environmental security (Tayeh et al. 2021).

4. RECOMMENDATIONS

In practice, all the above-discussed methods are employed in different sectors to minimize the waste, valorise the low-value materials/waste materials, and recover maximum energy for the promotion of circular economy with least environmental effects. However, their lone practice is difficult to achieve the objectives. Henceforth, most of the methods need to integrate for better performance in terms of energy, economy, and environmental safety.

The technologies such as MEC, and DLE have a great scope of advancements since they are less reported. Their applicability and efficiencies attract researchers to promote the technology and to enhance the processes to further extents. The major responsibility of WtE is to minimize its effect on environment. Currently, the by-products from various WtE technologies have an impact on environment by polluting air, water, or soil. Hence, developments in the technologies not only focus on economy and energy efficiencies but also must ensure environmental safety.

Additionally, applicability of these technologies for developing and under developing countries must be ensured so that they can adapt without experiencing economic burden. Environmental aspect of methane production must be considered as it can cause 20 times more global warming than carbon dioxide. Hence, the practices must ensure trade-off among several concerns.

5. CONCLUSION

This book chapter explains the effect of urbanization on the volume of MSW generated, and the environmental pollution caused due to solid waste. There is high need to convert waste to energy. This will not only help to minimize and control solid waste but will also help to meet the energy requirement of growing population. The waste to energy will prove an alternative to continuously depleting fossil fuel energy source. The primary and secondary solid waste collection process and need of segregation of solid waste at source collection is described. The classification (industrial, domestic, food waste, glass, paper metal) and characterization of solid waste based on the calorific value of solid waste is given. In the next section, the technical options to convert waste to energy and their proper disposal are explained with their operational procedure, volume handled, type of solid waste handled and energy that can be extracted from particular technology. The last section explains the techno-economic aspect of solid waste management in developing countries. Even though the technologies provide waste management and alternative energy resources, their employability still needs to answer following questions:

1. Energy output to input ratio
2. Environmental impact of the system
3. Time and labor required

REFERENCES

A Global Snapshot of Solid Waste Management to 2050, What a Waste 2.0, World Bank Group 2018. ISBN (electronic): 978-1-4648-1347-4.

Aleluia, J. and Ferrão, P. 2016. Characterization of urban waste management practices in developing Asian countries: A new analytical framework based on waste characteristics and urban dimension. Waste Manag. [Internet] 58: 415–429. Available from: https://linkinghub.elsevier.com/retrieve/pii/S0956053X16302409.

Azzaz, A.A., Khiari, B., Jellali, S. et al. 2020. Hydrochars production, characterization and application for wastewater treatment: A review. Renew. Sustain. Energy Rev. 127.

Chaaban, M.A. 2001. Hazardous waste source reduction in materials and processing technologies. J. Mater. Process Technol. [Internet] 119: 336–343. Available from: https://linkinghub.elsevier.com/retrieve/pii/S0924013601009207.

Chand Malav, L., Yadav, K.K., Gupta, N. et al. 2020. A review on municipal solid waste as a renewable source for waste-to-energy project in India: Current practices, challenges, and future opportunities. J. Clean Prod. [Internet] 277: 123227. Available from: https://doi.org/10.1016/j.jclepro.2020.123227.

Cheng, H., Zhang, Y., Meng, A. et al. 2007. Municipal solid waste fueled power generation in China: A case study of waste-to-energy in Changchun City. Environ. Sci. Technol. [Internet] 41: 7509–7515. Available from: https://pubs.acs.org/doi/10.1021/es071416g.

Fernández-González, J.M., Grindlay, A.L., Serrano-Bernardo, F. et al. 2017. Economic and environmental review of Waste-to-Energy systems for municipal solid waste management in medium and small municipalities. Waste Manag. 67: 360–374.

Gajalakshmi, S. and Abbasi, S.A. 2008. Solid Waste Management by composting: State of the art. Crit. Rev. Environ. Sci. Technol. [Internet] 38: 311–400. Available from: http://www.tandfonline.com/doi/abs/10.1080/10643380701413633.

Ge, S., Foong, S.Y., Ma, N.L. et al. 2020. Vacuum pyrolysis incorporating microwave heating and base mixture modification: An integrated approach to transform biowaste into eco-friendly bioenergy products. Renew. Sustain. Energy Rev. [Internet] 127: 109871. Available from: https://doi.org/10.1016/j.rser.2020.109871.

Huang, Y.F., Chiueh, P. Te and Lo, S.L. 2016. A review on microwave pyrolysis of lignocellulosic biomass. Sustain. Environ. Res. [Internet] 26: 103–109. Available from: http://dx.doi.org/10.1016/j.serj.2016.04.012.

Ischia, G. and Fiori, L. 2021. Hydrothermal carbonization of organic waste and biomass: A review on process, reactor, and plant modeling. Waste and Biomass Valorization [Internet] 12: 2797–2824. Available from: https://doi.org/10.1007/s12649-020-01255-3.

Jain, A.K. 2007. Sustainable development and waste management. Environews. Newsl ISEB India, 13.

Janajreh, I., Adeyemi, I., Raza, S.S. et al. 2021. A review of recent developments and future prospects in gasification systems and their modeling. Renew. Sustain. Energy Rev. [Internet] 138: 110505. Available from: https://doi.org/10.1016/j.rser.2020.110505.

Khoshnevisan, B., Duan, N., Tsapekos, P. et al. 2021. A critical review on livestock manure biorefinery technologies: Sustainability, challenges, and future perspectives. Renew. Sustain. Energy Rev. [Internet] 135: 110033. Available from: https://doi.org/10.1016/j.rser.2020.110033.

Kothari, R., Tyagi, V.V. and Pathak, A. 2010. Waste-to-energy: A way from renewable energy sources to sustainable development. Renew. Sustain. Energy Rev. [Internet] 14: 3164–3170. Available from: https://linkinghub.elsevier.com/retrieve/pii/S1364032110001437.

Kumar, A. and Samadder, S.R. 2017. A review on technological options of waste to energy for effective management of municipal solid waste. Waste Manag. [Internet] 69: 407–422. Available from: https://doi.org/10.1016/j.wasman.2017.08.046.

Logan, B.E., Call, D., Cheng, S. et al. 2008. Microbial electrolysis cells for high yield hydrogen gas production from organic matter. Environ. Sci. Technol. 42: 8630–8640.

Loizidou, M., Moustakas, K., Rehan, M. et al. 2021. New developments in sustainable waste-to-energy systems. Renew. Sustain. Energy Rev. 151.

Malinauskaite, J., Jouhara, H., Czajczyńska, D. et al. 2017. Municipal solid waste management and waste-to-energy in the context of a circular economy and energy recycling in Europe. Energy 141: 2013–2044.

Mehta, S., Kumar, A. and Lal, R. 2017. Soils and Waste Management in Urban India. Urban Soils. CRC Press, pp. 329–350.

Miandad, R., Barakat, M.A., Rehan, M. et al. 2017. Plastic waste to liquid oil through catalytic pyrolysis using natural and synthetic zeolite catalysts. Waste Manag. [Internet] 69: 66–78. Available from: https://doi.org/10.1016/j.wasman.2017.08.032.

Mohee, R. and Mudhoo, A. 2012. Energy from biomass in Mauritius: Overview of research and applications, 297–321. Available from: http://link.springer.com/10.1007/978-1-4471-2306-4_12.

Mukherjee, C., Denney, J., Mbonimpa, E.G. et al. 2020. A review on municipal solid waste-to-energy trends in the USA. Renew. Sustain. Energy Rev. [Internet] 119: 109512. Available from: https://doi.org/10.1016/j.rser.2019.109512.

Ouda, O.K.M., Raza, S.A., Nizami, A.S. et al. 2016. Waste to energy potential: A case study of Saudi Arabia. Renew. Sustain. Energy Rev. [Internet] 61: 328–340. Available from: https://linkinghub.elsevier.com/retrieve/pii/S1364032116300223.

Pattnaik, S. and Reddy, M.V. 2010. Assessment of Municipal Solid Waste management in Puducherry (Pondicherry), India. Resour. Conserv. Recycl. [Internet] 54: 512–520. Available from: https://linkinghub.elsevier.com/retrieve/pii/S0921344909002328.

Pujara, Y., Govani, J., Chabhadiya, K. et al. 2021. Waste-to-energy: Suitable approaches for developing countries. Handb. Environ. Chem. 99: 173–191.

Rajaeifar, M.A., Ghanavati, H., Dashti, B.B. et al. 2017. Electricity generation and GHG emission reduction potentials through different municipal solid waste management technologies: A comparative review. Renew. Sustain. Energy Rev. [Internet] 79: 414–439. Available from: http://dx.doi.org/10.1016/j.rser.2017.04.109.

Rajput, R., Prasad, G. and Chopra, A. 2009. Scenario of solid waste management in present Indian context.

Sevda, S., Garlapati, V.K., Naha, S. et al. 2020. Biosensing capabilities of bioelectrochemical systems towards sustainable water streams: Technological implications and future prospects. J. Biosci. Bioeng. [Internet] 129: 647–656. Available from: https://doi.org/10.1016/j.jbiosc.2020.01.003.

Sharholy, M., Ahmad, K., Vaishya, R.C. et al. 2007. Municipal solid waste characteristics and management in Allahabad, India. Waste Manag. 27: 490–496.

Sharma, B., Goswami, Y., Sharma, S. et al. 2021. Inherent roadmap of conversion of plastic waste into energy and its life cycle assessment: A frontrunner compendium. Renew. Sustain Energy Rev. [Internet] 146: 111070. Available from: https://linkinghub.elsevier.com/retrieve/pii/S1364032121003580.

Simatele, D.M., Dlamini, S. and Kubanza, N.S. 2017. From informality to formality: Perspectives on the challenges of integrating solid waste management into the urban development and planning policy in Johannesburg, South Africa. Habitat Internationa.

Tayeh, R.A., Alsayed, M.F. and Saleh, Y.A. 2021. The potential of sustainable municipal solid waste-to-energy management in the Palestinian Territories. J. Clean Prod. [Internet] 279: 123753. Available from: https://doi.org/10.1016/j.jclepro.2020.123753.

Troschinetz, A.M. and Mihelcic, J.R. 2009. Sustainable recycling of municipal solid waste in developing countries. Waste Manag [Internet] 29: 915–923. Available from: https://linkinghub.elsevier.com/retrieve/pii/S0956053X08001669.

Vij, D. 2012. Urbanization and Solid Waste Management in India: Present practices and future challenges. Procedia - Soc. Behav. Sci. [Internet] 37: 437–447. Available from: http://dx.doi.org/10.1016/j.sbspro.2012.03.309.

Ward, C., Goldstein, H., Maurer, R. et al. 2020. Making coal relevant for small scale applications: Modular gasification for syngas/engine CHP applications in challenging environments. Fuel [Internet] 267: 117303. Available from: https://doi.org/10.1016/j.fuel.2020.117303.

Wienchol, P., Szlęk, A. and Ditaranto, M. 2020. Waste-to-energy technology integrated with carbon capture – Challenges and opportunities. Energy 198.

Yang, E., Omar Mohamed, H., Park, S.G. et al. 2021. A review on self-sustainable microbial electrolysis cells for electro-biohydrogen production via coupling with carbon-neutral renewable energy technologies. Bioresour. Technol. 320.

Zhao, X., Jiang, G., Li, A. et al. 2016. Economic analysis of waste-to-energy industry in China. Waste Manag. [Internet] 48: 604–618. Available from: https://linkinghub.elsevier.com/retrieve/pii/S0956053X1530163X.

Zhou, H., Meng, A., Long, Y. et al. 2014. Classification and comparison of municipal solid waste based on thermochemical characteristics. J. Air Waste Manag. Assoc. [Internet] 64: 597–616. Available from: http://dx.doi.org/10.1080/10962247.2013.873094.

10

TREATMENT OF SOLID FOOD WASTE USING SOLID-STATE FERMENTATION
ITS CURRENT USE AND FUTURE PERSPECTIVES

Deepika Umrao,[1,#] *Saumya Singh,*[1,#] *Devendra Pratap Singh,*[1]
Shivani Maddirala,[2] *Surajbhan Sevda*[2] and *Meena Krishania*[1,*]

1. INTRODUCTION

With the rise in population, food production has also increased and so has food wastage. Different food and beverage industries have cropped up across the globe to fulfil this demand. However, the wastes generated from the food processing sector remain underutilized and are disposed of indiscriminately either by dumping burning or unplanned landfilling (Sadh et al. 2018). Several issues are being created by these untreated wastes such as an increase in greenhouse gases emission, and excess utilization of fossil fuels leads to the climate change. Furthermore, dumping of agro-industrial wastes in the soil promotes the growth of pathogens and their burning emits toxic gases into the atmosphere. Off lately, the impetus has shifted towards finding an alternative for the efficient valorisation of agro-industrial wastes leading to their use as cleaner and renewable bioenergy resources (Bos and Hamelinck 2014). Moreover, the presence of some anti-nutritional factors in agriculture residues such as caffeine, tannins, oxalate hydrogen, cyanide polyphenols, etc., interfere with the bioavailability and digestibility of the nutrients, hence limiting their use as feed for feedstock. Similarly, the waste produced from food industries contains high concentrations of BOD (Biological Oxygen Demand), COD (Chemical Oxygen Demand), and suspended solids. The lack of proper treatment not only causes an adverse effect on the environment but also on the health of humans. The agro-food wastes due to their rich

[1] Center of Innovative and Applied Bioprocessing (CIAB), Sector-81, Mohali-140306 India.
[2] Environmental Bioprocess Laboratory, Department of Biotechnology, National Institute of Technology (NIT), Warangal, Telangana 506004 India.
[#] Contributed equally.
[*] Corresponding author: meena@ciab.res.in

nutritional composition can be utilized as cheap raw materials for value-added product generation by biotechnological routes (Kiran et al. 2014). A huge amount of carbohydrate-rich waste is generated every year from agriculture and food sector which is further classified as straw pulp seeds, etc., depending on the process used (Fig. 1). Solid state fermentation (SSF) is the most coveted method for the biodetoxification of these wastes. The consumption of high energy has become significant reason for lower production rate and product yield while running liquid state fermentation industry. Additionally, claimable for creating environmental pollution and requires high budget for fulfilling raw material accessibility; these situations have become more prominent and are significantly restraining the sustainable growth of fermentation engineering. The present genera significantly employed SSF to formulate a broad category of food, for instance, traditional foodstuffs (vinegar flavored spices soy sauce); enzymes isolated from microbial species (amylase cellulase glucosidase); microbial cells indulge (single-cell proteins palatable fungus spirulina); microbial species derived metabolites (nucleotide vitamins lipids and amino acids) (Chen 2013). To retrieve efficient profit and products under sustainable development guidelines, the fermentation industry has essentially shifted towards solid-state fermentation because of energy-redeemable and affordability properties (Yafetto et al. 2022). Additionally, the agro-industrial residues serve as nutrition-rich substrates not only as solid supports but also as the source for producing fermentable sugars, thereby facilitating low-cost process operation. Various studies have been conducted to bio-convert food and agriculture wastes to produce several products by using different micro-organisms and reactor designs, the details of which have been discussed here. Thus, the present chapter describes the uses of agro-industrial wastes by SSF processes.

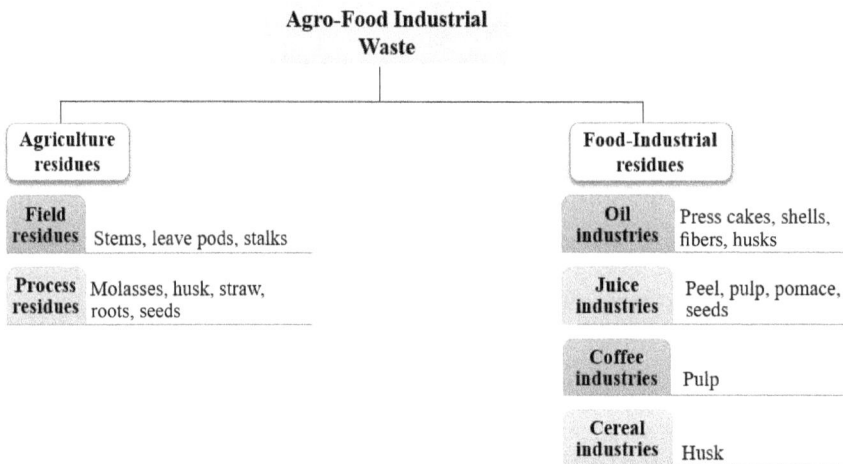

Figure 1. Schematic representation of agro-food industrial wastes.

2. SOLID-STATE FERMENTATION

2.1 Working principle

The modern era of SSF could employ the setup data of the liquid stage fermentation to overawe the complications of outmoded SSF. Researchers are now more focused on encountering the specific requirements of the production process and a complete understanding of the internal metabolic regulation of significant individual microorganisms for the fermentation process under the desired circumstances. The exceptional feature of SSF system is that there is no free water existing in the solid substrate (Chen 2013). This system has three-phases, consisting of the continuous gas stage, the liquid segment, and the solid phase. Moreover, substrate can be divided into two classes focused on its digestibility. Firstly, the inert carrier substrate works as a porous substrate that is chemically inert and microbial decomposition is difficult, for instance, polyurethane foam perlite microporous resin and vermiculite. The second one is the nutritional carrier substrate, for instance, crops; wheat bran and soybean meal; agriculture and forestry wastes indulges straw, sawdust and bagasse. These substrates are responsible for providing physical structure for microbial growth and work as a carbon nitrogen and growth factors for specific microorganisms (Singhania et al. 2010). The mentioned substrate plays critical role in the fermentation process and microorganisms can avail nutrition from the fluid culture that is present in the porous media gap (Wu 2006).

The SSF process regulation indulges different parameters based on biological and substrate characteristics. To achieve the goal of enhancing the production potential facility of the microbial strain, researchers are working to acquire complete knowledge of microbial metabolic engineering and fermentation process. Previously, fermentation strains were significantly obtained by random mutagenesis and natural selection process but recently, they were unable to meet fermentation industry requirements; therefore, the biochemical machinery of the synthesis should be discovered extensively. The specificity of the solid-substrate in SSF is based on metabolic characteristics and selection of specific microbial growth parameters for conducting experiments (Duan and Chen 2012).

Recent fermentation technology is broadly categorized as keen on four fragments: firstly, upstream engineering signifies the selection of specific microbial strains that are adapted for SSF, for instance, *Aspergillus oryzae*, *A. niger*, *A. awamori*, *Beauveria bassiana*, *Rhizopus oryzae*, *Trametes versicolor*, *Kluyveromyces marxianus*, *Trichoderma hazanium*, etc., and their preparation with the sterilization process. Afterward, midstream plays a role of hot spot for research ideas which includes designing and suitable equipment selection with fermentation process analysis. The process is followed by downstream and auxiliary technology as depicted in (Fig. 2), which works on extraction of desirable products and the purification process.

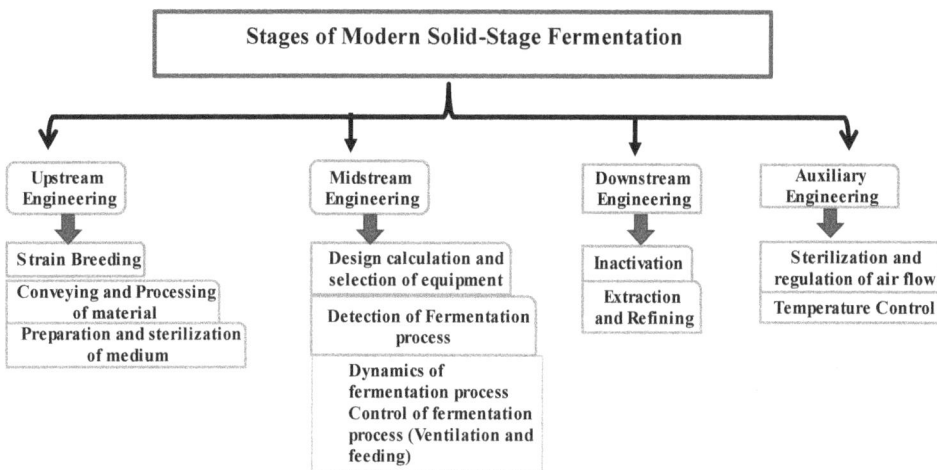

Figure 2. Phases of modern solid-state fermentation.

2.2 Bioreactor design for the modern solid-state fermentation (MSSF)

A bioreactor has a stainless-steel vessel that is responsible for supporting a biologically active environment by performing chemical process and which involves microorganisms or biochemically active substances extracted from specific organisms. Bioreactors exist in two forms according to the quantity of raw material treated. To employ dry solid medium from a range of few grams to kilograms, laboratory-scale fermenter is utilized, whereas to handle several tons of raw material, the pilot and industrial scale fermenters are significantly utilized. The broad classification of bioreactor for SSF is based on selection of microorganism based on producing primary and secondary metabolites as per industry requirements, for instance, column reactor, sterile reactor, rotating/perforated drum bioreactor and horizontal paddle mixer are used to perform solid-state fermentation process. The bioreactors are also classified into four classes centered on their mode of actions, namely (1) Tray bioreactor (2) Packed bed bioreactor (3) Air Pressure bioreactor and (4) Intermittent/continuously mixed SSF bioreactors. Hence, these modern engineering parameters are different from traditional SSF in modified fermentation process that have accomplished mechanisation, automation and intensification.

2.2.1 Tray bioreactor

Traditionally, tray bioreactors (TB) have been used for producing fermented food, for instance, tempeh/ miso/soy sauce/koji in various parts of Asian countries (Zhu and Tramper 2013). Trays comprise of wood/plastic or metal/perforations/filled with substrate-facility and arranged in a stacked manner with a well-controlled environment parameter. The use of tray bioreactor for scale-up is commonly accomplished by increasing the trays' number and surface area. This bioreactor has great efficacy to produce enzymes and is focused on regulating bed heights, substrate moisture at initial stage and their impact on yield. The traditional tray bioreactor (TB) has been modified for spore fabrication by (CRM-16) *Clonostachys rosea* mutant species (Zhang et al. 2014). Changes have been made in bioreactor sporulation area and this area was increased by two times compared to the traditional TB. The ratio of (3:1) wheat bran and maize meal was placed both on top and bottom and concealed by a permeable nature polyethylene membrane (responsible for lowering contamination risk). After two mixing procedures, sporulation was testified and achieved ten times better than the traditional tray bioreactor. Another significant application of modern TB is the fumaric acid formulation by consuming *Rhizopus oryzae* 1526 and here ultra-filtered apple derived pomace are used as substrate. Consequently, fumaric acid shows greater productivity as compared to submerged fermentation process (Das et al. 2015). The innovative trickling tray bioreactor comprises of a medium storage tank bio-reaction compartments and a tank for collecting extracted product. In this model, perforated trays have measurements of $(17 \times 11 \times 54.2)$ cm occupied as working site for fermentation and crude enzymatic clarifications that was passed from medium storage tank in batch/semi-batch pattern proposed by (Fath and Fazaelipoor 2015). The rice bran with wheat bran has shown significant results for providing greater porosity area; soybean meal and wheat flour works as nitrogen sources and protease inducer, respectively. The mentioned arrangement was held in the wood chamber at 30°C and sustained gases' exchange throughout the fermentation process. Consequently, on the 3rd day of the enzyme cultivation, protease shows higher activity (748U g-ds^{-1}) as compared to the liquid flask experimental studies (530 U g-ds^{-1}). (Ruiz et al. 2012) recommended a column tray bioreactor made from perforated 8 trays in the cylindrical support during lemon peel pomace fermentation for producing pectinase and resulted in a better protease action (2181U L^{-1}) with the help of perforated surface that allows circulation of sterile moist air and results in enhancing oxygen accessibility and improves bed drying process. Bioreactor tray measurements of $(32 \times 40 \times 5)$ cm were utilized for the *Phlebiopsis gigantia* spores' production and it has biofungicide potential (Virtanen et al. 2008). Other important parameters for tray bioreactor are to examine heat generated by microbial action. The studied work advised an ideal temperature regulator and the finest step was to formulate bed heights ranging from (1–2) cm and the airflow area should be in close proximity for microbial growth. A surge in biofilm width resulted in lesser oxygen percentage in bed and this condition is inevitable in tray bioreactor. Tray bioreactors with modifications have various applications and greater advantages over the traditional process to produce products such as conidia spores fumaric acid used for lignin degradation and other examples like this are shown in (Table 1).

Table 1. Products produced by modern tray bioreactor-solid-state fermentation engineering.

Substrate employed	Microorganism utilized	Product	Production value	References
Rice bran/soybean meal/wheat flour/ wheat bran	*Aspergillus oryzae*	Alkaline Protease	74 U gd/s	Fath and Fazaelipoor 2015
Wheat bran	*Aspergillus oryzae*	Polygalactouronase	298 U/g	Demir and Tari 2016
Sifted pine sawdust/ soybean powder/rice straw	*Irpex lacteus*	Manganese peroxidase	950 U/L	Zhao et al. 2015
Palm kernel cake and pressed fiber	*Aspergillus oryzae*	Protease	319.3 U/g	Tsouko et al. 2017
Wheat bran/rapeseed meal/soybean hulls	*Aspergillus oryzae* and *A. awamori*	Glucoamylase/protease/ xylanase pectinase/ cellulase	Specific substrate enzymatic activity increased	Manan and Webb 2016
Apple pomace	*A. niger*	Beta-glucosidase	64.181 U gd/s	Dhillon et al. 2011
Tomato pomace	*A. awamori*	Xylanase	195 IU gd/s	Umsza-Guez et al. 2011
Grape pomace/ orange peel	*A. awamori*	Xylanase	42.64 IU gd/s	Diaz et al. 2013
Rice	*Beauveria bassiana*	Conidia	3.92×10^{12} conidia/kg rice	Xie et al. 2013
Apple Pomace	*Rhizopus oryzae*	Fumaric	After 21 days 52g kg-ds^{-1}	Das et al. 2015
Corn forage	*Trametes versicolor*	Lignin degradation	71% lignin degradation after 7 days	Planinic et al. 2016

2.2.2 Packed bed bioreactor

The specific property of a packed bed reactor is the enforced aeration system over a static bed that results in the renewal of oxygen/moisture; moreover, accumulation of heat and carbon dioxide eases. PBR is significantly utilized under not required mixing condition. The heat transfer capability of PBR serves with conductivity/convection/evaporation but during scale-up, the procedure is frequently challenged with the compaction of bed and air channelling. Maintaining ideal bed moisture/temperature is a greater challenge in PBR. The structure of PBR generally consists of cylindrical glass metal tube or drum responsible for providing space to substrate and the cylinder wall has jackets for providing cooling effects to the plates to ease the heat allocation process.

During the aeration process in the PBR, substrate compaction increases, which leads to a decrease in the pressure that is exerted over a bed as a result of air channelling disturbances. Additionally, it shows heat/mass heterogeneity. These complications are developed during the formulation of Iturin A (antifungal compound) formed by *Bacillus subtilis* species (Piedrahta-Aguirre et al. 2014). Air flow rates stimulated the production of poly-gamma-glutamic acid which is responsible for creating inversely proportional relationship between oxygen uptake and bed viscosity air channeling. This issue seems relevant during production of pectinase by utilizing *A. niger* in a pilot study (PBR:200L) and results depicted that the pectinase yield (1350 U/kg/h) was lesser than the lab scale bioreactor (1930 U/kg/h) because of bed and wall shrinkage in bioreactors that is responsible for increasing temperature up to 37°C and air channelling (Finkler et al. 2017).

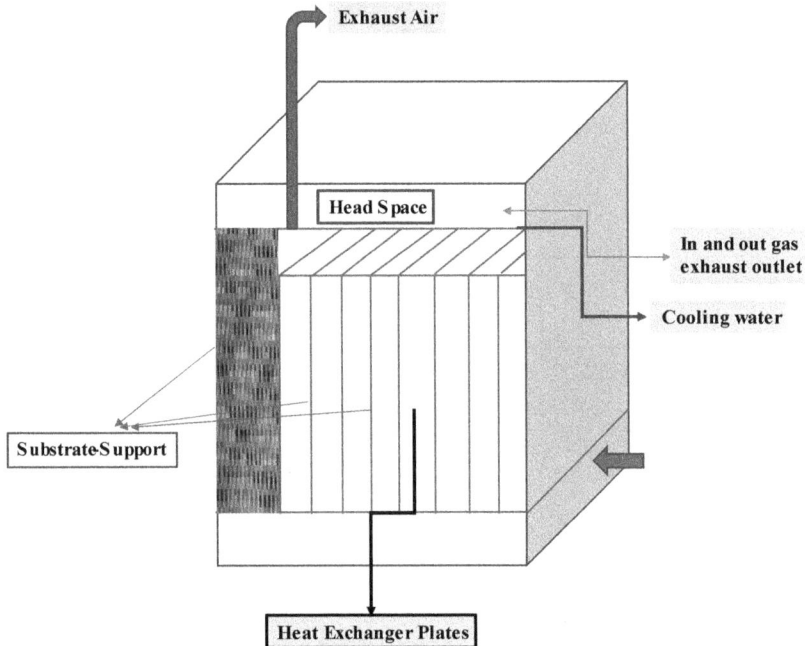

Figure 3. Zymotis packed bed reactor (Roussos et al. 1993).

To overcome this issue, sugarcane bagasse was utilized for increasing the porosity and allows circulation of saturated air. Prior studies suggested that productivity will only increase by providing cooler air flow up to 24°C during heat generation cycle and results in productivity increased to 1840 U/Kg/h. Moreover, to overcome PBR overheating problem, the modern version of packed bed reactor was suggested and formulated by (Roussos et al.1993) known as 'Zymotis' as depicted in (Fig. 3). The bioreactor comprises of a four-sided shaped box covered with sheets of acrylic. The outer domed mould also indulges outlet and inlet for water circulation inlet-outlet channels for gases and a holder for opening drive. Additionally, it also has heat transfer machinery responsible for conduction generated from substrate bed in the direction of the heat exchanger. This model is currently employed by German company (Prophyta) for biopesticide production. Packed bed reactor has significant role in producing various products by using specific microbial strains as shown in (Table 2).

2.2.3 Air pressure and intermittent mixed solid-state fermentation

Bioreactor under this class employs intermittent pulsation of air circulation to augment the microbial action and alleviates heterogeneity procedure. It is also known as GDD-SFF (Gas double-dynamic solid-state fermentation) and comprises of forced air drive and pulsation event. Due to the presence of high pulsation and force of air, the partial pressure of oxygen increases in the gas segment between the process of compression and results in enhancing oxygen proportion in the bed but during decompressing process, it facilitates removal of carbon dioxide and heat. To ensure that the oxygen concentration does not restrict the internal pressure of the bioreactor, it is necessary to change the pressure including plates placed in a stainless steel chamber. The outcome of pulsation of air compression was studied on cellulase formulation by using *Trichoderma viride* and consequently showed elevation in cellulase productivity three times as compared to tray bioreactor (Tao et al. 1999). Fujian et al. (2002) examined the effect of GDD-SFF on production of cellulose by *penicillium decumbens* where the pulse is recirculated in units of 0.20 MPa and the air flow at 1.5 m/s has been shown to have the best effect, but if the pulse is increased, hyphae disturbance occurs due to the gas inflating. The temperature gradient was reduced

Table 2. Products produced by packed bed bioreactor.

Substrate employed	Microorganism utilized	Product	Production value	References
Sugarcane bagasse/ Wheat bran	*Aspergillus niger*	Pectinase	22 U/g	Finkler et al. 2017
Wheat bran	*Aspergillus niger*	Endoglucanase	50.2 IU gd/s	Farinas et al. 2011
Babassu cake	*Aspergillus awamari*	Exoamylase/protease/ cellulase/xylanases	73.4 U/g	Castro et al. 2015
Citrus pulp/ sugarcane bagasse	*Aspergillus oryzae*	Pectinase	37 U/g	Biz et al. 2016
Wheat bran/ sugarcane bagasse	*Aspergillus niger*	Pectinase	20 U/g	Pitol et al. 2016
Waste bread	*Aspergillus awamari*	Glucoamylase	130.8 U/g	Melikoglu et al. 2015
Wheat bran	*Bacillus* species (KR-8104)	Alpha-amylase	473.7 Ug/ds	Derakhti et al. 2012
Wheat bran	*Aspergillus ficuum*	Phytase	580 U/ds	Badmchi et al. 2013
Immobilized strain in orange peel	*Aspergillus niger* (URM 5162)	Endo- and Exo-Polygalactouronase	1.18 U/ml	Maciel et al. 2013
Polyurethane foam	*Aspergillus niger*	Tannase	7955 U/L	Rodriguez-Duran et al. 2011
Pressmud	*Kluyveromyces marxianus*	Inulinase	300.5 Ugd/s	Dilipkumar et al. 2013
Copra waste	*Penicillium regulosum*	Inulinase	239 Ugd/s	Dilipkumar et al. 2014

by placing trays in horizontal position (0.8°C/cm) as compared to vertical position (4.2°C/cm). Yang et al. (2011) significantly focused on the ranges of pulsation on xylanase production by *Thermomycetes lanuginosus*. Exposure to continued stages of high pressure undesirably affected fungal growth.

In intermittent mixed SSF, there is a requirement of moderate agitation with insisted aeration to increase the heat/mass transfer process and microbial species growth. Mixing or assimilation facilitates transport because it the surface of substrate open to air and cooling fluids. Traditionally, to enhance the productivity with minimum mixing steps is a challenging task because lesser mixing could deteriorate fungal mycelia and also utilize extensive energy. Currently, rotating drum bioreactors (RDB) are utilized and various parameters are examined to overcome traditional issues by studying consequence of mixing on heat and mass transferences microbial activity yield efficiency. Recently, RDB is significantly employed for treating waste. RDB consisted of a drum-shaped vessel that is straddling on a rotating device and has three-subsystems for proper working, namely, drum wall/ headspace present above the substrate bed and the rotation of bioreactor that occurs in two patterns (i) intermittently, and (ii) continuously (Alam et al. 2002). Literature suggested that cellulase production by using 4kg of fruit bunch as a substrate shows two-fold increase in productivity. RDB has been widely employed for producing several products for human wellbeing, for instance, citric acid, ethanol, lactic acid, biosorbent, red 40 dye degradation enzyme and many more as mentioned in (Table 3).

2.3 Substrates used in solid-state fermentation

The features of the SSF substrate can be identified from micro and macro characteristics. In macro category, weight of dry and substrate air permeability is dynamic in nature during the fermentation procedures, whereas micro-class deals with the microbial growth. Another significant property of nutritional substrate carrier for fermentation is the pressure drop of substrate which is responsible

Table 3. Application of rotary drum bioreactor under SSF engineering.

Substrate employed	Microorganism utilized	Product	Production value	References
Soyabean meal	*Aspergillus niger*	Phytase	580 U gd/s	Sithia and Tongta 2016
Wheat bran	*Aspergillus niger*	Pectinase	Increase in 37% of pectinase recovery	Poletto et al. 2015
Defatted rice bran	*Aspergillus niger* LB-020SF	Amylogucosidase	886.2 U gd/m	Colla et al. 2017
Palm fruit bunch	*Penicillium verruculosum*	CM case	6.5 U CM case	Kim and Kim 2012
Palm fruit bunch	*Aspergillus niger*	Endoglycosidase	135 U/L	Noratiqah et al. 2013
Palm oil lignocellulosic biomass	*Trichoderma hazanium*	Cellulase	8.2 FPA gd/s	Alam et al. 2009
Soyabean meal	*Aspergillus niger*	Amylase and Protease	85000–110000 U g d/s	Sukumprasertsri 2013
Grape pomace and orange peels	*Aspergillus awamori*	Xylanase	54.4 IUg/ds	Diaz et al.2009
Grape pomace	*Aspergillus awamori*	Pectinase/xylanases /cellulases	8.77 IU g/ds	Diaz et al. 2009

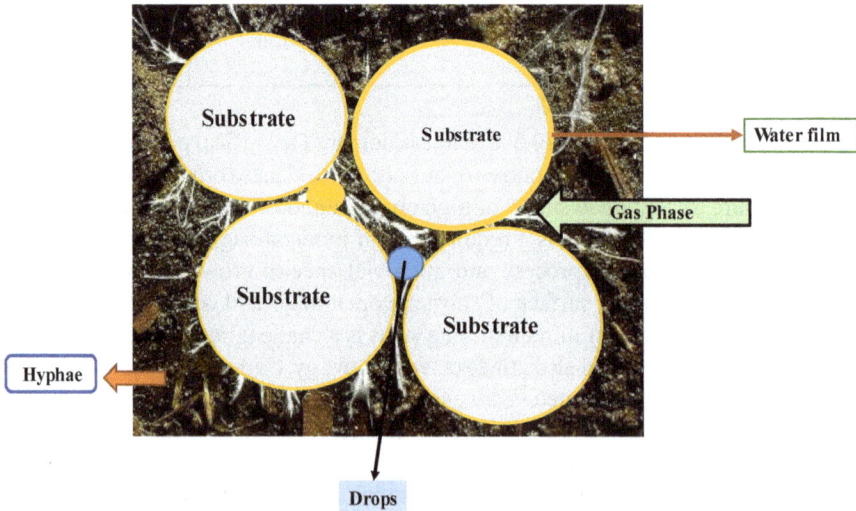

Figure 4. Internal cross-section view of the nutritional carrier substrate.

for determining the growth of cell trend. The substrate utilized for fermentation is solid in nature and gas is required as a continuous time with water-free environment in the substrate. Therefore, SSF comprises of a solid-liquid-gas phase system as depicted in (Fig. 4). The substrate proximities, porosity, mass, and heat allocation are always changing throughout the fermentation phenomena. SSF has proven for showing its classical application in different area by using nutritional rich substrate. Various products - for instance, enzymes, antibiotics, organic acids, biological control agents, and lipids-biodiesel - are successfully produced by using specific nutrition rich substrate as mentioned in (Table 4).

Naringinase enzyme production from grapefruit and orange substrates under SSF conditions by filamentous fungi. The composition of naringin hydrolyzed from grapefruit skin by *Aspergillus niger*,

Table 4. List of nutritional substrates utilized for solid state fermentation.

Substrate utilized	Product	References
Enzyme's production		
Orange/grapefruit rind	Naringinase	Mendoza-Cal et al. 2010
Apple bagasse/wheat bran	Polygalactouronase	Abbasi et al. 2011
Rice-husk/banana husk/millet/wheat bran	Alpha-amylase	Ozdemir et al. 2012
Sunflower seed/sugarcane bagasse	Lipase	Lin et al. 2013
Wheat bran/soybean meal	Protease	Imtiaz and Muhtar 2013
Corn straw/grass powder/sugarcane barbojo	Cellulose/hemicellulose	Saratale et al. 2014
Wheat bran	Phytase	McKinney et al. 2015
Popular saw dust	Laccase	Kuhar et al. 2015
Organic acid production		
Pineapple waste/valenica orange peel/sugarcane bagasse	Citric acid	Kumar et al. 2010, Torrado et al. 2011, Yadegary et al. 2013
Sugarcane bagasse/cassava bagasse for carbon source/rice straw/wheat bran	Lactic acid	John et al. 2006, Qi and Yao 2007, Naveena et al. 2005
Tea waste for support/molasses for carbon sources	Gluconic acid	Sharma et al. 2008
Sugarcane bagasse/pomegranate seeds/husk/ pomegranate peel	Ellagic acid	Singh et al. 2003, Robledo et al. 2008, Sepulveda et al. 2012
Bioactive compounds production		
Rice bran	Phenolic compounds (Antioxidant activity)	Schmidt et al. 2014
Wheat	Phenolic compounds (Antioxidant activity)	Dey and Kuhad 2014
Chick Pea	Phenolic compounds	Xiao et al. 2019
Lentils	Phenolic peptides	Torino et al. 2013
Apple Pomace	Phenolic compounds	Salar et al. 2012
Lipids Production		
Pear Pomace/rice bran/soy bean meal	Gamma-Linolenic acid	Fakas et al.2009, Jangbua et al. 2009
Sorghum	Oleic acid	Economou et al. 2011
Wheat straw-bran	Palmitic acid	Hui et al. 2010
Lipids	Soybean hull/ palm blank fruit bunches	Zhang and Hu 2012, Chiersilp and Kitcha 2015

and *Aspergillus foetidus* was 81% and 80%, respectively. The specific activity of naringinase and volumetric parameter of an individual strain was influenced by pH, temperature and water activity. *Aspergillus foetidus* shows highest volumetric activity with 2.58 U ml^{-1} (Mendoza-Cal et al. 2010). Apple bagasse are also employed as substrate in SSF to produce polygalacturonates with microbial strain of *Aspergillus niger* CCT0916. During optimization process, the impact of moisture content, ammonium sulphate percentage, spore formation and fermentation temperature on polygalactouronase activity was observed. Consequently, moisture become the limiting factor in the procedure. The parameter to produce maximum yield of polygalactouronase activity with 33 U/g, 50% moisture content, 106 spores/g and 1.5% w/w ammonium sulphate at 35°C temperature (Abbasi et al. 2011). Cellulase derived from sugarcane bagasse was used for producing lactic acid under SSF in media containing a cellulase enzyme derived from *penicillium janthinellum mutant EU1* and cellobiose utilizing *lactobacillus delbrueckii* mutant responsible for resulting in maximum yield of lactic acid (67 g/l) from bagasses (John et al. 2006, Qi and Yao 2007, Naveena et al. 2005).

2.4 *Microorganism utilized for solid state fermentation*

Microbes are the important participants during SSF process and according to the microorganism requirement, SSF can be categorised into two parts: (i) pure culture and (ii) non-pure culture phenomena. Non-pure culture includes (composting/koji formulation/silage) responsible for managing microbial population by environmental control parameters, for instance, Temperature, oxygen, humidity. The different group of microbial species present in SSF has different structures in solid matrix because different microbes have different functions. Consequently, it becomes difficult to acquire a specific characteristic of substrate for specific microbial species. The biofilm formed by microbial species shows that growth on substrate surface is not homogenous in nature for microbes. Microorganism has capacity to live under extreme environmental condition and according to the microbial taxonomy created for SSF for example; ability to survive at high concentrations of salt/ acid/heat/drought/pressure/radiation and alkalinity. Bacteria/yeast/fungi has greater efficiency to grow on the solid substrate and are responsible for producing significant products as mentioned in Fig. 5 (Raimbault 1998).

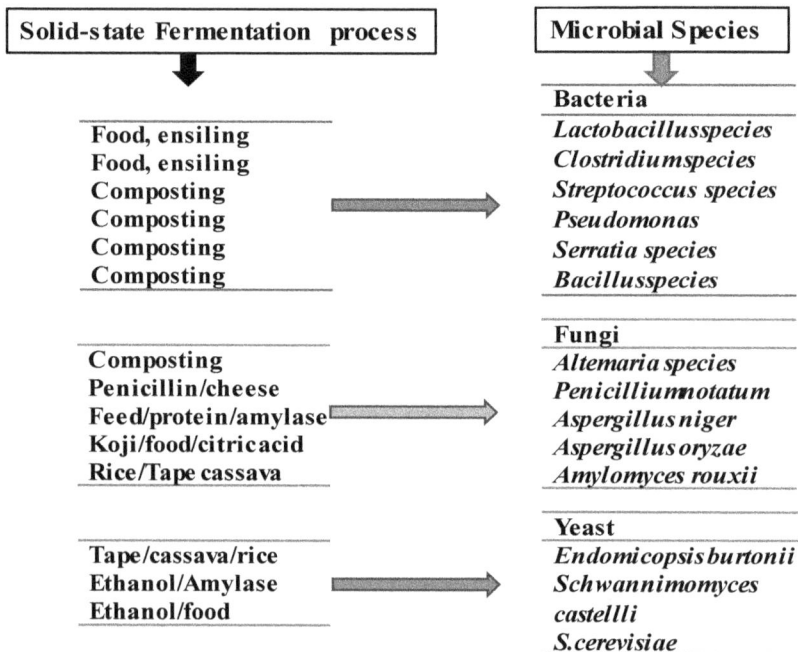

Solid-state Fermentation process		Microbial Species	
		Bacteria	
Food, ensiling		*Lactobacillus species*	
Food, ensiling		*Clostridium species*	
Composting		*Streptococcus species*	
Composting		*Pseudomonas*	
Composting		*Serratia species*	
Composting		*Bacillus species*	
Composting		**Fungi**	
Penicillin/cheese		*Altemaria species*	
Feed/protein/amylase		*Penicilliumnotatum*	
Koji/food/citricacid		*Aspergillus niger*	
Rice/Tape cassava		*Aspergillus oryzae*	
		Amylomyces rouxii	
Tape/cassava/rice		**Yeast**	
Ethanol/Amylase		*Endomicopsis burtonii*	
Ethanol/food		*Schwannimomyces*	
		castellli	
		S.cerevisiae	

Figure 5. Microbial species employed for SSF process (Raimbault 1998).

3. AGRO-FOOD INDUSTRIAL RESIDUES

Agriculture and food industries jointly generate huge amount of wastes each year which often go unused including the post-harvest field wastes and are termed as agro-food industrial residues. Agricultural residues are further grouped into those procured from crops fruits, vegetables, and food processing industries The production of agro-industrial residues has escalated with the rise in population worldwide, leading to drastic economic, environmental, and societal issues (soil, air, water pollution waste treatment and disposal) (Chilakamarry et al. 2022).

This has led to a growing interest in conversion of agriculture biomass into value added products such as feed, bio-fuels and bio-energy (Duque-Acevedo et al. 2020). One way to achieve this is by bio-valorization of these agro-food industrial wastes by solid state fermentation using microorganisms (Zepf and Jin 2013). Agro-industrial residues act as cheap but nutritious raw material enriched in starch, sugar, protein, cellulose (35–50 %w/w) hemicellulose (25–30 %w/w) or lignocellulosic (15–25 %w/w) contents which have an endless potential to replace costly carbon sources as alternative but cheap substrates for fermentation (Sadh et al. 2018, Salgadoa et al. 2012). Though some of the agro-food industrial wastes are employed as feed for livestock, their use remains limited due to the presence of anti-nutritional factors such as caffeine, oxalate, hydrogen cyanide, polyphenols, tannins, etc., that can meddle with the bioavailability and digestibility of the nutrients. The metabolism of these anti-nutritional factors in feed yields by-products that lower the presence of significant nutrients required by the livestock. SSF hence is the most effective way for biodetoxification of these agriculture as well as food industrial wastes. Moreover, agro-industrial wastes are suitable for use in SSF as they are abundant and easily available (Leite et al. 2021).

3.1 Agricultural residues

Agriculture is considered as one of the largest jobs providing sector worldwide with the maximum biomass generation. The amount of biomass generated by the agricultural activities has increased in the last 50 years accounting for the current production of 23.7 MT of food/day globally, thus creating a large pressure on the environment, impacting the quality of water, air, and soil due to indiscriminate disposal (Duque-Acevedo et al. 2020). Globally, about 5 billion MT of biomass is produced yearly from agriculture such as fruits, vegetable, wastes, rice bran, groundnut shells, rice straw, sugarcane bagasse, wheat bran, cotton leaf scraps, etc. Moreover, it's estimated that 709.2 and 673.3 million MT of wheat and rice straw is produced worldwide. Most of it is either utilized at farms as bedding or as feed for cattle. The unused agriculture waste is then either dumped as such or burnt either at farms or in the backyards of food-processing industries (Soccol et al. 2003, Yafetto 2022). For instance, the average rice production annually is 721.4 MT of which Asian countries hold 90.48% share in the total produce. Rice harvesting generates the most amount of residues (30 MMT). It is reported that only 20% of the rice straw produced globally is employed constructively. The chronological order of the major agricultural crop residues in India is in million metric ton (MMT): 112 rice straw; 22.4 rice husk; 109.9 wheat straw; 97.8 sugarcane tops and 101.3 bagasse (Sukumaran et al. 2010). Furthermore, agriculture wastes can be grouped into field residues and process residues. Field residues are residues remaining in the field after the crop has been harvested such as leaves, seed pods, stalk, and stems. On the contrary, process residues are ones generated during processing of the crop into alternate valuable resource such as molasses, husks, pulp, stubble, peels, leaves, straw, stalk, husks shells, roots bagasse seeds stem cereal bran, etc. (Oliveira et al. 2017). However, such wastes may differ in composition of protein content sugars and minerals, although some of the agriculture residues are used as animal feed for soil improvement as fertilizers, etc. Several other lignocellulosic residues like grasses, vine leafs, red grape stems, leafs, fruit pistachio and chestnut shells have been explored for value added product development by biotechnological route post suitable hydrolysis treatment. Owing to their high nutritional index (Table 5), agriculture residues are considered as potential raw materials for other product formation development by acting as solid support in SSF for making different beneficial products. In addition, wastes such as green walnut husks, lemon and pomegranate peels can be used as natural anti-microbial agents (Nguyen et al. 2010).

3.2 Food-industrial wastes

Food waste management is a major concern worldwide as it is difficult to handle because of high moisture content (approximately 80%). Food-industrial waste is broadly classified as the discarded

Table 5. Chemical composition of agriculture wastes.

Agriculture wastes	Chemical composition (% w/w)						
	Cellulose	Hemicellulose	Lignin	Ash (%)	Solids (%)	Moisture (%)	References
Sugarcane bagasse	30.2	56.70	13.4	1.9	91.66	4.8	Sadh et al. 2018
Rice straw	39.2	23.5	36.1	12.4	98.62	6.40	Sadh et al. 2018
Wheat straw	32.9	24.0	8.9	6.7	95.6	7	Sadh et al. 2018
Sunflower stalk	42.1	29.7	13.4	11.17	–	–	Motte et al. 2013
Cotton stalk	50.2	19.82	22.24	5.65	–	–	Ming et al. 2021
Barley straw	38	35	16	–	–	–	Kalita 2018
Sweet Sorghum bagasse	45	27	21	–	–	–	Kalita 2018
Reed stalk	33.1	25.3	19.7	2.86	–	–	Koullas et al. 1996

foodstuff which mainly contains the unsold food and the leftovers from food processing industry. Food processing industries engaged in manufacturing of chips, juice, confectionary and meat generates a huge quantity of organic residues per annum. The fruit-based industries generate process residues comprising of fruits peels (mango, apple, banana, and pomegranate) and seeds (black plum, tamarind, apple, mango) (Saeed et al. 2021). For instance, the mango juice processing industry produces tons of lignocellulosic waste annually of which pomace and peel accounts for 15–20% of the total waste produced from juice processing from mango (33–85% edible pulp) (Araya 2015). Of this, the mango kernel alone makes up to 45–75% of the seed weight. Similarly, the wine industry worldwide operating on a large scale produces 122,000 tons of grape pomace per annum which includes seeds and pulp in California alone (Zheng et al. 2012). The lignocellulosic food waste due to its chemical complexity, high moisture content, easy degradation, and nutrient rich composition can be employed as a valuable resource for higher value products' generation such as bio-ethanol and other commercial chemicals via fermentation (Table 6) (Kiran et al. 2014). As starch is also a major biopolymer present in fruit-industrial wastes (waste bread, potatoes savoury, cakes, fruits and vegetables), its saccharification using enzymes (e.g., glucoamylase) is imperative for bio-conversion into value-added products (Yang et al. 2006). The coffee producing industries also generate large number of wastes, particularly coffee pulp, husk, and skin, which make up half of the total coffee fruit production. Coffee pulp alone makes up to 40% mass of coffee fruit obtained post wet processing of coffee cherries and is regarded as anti-nutritional because of the presence of high content of caffeine polyphenols and tannins (Silveiraa et al. 2019). Similarly, the juice industries produce enormous wastes in the form of peels, cereal industries produce husks (Sadh et al. 2018) and oil industries produce oil cakes, which are residues obtained after oil extraction from seeds such as sunflower oil cake, coconut oil cake, rapeseed cake, cotton seed cake, sesame oil cake, palm kernel cake, soy bean cake, mustard oil cake, groundnut oil cake, canola oil cake, and olive oil cake. These untreated residues, when disposed of by the industries, cause air, water and soil pollution as these wastes contain high concentration of suspended solids, oil, fat, grease and dissolved solids (Rudra et al. 2015). At present, a major part of the food waste is either incinerated along with other municipal wastes to produce energy or is discarded off in landfills. However, incineration technique is facing regulations and environmental pressures as it causes severe air pollution and is an expensive waste conversion technique (Tian et al. 2018, Zhang et al. 2013a).

Table 6. Chemical composition of food-industrial wastes.

Food-Industrial wastes	Chemical composition (% w/w)						
	Cellulose	Hemicellulose	Lignin	Ash (%)	Solids (%)	Moisture (%)	References
Coffee pulp	53	11	16.85	–	–	5	Pudjiastuti et al. 2019
Rice husk	35	25	20	17	–	–	Abbas and Ansumali 2010
Nut shells	25-30	22-28	30-40	–	–	–	Adewuyi 2022
Sugarcane tops	35	32	14	–	–	–	Adewuyi 2022
Orange peel	9.21	10.5	0.84	3.5	-	11.86	Sadh et al. 2018
Pineapple peel	18.11	-	1.37	–	93.6	91	Sadh et al. 2018
Wheat bran	32.1	29.2	16.4	–	–	–	Xiao et al. 2019
Apple pomace	7-44	4-24	15-23	1.5-3.5	–	70-85	Costa et al. 2022

4. TREATMENT OF SOLID FOOD WASTE USING SOLID STATE FERMENTATION

Food waste is not only generated by kitchen but lot of food processing companies produce millions of tons of food waste. These food wastes cause severe environment health and economic problems; in order to tackle these problems, the best way is to produce commercial products from these food waste. Lot of commercial products are produced from these food waste like enzymes, bioactive compounds, various organic solvents (lactic acid, ethanol, butanol), biogas, biofertilizer, nutraceuticals and many other products as shown in Table 7 (Huang et al. 2015, Akpan et al. 2008, Song et al. 2022, Ballardo et al. 2016, O'Connor et al. 2021). The global food waste would be expected to grow around 2.2 billion tons in 2025 (O'Connor et al. 2021). Food waste is also used to produce some human consumable fungus spores like *Conidia* (Sala et al. 2022). A different strategy was used by (Moukamnerd et al. 2013) to utilize the food waste for bioethanol production; they consolidated continuous solid-state fermentation (CCSSF) method, which acts as an integrated system in which hydrolysis fermentation

Table 7. Food waste and their valorization using SSF.

Sr. no.	Substrate	Microorganism	Product	References
1.	Beer draff and Rice husk	*Beauveria bassiana*	Conidia spore as Biopesticide	Sala et al. 2022
2.	Cassava bagasse	*Lactobacillus delbrueckii*	L-Lactic Acid	John et al. 2006
3.	Bread crust, Potato chips and Rice Grain	*Super camellia* Dry Yeast	Bioethanol	Moukamnerd et al. 2013
4.	Sugarcane Bagasse	*Corynebacterium glutamicum*	L-Lysine	Anusree et al. 2015
5.	Pearl Oyster, Mushroom residues	*Streptomyces albulus* CICC 11022	Poly L-Lysine and Poly L-Diaminopropionic acid	Wang and Rong 2022
6.	Dairy manure	*Bacillus amyloliquefaciens*	poly-c-glutamic acid	Yong et al. 2011
7.	Rice bran and wheat straw	*Klebsiella pneumoniae*	Vitamin B12	El-sheekh et al. 2013
8.	Rice bran and wheat straw	*Citrobacter freundii*	Folic acid	El-sheekh et al. 2013

and ethanol recovery is done in a single process. This CCSSF system consists of a rotary drum reactor, condenser and humidifier. This CCSSF system helps to reduce cost of product, time and recovery processes. Wood waste is also used to produce various amino acids, lipids and many aromatic compounds (Aggelopoulos et al. 2014, Anusree et al. 2015, Wang and Rong 2022). Since food waste is a great source of protein, carbohydrate, lipid and minerals, there are various publications which adopt various strategies for the isolation of these biomolecules (Kamal et al. 2021, Ma et al. 2018); these extracted biomolecules can be used for further industrial application, e.g., extracted lipid can be used for the biodiesel production.

5. ENZYME PRODUCTION USING FOOD WASTE BY SSF

Enzymes are substrate specific, due to which they have commercial value in many sectors like pharmaceuticals (therapeutic enzymes), food, nutraceuticals, waste treatment, textile, beverages, biofuel, and many others (Londono-Hernandez et al. 2020, Binod et al. 2019, Sindhu et al. 2018, Tarafdar et al. 2021). The industrial enzyme market is approximately US \$4.5 billion which is expected to rise to \$7 billion by 2021 with a 4% annual growth rate. Novozymes DSM BASF and DuPont are the main enzyme producing companies that supply 75% of the overall commercial enzymes (Tarafdar et al. 2021). Developed countries like Western Europe, Canada, USA, and Japan share most global enzyme market. In current scenario, researchers are focusing on valorization of food waste for enzyme production because these food wastes have sufficient nutritional value for the growth of microbes to produce desired enzymes using SSF (Ravindran and Jaiswal 2016, Shukla and Kar 2006). SSF gets privilege over SmF (Submerged Fermentation) because SSF generates less waste, low requirement of water, less process control requirement, easy recovery of products, and many other factors which make SSF more efficient over SmF. Different types of food wastes produce various type of enzymes like proteases, cellulases, amylases, lipases, and many other enzymes using SSF (Uçkun et al. 2014).

Enzymes are produced either by directly using food waste or sometimes with adding some nutrient, e.g., phytase is produced from cassava dreg by adding ammonium nitrate (Hong et al. 2001). Food waste has different nutritional compositions like Total Solid Content, Total Volatile Content, Total Moisture Content, Carbohydrate, Lipid, and Protein Content in varying concentrations depending upon sources where these food wastes had been taken from (Chua et al. 2019). The food waste is treated prior to its use for solid state fermentation. Same enzyme shows different yield and activity depending upon food waste's source and microorganism used to produce it like α-amylase produced from kitchen waste by *Bacillus amyloliquefaciens* shows 10 U/g and same enzyme from potato peel by *Bacillus subtilis* shows 600 U/g on different optimized incubation time (Bhatt et al. 2020, Shukla and Kar 2006). Same pattern is followed by other microorganism with different food waste. A new term is used for enzyme production using food waste, i.e., garbage enzyme. This enzyme is a complex of organic substances, minerals, salts, and protein chains, and is produced from vegetables waste and fruits peel waste. This garbage enzyme has diverse applications, especially in bioremediation and insecticide. It contains mainly caseinate, protease, lipase, cellulose and amylase (Neupane and Khadka 2019).

Various statistical, numerical, and artificial neural network- genetic algorithm (ANN-AG) models are applied to enhance the yield and productivity of the enzymes using SSF (Luiz et al. 2021). This ANN-GA model is more efficient over RSM (Response surface model). Lipase enzyme produced from *Penicillium roqueforti ATCC 10110* shows 267.7% increase using ANN-AG model over RSM model (Luiz et al. 2021). Using these approaches, we can save our cost, time and work. The production of certain enzymes has been considered food waste by the SSF as shown in Table 8.

Table 8. Enzyme production from food waste by SSF and their industrial approach.

Sr. no.	Enzyme	Microorganism and substrate	Incubation time	Yield of enzyme	Respective industry	References
1.	Phytase	*Aspergillus niger* Cassava Dreg	8 days	6.73 U/g	Poultry feed	Hong et al. 2001
2.	α-amylase	*Bacillus lecheniformis* Potato Peel	3 Days	270 U/ml	Food industry	de Souza and e Magalhães 2010, Shukla and Kar 2006
3.	α-amylase	*Bacillus lecheniformis* Wheat Bran	3 Days	175 U/ml	Textile industry	de Souza and e Magalhães 2010, Shukla and Kar 2006
4.	α-amylase	*Bacillus subtilis* Potato Peel	3 Days	600 U/ml	Pharmaceutical	de Souza and e Magalhães 2010, Shukla and Kar 2006
5.	α-amylase	*Bacillus subtilis* Wheat Bran	1 Day	265 U/ml	Paper Industry	de Souza and e Magalhães 2010, Shukla and Kar 2006
6.	Laccase	*Pleurotus ostreatus* Apple peel Apple seed	3 Days	9 U/g	Textile waste water treatment	Iandolo et al. 2011
7.	Glucoamylase	*Aspergillus awamori* Waste cake	6 Days	108.47 U/g	Food Industry	Kiran et al. 2014
8.	(i) Endoglucanase (ii) Xylanase	*Aspergillus niger* mixed food waste	6 Days	(i) 17.37 U/g (ii) 189 U/g	Animal Feed Textile Pharmace-utical Industry	Tian et al. 2018
9.	n-demethylase	*Rhizopus oryzae* Sorghum and Coffee pulp mixture	4 Days	18.762 U/g	Decaffianation of Coffee and Tea Waste	Peña-Lucio et al. 2020
10.	Pectinase	*Aspergillus niger* Citrus Peel	5 Days	97.2 U/mg	Textile and Juice Industry	Ahmed et al. 2016
11.	Cellulase	Trichoderma atroviridae Vegetable waste	5 Days	77.39 U/g	Biofuel and food industry	
12.	Lipase	*Aspergillus niger* Palm oil waste	3 Days	20.7 U/g	Animal feed	Oliveira et al. 2017
13.	Protease	*Aspergillus awamori* Bread	6 Days	83.2 U/g	Pigment production PHB	Melikoglu et al. 2015
14.	Lipase	*Aspergillus niger* Copra waste	5 Days	170 U/ml	Polymer and Pharma Industry	Nurul et al. 2020

g-gram, mg-milligram, PHB-Polyhydroxybutyrate.

5.1 Global enzymes market

Constant growth is being observed in enzyme market because enzymes have diverse applications in many industrial sectors (Sarrouh et al. 2012). The sale of enzyme in US in year 2002 and 2009 was USD 1.3 billion and USD 5.1 billion, respectively; in this, 31% share was by food enzyme and 6%

Table 9. The global enzyme market and growth statistics.

Sr. no.	Enzyme	Global market size	CAGR (Compound Annual Growth Rate)	References
1.	Protease	USD 1.8 billion (2021)	7.3%	FMI Report
2.	Lipase	USD 425 million (in 2018)	6.8%	Chandra et al. 2020
3.	Amylase	USD 255 million (in 2015)	3.8%	Machado de Castro et al. 2018
4.	Phytase	USD 380 million (in 2017)	7.9%	Mrudula Vasudevan et al. 2019
5.	Cellulase	USD 5.9 billion (in 2020)	6.5%	Valerie et al. 2021

share was by feed enzyme (Table 9) (Tarafdar et al. 2021). The enzyme market is not only limited with food and feed but it has wide industrial applications in biofuel, detergent, pharmaceutical, and many other sectors.

6. FUTURE PERSPECTIVES

Although SSF is an important bioprocess whose cost of production can be reduced by using agriculture and food waste as the replacement of solid matrix, the large-scale feasibility and commercial viability is the main challenge. Besides this, operational parameters and reactor design also need to be studied extensively for maximization of product yield. Further study of downstream process for easy and efficient recovery of the product also needs to be done. Innovation in existing reactor designs and development of new bioreactors for proper mixing and heat removal needs to done in order to lessen the production costs. The optimization of inoculum size or careful curation of the microbial consortia of efficient wild or adapted strains is also crucial for the successful implementation of SSF process. Besides this, the use of SSF for production of other industrially important platform chemicals such as bio-plastics, bio-surfactants, and pigments need to be studied further.

The significant bioproducts produced by SSF is the most researched area in the bioprocess engineering scientific community and it is positively evolving. Despite all this, current scientific community, industry persons and researchers should explore different challenges of SSF technology. Firstly, to develop a novel bioreactor is a proven issue that requires to be studied further and in detail. To improve the mentioned issue, work should be done on improvising operational parameters' constrains such as heat removal, mixing properties and maintaining continuous producing pathway. The area that needs to be explored further is building a reactor design with an appropriate vaccination or feeding strategy, that has significant potential to produce a sustainable bioprocess inclined to provide simpler operation amendments leading to a fully optimized industrious process.

7. CONCLUSION

This chapter provided information on the principle of modern SSF and its four pillars, namely, upstream and midstream process, followed by downstream and auxiliary engineering. Various bioreactors have been employed to overcome products' subordinate yield generation by performing and adjusting different parameters in perforated tray, packed bed bioreactor, air-pressure and rotatory drum bioreactors. Surface area and selection of specific substrate have become key requirements for microbial growth at optimum temperature and pressure in stainless steel vessel. Moreover, the chapter also emphasized on application and efficacy of SSF to produce commercially important enzymes bioethanol, L-lysine, poly L-lysine, poly L-diamino, propionic acid, poly-c-glutamic acid, vitamin B12, folic acid, bioactive, lipids and organic acid production by utilizing agro-food industrial wastes as substrate.

REFERENCES

Abbas, A. and Ansumali, S. 2010. Global potential of rice husk as a renewable feedstock for ethanol biofuel production. Bioenergy Research 3(4): 328–334.

Abbasi, H., Mortazavipour, S.R. and Setudeh, M. 2011. Polygalacturonase (PG) production by fungal strains using agro-industrial bioproduct in solid state fermentation. Chemical Engineering Research Bulletin 15(1): 1–5.

Adewuyi, A. 2022. Underutilized lignocellulosic waste as sources of feedstock for biofuel production in developing countries. Frontiers in Energy Research 142(10): 1–21.

Alam, M.Z., Mamun, A.A., Qudsieh, I.Y. et al. 2009. Solid state bioconversion of oil palm empty fruit bunches for cellulase enzyme production using a rotary drum bioreactor. Biochemical Engineering Journal 46(1): 61–64.

Araya, A. 2015. Review of mango (Mangifera indica) seed-kernel waste as a diet for poultry. Journal of Biology Agriculture and Healthcare (5): 2224–3208.

Arora, S., Rani, R. and Ghosh, S. 2018. Bioreactors in solid state fermentation technology: Design, applications and engineering aspects. Journal of Biotechnology 269: 16–34.

Badamchi, M., Hamidi-Esfahani, Z. and Abbasi, S. 2013. Comparison of phytase production by Aspergillus ficuum under submerged and solid state fermentation conditions. Focusing on Modern Food Industry 2(3): 129–137.

Baruah, J., Nath, B.K., Sharma, R. et al. 2018. Recent trends in the pretreatment of lignocellulosic biomass for value-added products. Frontiers in Energy Research 6: 141.

Bos, A. and Hamelinck, C. 2014. Greenhouse gas impact of marginal fossil fuel use, 158.

Castro, A.M., Castilho, L.R. and Freire, D.M. 2015. Performance of a fixed-bed solid-state fermentation bioreactor with forced aeration for the production of hydrolases by Aspergillus awamori. Biochemical Engineering Journal 93: 303–308.

Cheirsilp, B. and Kitcha, S. 2015. Solid state fermentation by cellulolytic oleaginous fungi for direct conversion of lignocellulosic biomass into lipids: Fed-batch and repeated-batch fermentations. Industrial Crops and Products 66: 73–80.

Chen, H. 2013. Anaerobic solid-state fermentation. In Modern Solid State Fermentation, 199–242.

Chilakamarry, C.R., Sakinah, A.M., Zularisam, A.W. et al. 2022. Advances in solid-state fermentation for bioconversion of agricultural wastes to value-added products: Opportunities and challenges. Bioresource Technology 343: 126065.

Colla, E., Santos, L.O., Deamici, K. et al. 2017. Simultaneous production of amyloglucosidase and exo-polygalacturonase by Aspergillus niger in a rotating drum reactor. Applied Biochemistry and Biotechnology 181(2): 627–637.

Costa, J.M., Ampese, L.C., Ziero, H.D.D. et al. 2022. Apple pomace biorefinery: Integrated approaches for the production of bioenergy, biochemicals, and value-added products–an updated review. Journal of Environmental Chemical Engineering 108358.

Das, R.K., Brar, S.K. and Verma, M. 2015. A fermentative approach towards optimizing directed biosynthesis of fumaric acid by Rhizopus oryzae 1526 utilizing apple industry waste biomass. Fungal Biology 119(12): 1279–1290.

Demir, H. and Tari, C. 2016. Bioconversion of wheat bran for polygalacturonase production by Aspergillus sojae in tray type solid-state fermentation. International Biodeterioration & Biodegradation 106: 60–66.

Derakhti, S., Shojaosadati, S.A., Hashemi, M. et al. 2012. Process parameters study of α-amylase production in a packed-bed bioreactor under solid-state fermentation with possibility of temperature monitoring. Preparative Biochemistry and Biotechnology 42(3): 203–216.

Dey, T.B. and Kuhad, R.C. 2014. Enhanced production and extraction of phenolic compounds from wheat by solid-state fermentation with Rhizopus oryzae RCK2012. Biotechnology Reports 4: 120–127.

Dhillon, G.S., Brar, S.K., Valero, J.R. et al. 2011. Bioproduction of hydrolytic enzymes using apple pomace waste by A. niger: Applications in biocontrol formulations and hydrolysis of chitin/chitosan. Bioprocess and Biosystems Engineering 34(8): 1017–1026.

Díaz, A.B., Alvarado, O., de Ory, I. et al. 2013. Valorization of grape pomace and orange peels: Improved production of hydrolytic enzymes for the clarification of orange juice. Food and Bioproducts Processing 91(4): 580–586.

Díaz, A.B., De Ory, I., Caro, I. et al. 2009, May. Solid state fermentation in a rotating drum bioreactor for the production of hydrolytic enzymes. In The Ninth International Conference on Chemical and Process Engineering, Rome.

Dilipkumar, M., Rajamohan, N. and Rajasimman, M. 2013. Inulinase production in a packed bed reactor by solid state fermentation. Carbohydrate Polymers 96(1): 196–199.

Dilipkumar, M., Rajasimman, M. and Rajamohan, N. 2014. Utilization of copra waste for the solid state fermentative production of inulinase in batch and packed bed reactors. Carbohydrate Polymers 102: 662–668.

Duan, Y.Y. and Chen, H.Z. 2012. Effect of three-phase structure of solid-state fermentation substrates on its transfer properties. CIESC Journal 63(4): 1204–1210.

Duque-Acevedo, M., Belmonte-Urena, L.J., Cortés-García, F.J. et al. 2020. Agricultural waste: Review of the evolution, approaches and perspectives on alternative uses. Global Ecology and Conservation (22): e00902.

Duque-Acevedo, M., Belmonte-Ureña, L.J., Cortés-García, F.J. et al. 2020. Recovery of agricultural waste biomass: A sustainability strategy for moving towards a circular bioeconomy. Handbook of Solid Waste Management: Sustainability through Circular Economy, 1–30.

Economou, C.N., Aggelis, G., Pavlou, S. et al. 2011. Single cell oil production from rice hulls hydrolysate. Bioresource Technology 102(20): 9737–9742.

Elizabeth, O.V. 2010. Naringinase production from filamentous fungi using grapefruit rind in solid state fermentation. African Journal of Microbiology Research 4(19): 1964–1969.

Fakas, S., Makri, A., Mavromati, M. et al. 2009. Fatty acid composition in lipid fractions lengthwise the mycelium of Mortierella isabellina and lipid production by solid state fermentation. Bioresource Technology 100(23): 6118–6120.

Farinas, C.S., Vitcosque, G.L., Fonseca, R.F. et al. 2011. Modeling the effects of solid state fermentation operating conditions on endoglucanase production using an instrumented bioreactor. Industrial Crops and Products 34(1): 1186–1192.

Fath, M. and Fazaelipoor, M.H. 2015. Production of proteases in a novel trickling tray bioreactor. Waste and Biomass Valorization 6(4): 475–480.

Finkler, A.T.J., Biz, A., Pitol, L.O. et al. 2017. Intermittent agitation contributes to uniformity across the bed during pectinase production by Aspergillus niger grown in solid-state fermentation in a pilot-scale packed-bed bioreactor. Biochemical Engineering Journal 121: 1–12.

Fujian, X., Hongzhang, C. and Zuohu, L. 2002. Effect of periodically dynamic changes of air on cellulase production in solid-state fermentation. Enzyme and Microbial Technology 30(1): 45–48.

Hui, L., Wan, C., Hai-Tao, D. et al. 2010. Direct microbial conversion of wheat straw into lipid by a cellulolytic fungus of Aspergillus oryzae A-4 in solid-state fermentation. Bioresource Technology 101(19): 7556–7562.

Jangbua, P., Laoteng, K., Kitsubun, P. et al. 2009. Gamma-linolenic acid production of Mucor rouxii by solid-state fermentation using agricultural by-products. Letters in Applied Microbiology 49(1): 91–97.

John, R.P., Nampoothiri, K.M. and Pandey, A. 2006. Solid-state fermentation for L-lactic acid production from agro wastes using Lactobacillus delbrueckii. Process Biochemistry 41(4): 759–763.

Kim, S. and Kim, C.H. 2012. Production of cellulase enzymes during the solid-state fermentation of empty palm fruit bunch fiber. Bioprocess and Biosystems Engineering 35(1): 61–67.

Kiran, E.U., Trzcinski, A.P. and Liu, Y. 2014. Glucoamylase production from food waste by solid state fermentation and its evaluation in the hydrolysis of domestic food waste. Biofuel Research Journal 1(3): 98–105.

Koullas, D.P., Mamma, D., Kekos, D. et al. 1996. Innovative concepts in agricultural residues utilisation for sustainable development in developing countries (icarus). In Biomass for Energy and the Environment, 1498–1503.

Kuhar, F., Castiglia, V. and Levin, L. 2015. Enhancement of laccase production and malachite green decolorization by co-culturing Ganoderma lucidum and Trametes versicolor in solid-state fermentation. International Biodeterioration & Biodegradation 104: 238–243.

Kumar, D., Verma, R. and Bhalla, T.C. 2010. Citric acid production by Aspergillus niger van. Tieghem MTCC 281 using waste apple pomace as a substrate. Journal of Food Science and Technology 47(4): 458–460.

Leite, P., Sousa, D., Fernandes, H. et al. 2021. Recent advances in production of lignocellulolytic enzymes by solid-state fermentation of agro-industrial wastes. Current Opinion in Green and Sustainable Chemistry 27: 100407.

Li, M., He, B., Chen, Y. et al. 2021. Physicochemical properties of nanocellulose isolated from cotton stalk waste. ACS Omega 6(39): 25162–25169.

Liu, Y., Li, C., Meng, X. et al. 2013. Biodiesel synthesis directly catalyzed by the fermented solid of Burkholderia cenocepacia via solid state fermentation. Fuel Processing Technology 106: 303–309.

Machado De-Melo, A.A., Almeida-Muradian, L.B.D., Sancho, M.T. et al. 2018. Composition and properties of *Apis mellifera* honey: A review. Journal of Apicultural Research 57(1): 5–37.

Maciel, M., Ottoni, C., Santos, C. et al. 2013. Production of polygalacturonases by Aspergillus section Nigri strains in a fixed bed reactor. Molecules 18(2): 1660–1671.

Manan, M.A. and Webb, C. 2016. Multi-enzymes production studies in single tray solid state fermentation with opened and closed system. Journal of Life Sciences 10: 342–356.

McKinney, K., Combs, J., Becker, P. et al. 2015. Optimization of phytase production from Escherichia coli by altering solid-state fermentation conditions. Fermentation 1(1): 13–23.

Melikoglu, M., Lin, C.S.K. and Webb, C. 2015. Solid state fermentation of waste bread pieces by Aspergillus awamori: Analysing the effects of airflow rate on enzyme production in packed bed bioreactors. Food and Bioproducts Processing 95: 63–75.

Motte, J.C., Trably, E., Escudié, R. et al. 2013. Total solids content: A key parameter of metabolic pathways in dry anaerobic digestion. Biotechnology for Biofuels 6(1): 1–9.

Muhammed, M. and Nambisan, P. 2012. Optimization of lignin peroxidase, manganese peroxidase, and lac production from ganoderma lucidum under solid state fermentation of pineapple leaf. BioResources 8(1): 250–271.

Naveena, B.J., Altaf, M., Bhadriah, K. et al. 2005. Selection of medium components by Plackett–Burman design for production of L (+) lactic acid by Lactobacillus amylophilus GV6 in SSF using wheat bran. Bioresource Technology 96(4): 485–490.

Nguyen, T.A.D., Kim, K.R., Han, S.J. et al. 2010. Pretreatment of rice straw with ammonia and ionic liquid for lignocellulose conversion to fermentable sugars. Bioresource Technology 101(19): 7432–7438.

Noratiqah, K., Madihah, M.S., Aisyah, B.S. et al. 2013. Statistical optimization of enzymatic degradation process for oil palm empty fruit bunch (OPEFB) in rotary drum bioreactor using crude cellulase produced from Aspergillus niger EFB1. Biochemical Engineering Journal 75: 8–20.

Oliveira, F., Salgado, J.M., Abrunhosa, L. et al. 2017. Optimization of lipase production by solid-state fermentation of olive pomace: from flask to laboratory-scale packed-bed bioreactor. Bioprocess and Biosystems Engineering 40(7): 1123–1132.

Özdemir, S., Matpan, F., Okumus, V. et al. 2012. Isolation of a thermophilic Anoxybacillus flavithermus sp. nov. and production of thermostable α-amylase under solid-state fermentation (SSF). Annals of Microbiology 62(4): 1367–1375.

Pitol, L.O., Biz, A., Mallmann, E. et al. 2016. Production of pectinases by solid-state fermentation in a pilot-scale packed-bed bioreactor. Chemical Engineering Journal 283: 1009–1018.

Planinić, M., Zelić, B., Čubel, I. et al. 2016. Corn forage biological pretreatment by Trametes versicolor in a tray bioreactor. Waste Management & Research 34(8): 802–809.

Poletto, P., da Rocha Renosto, D., Baldasso, C. et al. 2015. Activated charcoal and microfiltration as pretreatment before ultrafiltration of pectinases produced by Aspergillus niger in solid-state cultivation. Separation and Purification Technology 151: 102–107.

Pudjiastuti, L., Iswanto, T., Altway, A. et al. 2019, June. Lignocellulosic properties of coffee pulp waste after alkaline hydrogen peroxide treatment. In IOP Conference Series: Materials Science and Engineering 543(1): 012081. IOP Publishing.

Qi, B. and Yao, R. 2007. L-Lactic acid production from Lactobacillus casei by solid state fermentation using rice straw. BioResources 2(3): 419–429.

Raimbault, M. 1998. General and microbiological aspects of solid substrate fermentation. Electronic Journal of Biotechnology 1(3): 26–27.

Robledo, A., Aguilera-Carbó, A., Rodriguez, R. et al. 2008. Ellagic acid production by Aspergillus niger in solid state fermentation of pomegranate residues. Journal of Industrial Microbiology and Biotechnology 35(6): 507–513.

Rodrigues, A., Leitao, A., Moreira, S. et al. 2010. Advancement and comparative profiles in the production technologies using solid-state and submerged fermentation for microbial cellulases. Enzyme and Microbial Technology 46: 541–549.

Rodriguez-Duran, L.V., Contreras-Esquivel, J.C., Rodríguez, R. et al. 2011. Optimization of tannase production by Aspergillus niger in solid-state packed-bed bioreactor. Journal of Microbiology and Biotechnology 21(9): 960–967.

Rudra, S.G., Nishad, J., Jakhar, N. et al. 2015. Food industry waste: Mine of nutraceuticals. Int. J. Sci. Environ. Technol., 4(1): 205–229.

Ruiz, H.A., Rodríguez-Jasso, R.M., Rodríguez, R. et al. 2012. Pectinase production from lemon peel pomace as support and carbon source in solid-state fermentation column-tray bioreactor. Biochemical Engineering Journal 65: 90–95.

Saba, I. and Hamid, M. 2013. Production of alkaline protease by Bacillus subtilis using solid state fermentation. African Journal of Microbiology Research 7(16): 1558–1568.

Sadh, P.K., Duhan, S. and Duhan, J.S. 2018. Agro-industrial wastes and their utilization using solid state fermentation: A review. Bioresources and Bioprocessing 5(1): 1–15.

Saeed, S., Aslam, S., Mehmood, T. et al. 2021. Production of gallic acid under solid-state fermentation by utilizing waste from food processing industries. Waste and Biomass Valorization 12(1): 155–163.

Saithi, S. and Tongta, A. 2016. Phytase production of Aspergillus niger on soybean meal by solid-state fermentation using a rotating drum bioreactor. Agriculture and Agricultural Science Procedia 11: 25–30.

Salar, R.K., Certik, M. and Brezova, V. 2012. Modulation of phenolic content and antioxidant activity of maize by solid state fermentation with Thamnidium elegans CCF 1456. Biotechnology and Bioprocess Engineering 17(1): 109–116.

Salgadoa, J.M., Maxa, B., Rodríguez-Solanaa, R. et al. 2012. Purification of ferulic acid solubilized from agroindustrial wastes and further conversion into 4-vinyl guaiacol by Streptomyces setonii using solid state fermentation. Industrial Crops and Products 39: 52–61.

Saratale, G.D., Kshirsagar, S.D., Sampange, V.T. et al. 2014. Cellulolytic enzymes production by utilizing agricultural wastes under solid state fermentation and its application for biohydrogen production. Applied Biochemistry and Biotechnology 174(8): 2801–2817.

Schmidt, C.G., Gonçalves, L.M., Prietto, L. et al. 2014. Antioxidant activity and enzyme inhibition of phenolic acids from fermented rice bran with fungus Rizhopus oryzae. Food Chemistry 146: 371–377.

Sepúlveda, L., Aguilera-Carbó, A., Ascacio-Valdés, J.A. et al. 2012. Optimization of ellagic acid accumulation by Aspergillus niger GH1 in solid state culture using pomegranate shell powder as a support. Process Biochemistry 47(12): 2199–2203.

Sharma, A., Vivekanand, V. and Singh, R.P. 2008. Solid-state fermentation for gluconic acid production from sugarcane molasses by Aspergillus niger ARNU-4 employing tea waste as the novel solid support. Bioresource Technology 99(9): 3444–3450.

Silveiraa, J.S., Duranda, N., Lacour, S. et al. 2019. Solid-state fermentation as a sustainable method for coffee pulp treatment and production of an extract rich in chlorogenic acids. Food and Bioproducts Processing 115: 175–184.

Singh, O.V., Jain, R.K. and Singh, R.P. 2003. Gluconic acid production under varying fermentation conditions by Aspergillus niger. Journal of Chemical Technology & Biotechnology: International Research in Process Environmental & Clean Technology 78(2-3): 208–212.

Soccol, C.R. and Vandenberghe, L.P.S. 2003. Overview of applied solid-state fermentation in Brazil. Biochemical Engineering Journal 13: 205–218.

Sukumaran, R.K., Surender, V.J., Sindhu, R. et al. 2010. Lignocellulosic ethanol in India: Prospects challenges and feedstock availability. Bioresourc. Technol. 1014826–33.

Sukumprasertsri, M., Unrean, P., Pimsamarn, J. et al. 2013. Fuzzy logic control of rotating drum bioreactor for improved production of amylase and protease enzymes by Aspergillus oryzae in solid-state fermentation. Journal of Microbiology and Biotechnology 23(3): 335–342.

Tao, S., Zuohu, L. and Deming, L. 1996. A novel design of solid state fermenter and its evaluation for cellulase production by Trichoderma viride SL-1. Biotechnology Techniques 10(11): 889–894.

Tian, M., Wai, A., Guha, T.K. et al. 2018. Production of endoglucanase and xylanase using food waste by solid-state fermentation. Waste and Biomass Valorization 9: 2391–2398.

Torino, M.I., Limón, R.I., Martínez-Villaluenga, C. et al. 2013. Antioxidant and antihypertensive properties of liquid and solid state fermented lentils. Food Chemistry 136(2): 1030–1037.

Torrado, A.M., Cortés, S., Salgado, J.M. et al. 2011. Citric acid production from orange peel wastes by solid-state fermentation. Brazilian Journal of Microbiology 42: 394–409.

Tsouko, E., Kachrimanidou, V., Dos Santos, A.F. et al. 2017. Valorization of by-products from palm oil mills for the production of generic fermentation media for microbial oil synthesis. Applied Biochemistry and Biotechnology 181(4): 1241–1256.

Umsza-Guez, M.A., Díaz, A.B., Ory, I.D. et al. 2011. Xylanase production by Aspergillus awamori under solid state fermentation conditions on tomato pomace. Brazilian Journal of Microbiology 42: 1585–1597.

Vasudevan, U.M., Jaiswal, A.K., Krishna, S. et al. 2019. Thermostable phytase in feed and fuel industries. Bioresource Technology 278: 400–407.

Virtanen, V., Nyyssölä, A., Vuolanto, A. et al. 2008. Bioreactor for solid-state cultivation of Phlebiopsis gigantea. Biotechnology Letters 30(2): 253–258.

Xiao, Y., Liu, Y., Wang, X. et al. 2019. Cellulose nanocrystals prepared from wheat bran: Characterization and cytotoxicity assessment. International Journal of Biological Macromolecules 140: 225–233.

Xie, L., Chen, H.M. and Yang, J.B. 2013. Conidia production by Beauveria bassiana on rice in solid-state fermentation using tray bioreactor. In Advanced Materials Research 610: 3478–3482.

Yadegary, M., Hamidi, A., Alavi, S.A. et al. 2013. Citric acid production from sugarcane bagasse through solid state fermentation method using Aspergillus niger mold and optimization of citric acid production by Taguchi method. Jundishapur Journal of Microbiology 6(9): 1–6.

Yafetto, L. 2022. Application of solid-state fermentation by microbial biotechnology for bioprocessing of agro-industrial wastes from 1970 to 2020: A review and bibliometric analysis. Heliyon. 8(3): e09173: 1–17.

Yang, G., Hou, L.L. and Zhang, F.L. 2011. Study on the solid-state fermentation conditions for producing thermostable xylanase feed in a pressure pulsation bioreactor. In Advanced Materials Research 236: 72–76.

Yang, S.Y., Ji, K.S., Baik, Y.H. et al. 2006. Lactic acid fermentation of food waste for swine feed. Bioresour. Technol. 97: 1858–1864.

Zepf, F. and Jin, B. 2013. Bioconversion of grape marc into protein rich animal feed by microbial fungi. Chem. Eng. Proc. Techniq. 1: 1011.

Zhang, C., Xiao, G., Peng, L. et al. 2013. The anaerobic co-digestion of food waste and cattle manure. Bioresour. Technol. 129: 170–176.

Zhang, J. and Hu, B. 2012. Solid-state fermentation of Mortierella isabellina for lipid production from soybean hull. Applied Biochemistry and Biotechnology 166(4): 1034–1046.

Zhang, Y., Liu, J., Zhou, Y. et al. 2014. Spore production of Clonostachys rosea in a new solid-state fermentation reactor. Applied Biochemistry and Biotechnology 174(8): 2951–2959.

Zhao, X., Huang, X., Yao, J. et al. 2015. Fungal growth and manganese peroxidase production in a deep tray solid-state bioreactor and *in vitro* decolorization of poly R-478 by MnP. Journal of Microbiology and Biotechnology 25(6): 803–813.

Zheng, Y., Lee, C., Yu, C. et al. 2012. Ensilage and bioconversion of grape pomace into fuel ethanol. Journal of Agricultural and Food Chemistry 6: 11128–11134.

Zhu, Y. and Tramper, J. 2013. Koji–where East meets West in fermentation. Biotechnology Advances 31(8): 1448–1457.

11

SOLID WASTE TREATMENT USING VERMICOMPOSTING
ITS CURRENT USE AND FUTURE PERSPECTIVES

Vaishnavi Sharma,[1] *Shivani Maddirala,*[2] *Sudipa Bhadra,*[2]
P. Venkateswara Rao[1] **and** *Surajbhan Sevda*[2,*]

1. INTRODUCTION

Solid waste management is a very crucial challenge for everyone around the world. Due to increasing population, industrialisation, and urbanisation, municipal solid waste has emerged as a serious environmental issue (Alshehrei and Ameen 2021). According to the World Bank report, waste generated per person per day averages 0.74 kilograms. High-income countries are found to produce about 650 million tonnes of the world's waste. According to a 2018 research, worldwide garbage is predicted to increase to 3.4 billion tonnes by 2050. The output of garbage per capita is closely correlated with increases in economic growth and fast urbanisation. Municipal trash management is therefore significantly more expensive in cities. Waste management is the largest budget item in low-income countries, accounting for over 20% of municipal spending, more than 10% in middle-income countries, and just 4% in high-income nations (Alshehrei and Ameen 2021). India is a country of farmers, producing more biodegradable waste whose 50–60% can be easily converted to biogas for various uses and the remaining can be converted into manure.

Composting may be done in a variety of methods, with aerobic and anaerobic composting being the two main forms. In contrast to anaerobic composting, which lacks oxygen, aerobic composting uses bacteria to break down organic waste in the presence of oxygen. However, these more straightforward

---

[1] Department of Civil Engineering, National Institute of Technology Warangal, Warangal-506004, India.
[2] Environmental Bioprocess Laboratory, Department of Biotechnology, National Institute of Technology Warangal, Warangal-506004, India.
* Corresponding author: sevdasuraj@gmail..com, sevdasuraj@nitw.ac.in

methods of composting are transformed into the right automated approach for better and more effective results (Chand Malav et al. 2020):

- Bangalore technique (waste is placed in designated pits and layered at the end with soil material under anaerobic conditions, taking 4–6 months for the trash to breakdown).
- The Indore technique, which involves filling a pit with alternate layers of garbage and dirt and stirring it to maintain the aerobic state.
- Windrow technique (each windrow is a particular size of 3 m 2 m 1.5 m and can be used to transform MSW into manure in an aerobic environment).
- Vermicomposting (final stabilisation of organic waste into end product using several kinds of earthworms) (Chand Malav et al. 2020).

Vermicomposting is an age-long method, which has been in practice since the start of the 1970s. In recent years, vermitechnology has emerged as a leading contender to replace conventional composting as a rapid, simple, energy-efficient, economical, waste-free, and environmentally friendly method of producing natural fertiliser from organic solid waste (Soobhany 2019). Composting method involving worms (i.e., vermicomposting) produces manure of high quality. Worms are introduced to the pit where the waste is collected and a proper atmosphere is maintained for them. The compost takes about two months to two years to get ready depending on the composites and ingredients, the atmosphere, and the procedure followed to let the worms work. Vermicomposting frequently yields a better-quality homogeneous product than traditional composting, with less mass, a shorter processing time, a high humus content, less phytotoxicity, and a higher fertiliser value in the microbiologically active manure, or vermicompost (Soobhany 2019). Greenhouse gas (GHG) emissions are comparatively less in vermicomposting than in the traditional composting method. Hence, it is an advanced way of composting and is eco-friendlier.

Vermicomposting provides a number of advantages over other waste management techniques, including the ability to be done both indoors and outdoors and year-round composting. When compared to other composts, this method makes it possible to quickly get organic nutrient sources for crops that are effective from a nutritional, physical, and biochemical standpoint (Alshehrei and Ameen 2021). Composting using worms is a well-known low-cost technological approach for handling or treating organic waste. Vermicomposting produced an enriched compost with high levels of N, K, and P, which in turn led to a drop in the concentration of heavy metals, according to a comparative study between traditional composting and vermicomposting (Figure 1). It is a method that turns solid waste into usable compost, thereby replacing chemical fertilisers and lowering pollutants in the process. As a result, it is crucial to adopt cutting-edge waste to energy and vermicomposting

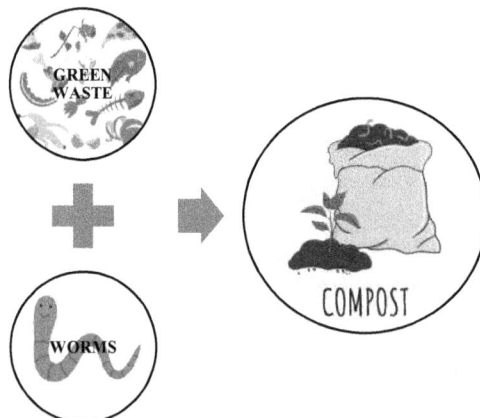

Figure 1. Illustration of vermicomposting to form manure for plants.

processes rather than these widely used waste management techniques (Alshehrei and Ameen 2021, Soobhany 2019).

The possible advancements are later suggested at the end of the study. The main objectives of this are as follows:

1. To understand the mechanism of vermicomposting in different applications.
2. To use that knowledge to suggest more ways of efficient vermicomposting.
3. To explore the application of vermicomposting in solid waste management.

The initial study involved the reading of different works involving vermicomposting, e.g., vermicomposting carried out on corn waste, sludge being treated using worms, etc. In addition, all the factors influencing the process of compost formation were studied. Later, a comparative study was carried forward to analyse the differences between different processes and the knowledge that was used to apply vermicomposting in new industries.

The following limitations of the current study can be listed:

1. The study was intended as a preliminary investigation of multiple factors affecting vermicomposting to obtain good compost. Therefore, this was done by studying the work done to date. The results of the analysis are aimed to be interpreted, at this stage, as qualitative, not quantitative.
2. To find out the actual microbial process, the study relies on the already available results from various research.

In this paper, a comparative study was carried out on vermicomposting practices. This critical review explains the scope of the application of vermicomposting on large scale in different sectors. Even though the concept has largely evolved with time, with the push of sustainability practices, it has got a new pace.

2. MECHANISM OF VERMICOMPOSTING

2.1 Vermiculture

In the process of composting, organic solid wastes break down into humus, sometimes referred to as compost, in the presence of air. This humus-like substance works wonders for plants as fertiliser. Worm composting, sometimes referred to as "vermicomposting," is a significant form of composting that involves an improved humification (decomposition) process carried out in the company of earthworms under non-thermophilic circumstances. In this vermicomposting process, organic solid waste is stabilised into a dark-coloured soil conditioner compost that is nutrient-rich and earthy-smelling and contains a large amount of both major and micronutrients (Alshehrei and Ameen 2021).

The fecundity of countless earthworms in waste material is called vermiculture and the process that occurs because of their presence is called vermicomposting. It is a fast-growing industry as both the raw materials required are costless and the products thus obtained are of great value. Vermicomposting is a non-thermophilic decomposition of organic waste carried out by earthworms and microorganisms (Bhata et al. 2018). It can also make the substrate free from parasite eggs and faecal coliforms (Parseh et al. 2021).

Vermicomposting is carried out in any open space, containers, or in any space, containers, or vermireactor (Figure 2). The waste is collected and placed in the reactor and earthworms are periodically introduced to fasten the process of vermicomposting (Yuvaraj et al. 2020).

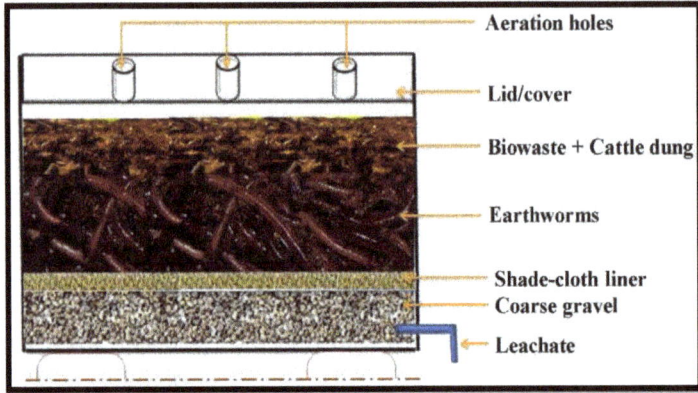

Figure 2. Traditional Vermireactor (Yuvaraj et al. 2020).

Vermicomposting smart closed reactor is a recent study describing an innovation in the traditional vermicomposting method, which brings forth various advantages like odour control, units can be set in public places, and the maturity time is faster for the substrate (Ghorbani et al. 2021).

2.2 Microbial process

In vermicomposting, the earthworms eat up the waste and break them down to smaller pieces, hereby increasing its surface area which eventually lets the microbes carry out the decomposition faster. The entire process depends on the time taken by the earth worms in their gut transit process (Figure 3) (Yuvaraj et al. 2020). The earthworms ventilate the waste and shred the waste to accelerate the microbial activity. The loss of organic matter in the substrate is attributed to the synergistic effect of microbes with worms (Chea et al. 2020).

The waste materials often get short of minerals like nitrogen, phosphorus, potassium, etc., which are formed after the earthworm excretes the vermicast (Figure 4). The earthworm secretes various enzymes in its guts like lipases, proteases, cellulases, amylases, chitinases, etc., which helps to convert the organics into plant-useable forms (Bhata et al. 2018).

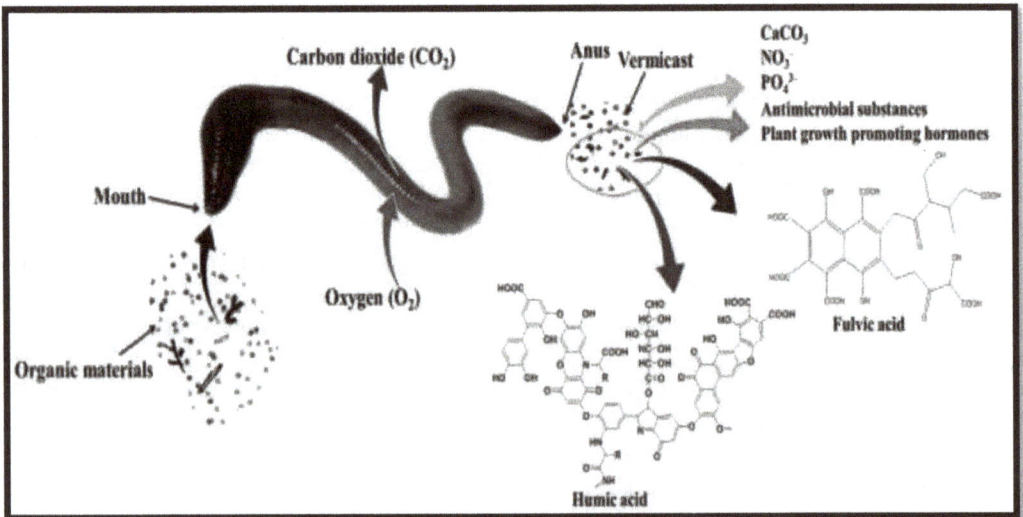

Figure 3. The mechanistic way of organic waste materials' consumption and decomposition by earthworms (Yuvaraj et al. 2020).

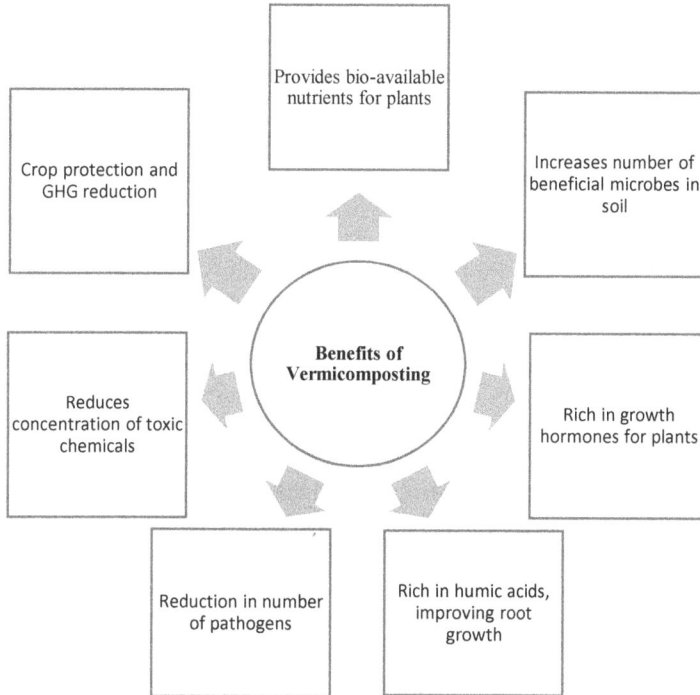

Figure 4. Benefits of Vermicomposting.

The decomposition process is mainly divided into two stages, i.e., the active phase and the maturation phase. The active phase involves the enhancement of the mineralization process by earthworms and organic waste. The maturation phase deals with the exertion of vermicast which gives stratification to the waste (Yuvaraj et al. 2020).

The final product formed as vermicompost is rich in nutrients, microbe-free and odourless. This is possible because the earthworms release coelomic fluids which kill the pathogens (Bhata et al. 2018).

3. MICROBIAL COMMUNITIES IN VERMICOMPOSTING

3.1 Earthworm genre

Vermicomposting is a kind of composting in which earthworms and aerobic microorganisms work together in a synergistic manner to break down and stabilise organic waste. Biodegradable wastes are mostly broken down by extracellular enzymatic activities carried out by microorganisms. Waste is secondary, degraded by earthworms, who consume partially decomposed materials, break them down in their guts, and then excrete vermicast. Particle size reduction and organic matter digestion in the worms' intestines produce the vermicast. It is a granular, typically odourless product with reduced levels of pollutants that is enriched with vital water-soluble soil nutrients. The vermicast may be used in agriculture as a green manure and organic fertiliser (Nanda and Berruti 2021).

The first stage in vermicomposting is to choose the right kind of earthworm and to choose them correctly since they determine how quickly waste stabilises. Many different kinds of earthworms have the potential to be utilised in sludge stabilisation and waste treatment procedures. The species of earthworms that are capable of naturally colonising organic waste, having high rates of organic matter consumption, digestion, and assimilation, being able to withstand a wide range of environmental stress, having high reproductive rates by producing numerous cocoons with quick hatch times, and having rapid growth and maturation rates of hatchlings to adults are suitable for use in the vermicomposting process (Mohee and Soobhany 2014).

Microbial process efficiency is dependent on the type of earthworm used and the kind of gut fluids it secrets throughout the process of decomposition. The overall time required for the active phase and maturity phase to get over depends on the number of earthworms present in the waste, the type of earthworms used (epigeic, anecic, endogeic), and the quality, and quantity of the waste collected (Yuvaraj et al. 2020).

With their high rates of intake, digestion, and assimilation of organic matter, tolerance to a variety of environmental conditions, short life cycles, high reproduction rates, endurance, and tolerance of handling, epigeic species of earthworms have strong potential for vermicomposting. Only five earthworm species, *Eisenia andrei* (Bouché), *Eisenia fetida* (Savigny), *Dendrobaena veneta* (Savigny), and, to a lesser extent, *Perionyx excavatus* (Perrier), and *Eudrilus eugeniae*, exhibit all of these traits (Kinberg). The ability of certain epigeic species of earthworms to stabilise organic waste and create vermicompost has been studied. The enormous populations of microorganisms required for the growth and reproduction of species of earthworms belong to the genera *Eisenia*, *Eudrilus*, *Dendrobaena*, *Perionyx*, and *Pheretima* (Mohee and Soobhany 2014). In Asian nations, where indigenous earthworm species including *Eisenia fetida*, *Eudrilus eugeniae*, *Lampito mauritii*, *Perionyx ceylanensis*, and *Perionyx excavatus* have been employed, vermicomposting has attracted a lot of interest (Nanda and Berruti 2021).

Eisenia Andrei is an institutionalized earthworm species used for vermicomposting since it has high utilization, maceration of organic matter, is immune to many environmental aspects, ephemeral, and has high reproductive rates (Gómez-Brandón 2021). When it comes to turning organic materials like food scraps, yard trash, agricultural waste, MSW, etc., into high-quality compost, Eisenia fetida excels. Additionally, it plays a bigger part in securing heavy metals and lessens their bioavailability through the bio-mineralization process. It is important to highlight that the earthworm's skin tissues release extracellular polymeric material (EPS) in response to metal stress, which is in charge of binding with the heavy metals and reducing their mobility by immobilising them on the earthworm's skin (Alshehrei and Ameen 2021). This EPS also functions as an enzyme for the bacterial population growth and waste decomposition. Additionally, it has been discovered that such EPS, which mostly consists of carbs and protein, increase the compost's quality and nutritional value. Earthworms now play a very important part in the vermicomposting process and produce high-quality compost as a result (Alshehrei and Ameen 2021).

3.2 Microbes involved in vermicomposting

It is well recognised that microbes play a crucial part in the biodegradation of complex chemical compounds into simpler ones. The pH and rearing substrate have an impact on the microbial community, according to functional profiling of the microbiota in casts of Eisenia fetida during vermicomposting (Alshehrei and Ameen 2021). Higher activities of β-galactosidase, dehydrogenases, proteases, and phosphatases were observed during the vermicomposting of duckweed, according to Gusain and Suthar (2020). Additionally, it has been discovered that adding nano-carbon or biochar to an amendment dramatically increased microbial activity, resulting in a greater rate of breakdown and enzymatic activity (Alshehrei and Ameen 2021).

Earthworms improve the respiration of the organic mass, which provides it with more oxygen which results in effective microbial activities. Fragmented mass (food for microbes) and availability of nutrients increases the number of microbes rapidly (Yuvaraj et al. 2020). Functional groups present in the substrate are broken down into smaller pieces by microbes thus enhancing the vermicomposting process (Sun et al. 2020). When exposed to metal stress, earthworms release a mucus-like material that helps to reduce the bioavailability of metals by forming organic-metallic complexes and bioaccumulates them on the surface of their tissue. It's also important to note that the EPS released by earthworms and the microbial communities they connect with during vermicomposting have been shown to slow metal mobility through enzymatic activity in processes related to the gut and cast (Alshehrei and Ameen 2021).

The inner side of the gut of earthworms secrets some enzymes which instigate the growth of microorganisms. These microorganisms can be aerobic or anaerobic. Through complex internal processes, these microbes are eliminated and lastly, an economically feasible vermicast is obtained that contributes to the compost. The studies have also confirmed that earthworms can increase the population of microbes by eight times (Yuvaraj et al. 2020).

The quality of compost depends on the type of defecation excreted out by the worms used in vermicomposting. The stability of the compost is decided by the excreta of the worms. In general, earthworms produce four types of faecal matter, i.e., (i) andogeic, anecic earthworms have flattened or globular vermicasts, (ii) few of them even produce the irregular shape of vermicast, (iii) many earthworms have cylindrical vermicast, (iv) few earthworm species including Andogeic, Anecic, Epigeic have their casts in pellets form. This casting material affects the soil porosity, water holding capacity, affects microbial activities, and also makes available nutrients for plant growth (Yuvaraj et al. 2020).

4. EFFECTS OF SUBSTRATE TYPES OF VERMICOMPOSTING

Due to its technological simplicity, efficiency, and ease of integration into farming and waste management techniques, vermicomposting has been used in many nations. Vermicomposting has been suggested as a practical way to turn animal dung (such as pig or bovine manure) into nutrient-rich bio-fertilizer using a variety of earthworm species. Earthworms in various livestock manures acquired net biomass in the following order: sheep, donkey, buffalo, goat, cow, horse, and camel (Khoshnevisan et al. 2021). Vermicomposting is used as a suitable bioprocess to stabilise manure combined with crops' straws and husks, or combined with traditional composting to form an integrated composting-vermicomposting process for treating livestock manure and other agricultural wastes, in addition to the direct use of livestock manure for the production of bedding materials (Khoshnevisan et al. 2021).

4.1 Effect of biomass on earthworm activities

The physio-chemical parameters and biological parameters of the biomaterial used as a substrate can alter the inner digestive processes of earthworms, i.e., homogenization, metabolism, catabolism, microbial activities, etc. The type of cast can be granular, globular, etc., which depends on the genre of the earthworm being used (Yuvaraj et al. 2020).

4.2 Sewage sludge as a substrate

To incorporate the circular economy concept in agricultural fertilization, the use of sewage sludge is a new growing practice. The treatment of sewage sludge poses different environmental issues like leachate formation, methane production in landfill, uncontrolled fire in landfills, and different impacts on soil, air, water, etc. (Dermendzhieva et al. 2021).

Sewage sludge is a rich mixture of organic matter and microorganisms; hence, to gain maximum economic output, vermicomposting is practiced on SS. The gut fluids from earthworm break and decompose the sludge and forms the best compost rich in nutrients that are good for plant growth. The produced vermicompost is light in weight, and its weight loose gradually with less moisture content in it, and better smelling than sludge itself. The compost thus formed from SS is rich in N,

P, K, Mn, Cu, Zn, Ni, Cr, and Pb. The vermicompost does not alter the presence of heavy metals in the SS; however, it eliminates *E. Coli*. Though, it cannot be used on a large scale unless the soil is tested properly (Dermendzhieva et al. 2021).

Changes in the humidity level of the substrate plays a vital role in governing the substrate's microbial activities. Moisture content orders the biological changes in the substrate and also changes a lot of physiochemical characteristics of the compost. However, the moisture content selection criteria are not presently clear (Qin et al. 2021). Due to the heterogeneous nature of sewage sludge and different wastewater treatment carried on it, varied results are produced when vermicomposting is carried on different substrates (Domínguez et al. 2021).

4.3 Grape Marc as substrate

Grape Marc proves to be the finest substrate to perform vermicomposting. It provides optimal conditions for the growth and reproduction of microbes, to assure enough food and energy to hold a large population of worms. The compost formed after about 4 months is found to contain a high amount of micro and macronutrients, lower phenol content, and the compost also had low microbial activity at the end of the process (Gómez-Brandón 2020). Using grape marc as a substrate contributed to treating the highest quantity of waste produced in the wine industry and also prevented pollution of the environment (Gómez-Brandón 2021).

4.4 Dairy farm wastewater as substrate

The wastewater from the dairy farm contains large quantities of nitrogen and carbon content; it is continuously added to the system to ensure proper water content and to recover the water loss which would facilitate the growth of worms. To judiciously utilize the resources and to keep in check the zero release of rural organic waste, it is important to incorporate vermicomposting as the manure produced has high value (Liu et al. 2021). The earthworms affect the biochemistry of dairy water and thus carry out the decomposition process by even altering the microbe's population (Liu et al. 2021).

4.5 Lignocellulosic waste as substrate

The lignocellulosic wastes (flower waste, kitchen waste, food waste, agricultural crop residue) can also form good compost through vermicomposting. The production of compost from this substrate has numerous enzymatic production and microbial activities present in the final stage (Pandita et al. 2020).

4.6 Biogas residue as substrate

After biogas production, the organic material given out as sludge can be treated with the vermicomposting method to provide better yields as fertilizers. Biogas residue contains a high level of heavy metals like Cd which can be altered correctly with the help of worms (Sun et al. 2020). Mature earthworms have high strength to store heavy metals in harmless forms and thus can be treated well by a vermicomposting process (Sun et al. 2020).

4.7 Industrial waste as substrate

Industrial sludge could include a lot of putrescible material and pose a threat from pathogens. Conditioning and stabilising the industrial sludge through a pretreatment procedure is one potential technique to guarantee that it might be utilised on agricultural land again. Vermicomposting is one of the pretreatment techniques that might be used in this situation. With the inclusion of earthworms, vermicomposting is an option for biologically stabilising organic wastes. Industrial sewage might be quickly converted into mature organic fertiliser or vermicompost by vermicomposting. In order to determine the viability of this approach, this investigation analysed recent literature on using the vermicomposting process to handle industrial sludge (Lee et al. 2018). The industrial waste can be hazardous or non-hazardous. It comprises heavy metals, hydrocarbons (benzene, vinyl chloride), pesticides, organic chemicals (dioxins, PCBs), fly ash, etc. To deal with such toxic waste, incorporating vermicomposting method is an improved technique (Bhata et al. 2018).

The waste toxicity is decreased and high-quality of micro-, macro- nutrient-rich compost is formed through this substrate. The disposal problems of such waste can be sorted to a large extent through vermicomposting technique. To fasten the composting process of such waste, cow dung and other organic waste can be mixed with the substrate (Bhata et al. 2018). Cow dung is mixed with the substrate as it fastens the palatability of waste, increases the mineralization process, and also creates a healthy atmosphere for microbes and earthworms to carry out vermicomposting easily (Yuvaraj et al. 2020).

4.8 Paper mill sludge and tea waste as substrate

To assimilate the concept of circular economy, vermicomposting is carried out on this substrate. The lightening of the substrate is largely governed by the amount of cow dung mixed with the sample. The entire decomposition process is governed by factors like substrate composition, habitat, worm species and the enzymatic fluids it secretes, and the microbes involved in the process. Ash content also commands the stabilization and mineralization process of the substrate (Badhwar et al. 2020). Paper mill sludge mixed with good amendments like cow dung also create good-quality compost. If not given proper amendments, the degradation of polysaccharides in the substrate takes longer (Negi and Suthar 2018).

4.9 Spent mushroom as substrate

This substrate can be converted into compost using this method, but when analysed in terms of total and available nutrients, it is not of considerable quality. So, a better method to treat this kind of substrate is required. Involving worms with this substrate resulted in lowering the conductivity of the substrate which is not desirable. This proved that earthworms can be easily used for spent mushroom substrate, not for the original raw material used (Hřebečkováa et al. 2020).

4.10 Poultry waste as substrate

Involving vermicomposting in managing poultry waste not only generates a clean source of energy but also proves to be a better compost with high availability of nutrients for plant growth. Composting followed by hydration of poultry waste is carried as a pre-treatment for vermicomposting to be carried on. By doing this, the toxicity of the waste is reduced for earthworms to carry on the vermicomposting process easily. Carrying the pre-treatment before actual vermicomposting reduces the electrical conductivity of the substrate, and nitrogen level in the matter which helps the worms to survive better and decompose the matter quickly and effectively (Niedzialkoski et al. 2021).

4.11 Fly ash as substrate

Fly ash is the end product of coal-based industries. It is rich in heavy metals and its disposal poses a huge challenge for the consumers of coal. Its usual disposal is carried by wet process on land which is extremely harmful both for humans and the environment. Fly ash is found in an unsteady state and can be improved through vermicomposting. Vermicomposting when carried on fly ash converts the heavy toxic metals to nutrients in non-toxic forms which can easily be utilized by plants (Sohal et al. 2021).

4.12 Toxic weed as substrate

A sustainable approach to dealing with toxic weeds is vermicomposting. The agricultural waste when mixed with cow dung served as a feedstock and then using it for vermicomposting gives positive results. Vermicomposting when carried on toxic weed reduces its toxicity and converts it into nutrient-rich compost which has high quality. Vermicomposting is a more feasible and non-polluting sustainable approach to tackling weed waste management challenges (Sharma and Garg 2020).

4.13 Spent coffee grounds as substrate

Spent coffee grounds (SCG) is highly humid and has a small particle size. SCG when mixed with straw pallets forms compost by vermicomposting, and is found to reduce the caffeine content in the substrate to a large extent. If not dealt with SCG, it is disposed of in landfills or drains which has its effects on the environment (Hanc et al. 2021). Straw pellets also prove to be a good additive to deal with malt house waste and turn it into a value-added product (Hanc et al. 2020).

    The compost formed through this waste is enzymatically active and is also ready to use in less time. SCG when amended with different residues also gives effective results like loss of pH which is also very useful in practice. SCG in all is not a suitable substrate for vermicomposting because of its high toxic content although, when mixed with different amendments like straw pallets, which increase the aeration of the substrate, decreases the bulk density, and provides a good breeding environment for worms give fine results (Hanc et al. 2021).

4.14 Fresh human excreta as substrate

To manage the poor sanitation problems in densely populated urban areas, vermicomposting can play a major role. However, there has been less research on this topic, but the results so far produced show a high potential for the application. Worms ease the process of mass reduction and improve the degradation of substrates (Nyame et al. 2021).

4.15 Food waste as substrate

Vermicompost formed from food waste is high-quality manure, it promotes elongated root growth in plants, has higher germination index, and high cation exchange capacity, which can be utilized on a large scale. Promoting a circular economy in managing food waste from hotels, institutions, etc., will reduce their waste generation and will also produce high-quality compost which holds good value in the market (Hashimoto et al. 2021).

4.16 Rice straw and kitchen waste as substrate

One of the simplest methods to get rid of the excessive amount of biomass left in the fields is to make compost using rice straw. The key factors in producing compost are temperature, pH, the carbon to nitrogen ratio, and moisture content. The ideal parameters for temperature, pH, C-N ratio, and moisture content needed to produce compost are 40 to 60°C, 6.9 to 8.3, 25 to 30, and 50 to 65%, respectively. But in recent years, vermicomposting of rice straw has also drawn a lot of interest (Singh and Patel 2022). The growth of worms in the substrate is highly based on the composition of matter in the substrate. The studies showed that the increase in the percentage of rice straw in the substrate results in earthworm reproduction and growth in many aspects (Song et al. 2019).

The compost made from rice straw has a range of bacteria that improve the qualities of the soil and crops. Since the nutrients in the compost are released gradually, the amount of nutrients lost due to leaching is essentially non-existent. However, tasks like collecting rice straw, rotating compost piles, and distributing compost in agricultural fields may need a significant amount of labour and equipment, raising the overall cost. Compared to inorganic fertiliser, the compost may not be able to provide the crop or soil with the precise amount of nutrients. Challenge is that the majority of farmers aren't aware that rice straw may be turned into compost. This is due to the fact that many nations' farmers lack access to basic infrastructure and training through seminars, lectures, and hands-on demonstrations (Singh and Patel 2022).

4.17 Municipal solid waste as substrate

Elango et al. (2009) used *Bacillus megaterium* and *Pseudomonas fluorescens* to study the thermophilic composting of MSW. The composting of MSW was hastened by thermophilic bacteria, resulting in a 78% volume decrease in 40 days. The need for a high C/N ratio and moisture content was decreased due to proper substrate mixing throughout the composting process. Early composting phases cause the pH of the substrate to decrease because microbial hydrolysis produces organic acids. The pH was neutralised as a result of microbial fermentation, which changed the acids into CO_2. Additionally, the maturation process led to the formation of NH_3, which raised the compost's pH (Nanda and Berruti 2021).

Paul et al. (2011) used *Perionyx ceylanensis* earthworms to do vermicomposting of MSW with cow dung for 50 days. With a high breakdown rate of 78% and a decreased C/N ratio of 15.4, cow dung improved the vermicomposting of MSW. Additionally, vermicomposting enhanced the population of vital bacteria, fungi, and actinomycetes as well as the availability of NPK (nitrogen, phosphorous, and potassium), which is a nutrient important to plants. Earthworm assistance in the composting of MSW and cow dung also increased the rate of breakdown of cellulose and lignin to 37% and 12%, respectively (Nanda and Berruti 2021).

Environmental impact assessments and lifecycle analyses of several MSW streams and treatment techniques with potential energy recovery were carried out by Arafat et al. (2015). The MSW under investigation comprises food waste, yard waste, waste paper, wood, plastic, and textile waste. Anaerobic digestion, bio-landfills, composting, incineration, and gasification are the treatment procedures under investigation. This evaluation finds that recycling paper, wood, and plastics using thermochemical conversion technology is sustainable. Contrarily, it is best to anaerobically digest food and yard wastes, whereas cremation of textile wastes is best. Anaerobic digestion and gasification showed more environmental advantages than composting, which had the least favourable effects on the environment (Nanda and Berruti 2021).

4.18 Livestock solid waste as substrate

The study looked at nutrient recovery and vermicompost generation from native and foreign cow breeds' solid wastes using epigeic earthworms. Pre-decomposed faeces from Vechur native and exotic Jersey breeds was vermicomposted for 45 days (Cycle I) and 90 days (Cycle II) with *Perionyx excavatus* and *Eudrilus eugeniae*, as well as the corresponding controls without earthworms.

NPK, Ca, and micronutrients in vermicomposts from both substrates over Cycle I and II increased (P 0.05); however, pH, total organic carbon, C/N, and C/P ratios revealed decreases (P 0.05) above starting values for both earthworms. Vechur native faeces produced nutrient-rich vermicompost, and E. eugeniae was proven to be effective in vermicomposting cow solid wastes in 45 days. The study came to the conclusion that Cycle I is suitable for earthworm mass reproduction in Cycle II, nutrient recovery, and vermicompost production (Rini et al. 2020).

4.19 Corn waste as substrate

In semi-arid areas, where high temperature prevails, it is hard to provide proper moisture to the substrate throughout the process of vermicomposting. But when corn cob and straw were mixed as a substrate for worms, they produced good results. The compost formed can be effectively used as a fertilizer on field generating economy for the producer (Silva et al. 2021). To prevent such farm waste from being burnt in open when in excess, vermicomposting can be carried out to promote eco-friendly practices. The corn waste is mixed with amendments like goat manure and sawdust in the form of an organic substrate which gave a high quality composed in the maturity phase (Aili et al. 2021, Silva et al. 2021).

Various instrumental techniques, such as elemental analysis (CHNSO), ultraviolet-visible spectroscopy (UV/Vis), Fourier-transform infrared spectroscopy (FTIR), fluorescence spectroscopy, or thermogravimetric analysis, must be used to precisely analyse the stability, particularly the humification, of the biological maturity, which has been extensively investigated (TGA). In addition, the spectroscopic characterisation will be a rapid approach to learn more about and comprehend the vermicomposting process better. Thus, the emphasis should be retained on enhancing stability and maturity as well as innovative technologies that can ease the production process in order to improve the vermicomposting process (for example, pretreatment methods, short composting time) (Khoshnevisan et al. 2021).

Vermicompost is a more effective supplement to enhance and stimulate plant development than traditional compost and inorganic fertiliser because it has a usually higher availability of macronutrients and micronutrients. So, the potential for using vermicompost on agricultural crops is enormous. Vermicomposts should be used and extensively applied to agricultural land in the near future to replace or supplement inorganic fertilisers (Lim et al. 2015).

Table 1 shows the nutrient % of vermicompost (Erick et al. 2008). Regarding the use of vermicompost, it can reduce the mobility and availability of heavy metals present in livestock manure; however, there are few reports on how vermicompost affects the sustainability of crops

Table 1. Nutrient percentage of vermicompost (Erick et al. 2008).

Nutrient element	Vermicompost (%)
Organic Carbon	9.8–13.4
Nitrogen	0.51–1.61
Phosphorus	0.19–1.02
Potassium	0.15–0.74
Calcium	1.18–7.61
Magnesium	0.093–0.568
Sodium	0.058–0.158
Zinc	0.0042–0.110
Copper	0.0026–0.0048
Iron	0.2050–1.3313
Manganese	0.0105–0.2038

and nutrient management under intensive agricultural practises, and this area still requires further research. Vermicompost's commercial use is still not widely accepted. As a result, it is necessary to calculate vermicompost concentrations under typical soil-water plant micrometeorological regimes as well as to get a deeper knowledge of the composition of immature vermicompost and the reasons why its application has failed (Khoshnevisan et al. 2021).

5. EFFECT OF NANOPARTICLES ON VERMICOMPOSTING

Nanoparticles [NPs] when present in the substrate pose a greater risk to the environment. The interaction of worms with such substrates depends on the type and concentration of nanoparticles, thus defining the quality of compost formed. Nanoparticles are mostly toxic, and are not only hazardous to human health but also affect the microbial community taking part in the vermicomposting process of such NP mixed substrate. The presence of nanoparticles reduces the number of both alive and dead worms and microbes in the substrate. Nanoparticles when present even in small concentrations hinder the decomposition process of microbes, but because of the cost-effectiveness of vermicomposting treatment of substrate having nanoparticles is carried out (Hui and Kui 2021).

Eisenia fetida has a high immortality rate in substrates having high nanoparticle concentration, humidity, and ammonia, and therefore is generally used to carry on the vermicomposting process (Hui and Kui 2021). It has also got the immense ability to reduce the pathogens in the substrate without any requirement for temperature change (Parseh et al. 2021). Vermicomposting when carried on sludge with nanoparticles formulates the interaction of these particles and organic matter. Even in the presence of varied pollutants, worms ease the formation of complexes of organic waste and NPs. Earthworms are even found to absorb some NPs from the substrate. The quality of compost formed from such substrate is affected by the concentration of NP, its toxicity, intrinsic properties, bacterial interactions, and different environmental conditions. Vermicomposting hardly alters the concentration of nanoparticles in the final mature compost formed, so this compost before being used as bio-fertilizer should be properly analysed as it holds higher environmental risk due to the presence of heavy metals (Hui and Kui 2021).

Biochar in combination with nanocarbon produces high vermicompost quality index (VQI), even when amendments like straw mixed with nanocarbon produce considerable agronomic value (Cao et al. 2021). The presence of nanocarbon enhances microbial activities but does not affect the worm's performance. So instead of vermicomposting, C-rich amendments can be used for composting as they prove unhealthy for worms (Cao et al. 2021).

6. FACTORS AFFECTING THE VERMICOMPOSTING TREATMENT OF SOLID WASTE

6.1 General

Researchers have been concentrating on the study of the intricate interaction between physical, chemical, and biological elements that occurs during natural composting under controlled settings over the past few decades (Wainaina et al. 2020). The qualities of compost are influenced by a number of things. Inoculum type, particle size, pH, moisture content, aeration duration, temperature, and heat insulation are a few of these. Dust, odour, and maintaining the proper levels of N, K, P, and C are all part of proper composting (Awasthi et al. 2022). Since they define the ideal circumstances for organic matter decomposition and microbial development, the regulation of these parameters, including porosity, nutrient content, C/N ratio, temperature, pH, moisture, and oxygen supply, has established itself as a critical component for optimising the composting process (Wainaina et al. 2020).

Composting improvements have concentrated on ammonia volatilization, greenhouse gas emissions, and composting time. Composting takes longer than procedures like pyrolysis and incineration. Utilizing microorganisms with accelerated maturation and breakdown can improve the composting process (Figure 5). Biochar and the usage of earthworms are further methods. The strategies will result in lower greenhouse gas emissions. To reduce nitrogen loss, mineral additives like zeolite can be added to the MSW organic portion (Awasthi et al. 2022).

The substrate plays an important role in deciding the quality of the compost formed. The physiochemical composition of the organic matter, the earthworm species involved and the interaction of worms with the microbe present decide the quality of the compost formed (Liu et al. 2021). The population of microbes during vermicomposting is highly affected by the kind of earthworm used for the process. The loss of lignin, cellulose, and hemicellulose content in the substrate is highly affected by the enzyme production in the gut of the worms by microbes (Pandita et al. 2020).

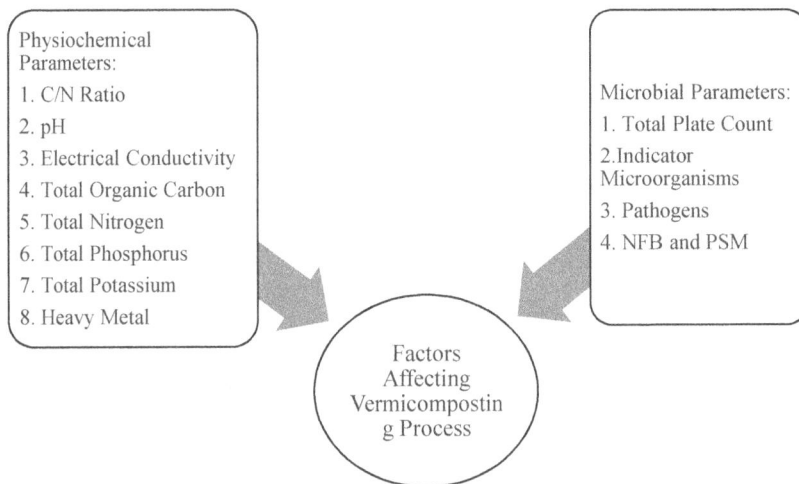

Figure 5. Factors affecting vermicomposting process in waste treatment.

There is a loss of microorganisms as the humidity level decreases. Also, the higher the moisture content, the faster the N mineralization process which is involved in the decomposition process (Qin et al. 2021).

6.2 Physiochemical parameters affecting vermicomposting

The mix of the feed materials determines the vermicompost's overall nutritional status, and some of these different feed components are used as an earthworm food source. The majority of vermicomposting techniques for various organic solid waste, whether from earlier or more contemporary study, showed an increase in nutrient concentrations. Few vermicomposting methods from earlier studies, meanwhile, demonstrated a drop in nutrient contents (Soobhany 2019).

6.2.1 C/N ratio

The C/N ratio is a very important factor in the vermicomposting process. It initiates the mineralization of organics present in the substrates. It also alters the reproduction of the worms and changes their weight. Throughout the process, the nutrient content increases in the compost while the C/N ratio and microbial activities reduce. If the nitrogen level is high in the substrate, it can result in a time lag in completing the process as it can result in the death of worms and microbes (Liu et al. 2021).

The parameter to evaluate the maturity and quality of compost comes from the C/N ratio, as the amount of carbon reduced in the substrate is equivalent to the earthworms consuming it as a source of energy (Niedzialkoski et al. 2021). If its value is below 15, it can be easily used for agricultural purposes but if its value is below 20, it qualifies as superior in quality (Badhwar et al. 2020).

6.2.2 *pH*

During vermicomposting, the pH of the substrate is highly affected by the decomposition and consumption activities of the worms. Ammonia vitalization plays a key role in changing the pH of the compost (Liu et al. 2021). pH also influences enzymatic processes and nutrient availability in the substrate (Gómez-Brandón 2020).

Studies show that when the pH of the soil varies from weakly acidic to weak alkaline, plant growth is best favoured. Microbes cannot work in acidic pH conditions, vermicomposting process helps neutralize the acidic substrate (María Gómez-Brandón 2020). The lower pH is because of organic acid production during the vermicomposting process which are raw materials for biogas production energy (Pandita et al. 2020, Niedzialkoski et al. 2021).

6.2.3 *Electrical conductivity*

The change in the electrical conductivity is defined by the change in the pH of the substrate. Acidification of the waste leads to an increase in the readily available nutrient content and ions in the sample, which can easily get mixed with the fluid of the sample. Although the dehydration process of these minerals increases the salt content in the substrate (Dermendzhieva et al. 2021).

6.2.4 *Total organic carbon*

The carbon present in the waste is utilized as an energy source by the worms and microorganisms throughout vermicomposting and as a by-product CO_2 is released. The transformation process of the carbon is similar to all substrates (Diyana Dermendzhieva et al. 2021, Pandita et al. 2020). The CO_2 emissions contribute to the GHG emissions (Lv et al. 2018). The finished compost's nitrogen concentration increases when TOC is reduced (Table 2) (Yuvaraj et al. 2020).

In the case of SCG, the carbon content plays a vital role. Due to the unavailability of carbon as a food source, the number of worms is less which can survive in the substrate and hence govern the process. The quality of compost formed is ordered by this factor (Hanc et al. 2021).

6.2.5 *Total Kieldahal Nitrogen (TKN), N-NH$_4^+$ AND N-NO$_3^-$*

Microorganisms carry the nitrification of $N-NH_4^+$ and convert it into $N-NO_3^-$. N_2 is lost as volatile ammonia during the acidification of the substrate at high pH. Earthworms alter the nitrogen level in the waste by nitrogen transformation, mucus formation, and the nitrogenous faeces production (Dermendzhieva et al. 2021). The reduction in the carbon level, pH, and dry biomass all contribute to the increase in the nitrogen level of the final compost (Pandita et al. 2020).

Various enzymes like amylase, invertase, phosphatase, protease, urease, etc., generated by worms and microbes during vermicomposting play a vital role in N-mineralization. The nitrogen content is also ordered by the dead tissues of the earthworms. Earthworm activities improve aeration, nitrification, and mineralization process which decreases the concentration of ammonical nitrogen and increases the concentration of nitrate N in the final product (Yuvaraj et al. 2020). N_2O is generated more as a by-product which contributes more to the overall GHG emission of a substrate as compared to the CO_2 emissions (Lv et al. 2018).

6.2.6 *Total P and P₂O₅*

Mineralization of phosphorous content in the substrate is more when in the presence of worms. The phosphorous which is not readily available for the plants can be made easily available by this process because the faecal matter excreted by the worms is rich in phosphorous (Dermendzhieva et al. 2021, Pandita et al. 2020).

Augmentation of phosphorus content is facilitated by the phytase and acid phosphatase enzyme activity carried on in the intestine of worms. These internal activities of worms make phosphorus available to the plants at the end of the process (Sharma and Garg 2020).

6.2.7 *Total K (TK) and K₂O*

Organic potassium gets easily mineralized in the vermicomposting method (Dermendzhieva et al. 2021). The gradual rise in the potassium level is contributed by the lowering of mass and volume of the substrate used (Pandita et al. 2020). At the end of vermicomposting, potassium content is found more and this observation can be credited to the decomposition, mineralization processes and also to the excreted matter by earthworms (Sohal et al. 2021).

6.2.8 *Heavy metals (Fe, Mn, Cu, Zn, Cr, Ni, Pb, Cd)*

Aspects of the impact of metals on earthworms at many levels of biological organisation, from molecules and cells to individual performance and population dynamics, are described in the literature (Table 2). These investigations are related by the fact that direct or indirect harmful effects progress from initial molecular-genetic targets to actual disturbances in earthworm populations and communities (Mohee and Soobhany 2014). The observed decrease in earthworm reproductive production in feedstocks polluted with metals may limit vermicomposting's effectiveness. Earthworms can develop tolerance after being exposed to metal- or metalloid-contaminated soil for several generations. The theoretical debate over whether it makes sense to find metal-resistant clones of vermicomposting species to process severely metal-contaminated feedstocks is fascinating, but its practical applicability may be in doubt given that the metal-enriched vermicomposts will have little intrinsic value

Table 2. Different physiochemical parameters reported by different researchers (Yuvaraj et al. 2020).

Physiochemical parameters	Yuvraj et al. (2020a)	Gong et al. (2019)	Balachandar et al. (2020)	Bhat et al. (2016)	Gusain and Suthar (2020a)	Devi and Khwariakpm (2020a)	Rini et al. (2020)
pH	8.56	8.8	7.73	8.35	8.2	6.5	8.84
EC (dS/m)	0.72	1.7	1.3	4.13	3.85	3.10	–
TOC (%)	42.90	31.2	35.3	46.28	59.57	28.56	33.0
TKN (%)	0.8	2.24	1.12	1.34	1.35	1.2	1.72
TP (%)	0.24	0.87	0.78	0.59	0.25	–	0.76
TK (%)	0.2	0.91	1.32	2.23	8.56	–	0.43
C/N Ratio	53.75	13.93	31.54	34.53	44.04	23.80	19.15

EC= Electrical Conductivity; **TOC** = Total Organic Carbon; **TK** = Total Potassium; **TP** = Total Phosphorus; **C/N Ratio** = Carbon to Nitrogen Ratio.

(Mohee and Soobhany 2014). Despite the fact that regulation limitations are based on individual metal concentrations, contaminated feedstocks and the composts made from them frequently contain combinations of metals that might possibly interact to modify the availability and possible harmful effects of each individual ingredient (Mohee and Soobhany 2014).

The heavy metals' removal is not observed when there is no leaching in the water drainage system. On the contrary, there are previous studies showing that heavy metal removal takes place in the vermicomposting process. The heavy metal content in the vermicompost is influenced by factors like substrate type, worms used, process methodology used, and climatic conditions (Dermendzhieva et al. 2021).

6.3 Microbial parameters

6.3.1 Total plate count

Total plate count depends on the substrate composition, worm genre used and the temporal effect, which is a significant aspect as well (Dermendzhieva et al. 2021).

6.3.2 Indicator microorganisms

Coliforms, *E. Coli* numbers cannot be clearly quantified as it varies from substrate to substrate, worm's species and the environmental aspect is prominent too (Dermendzhieva et al. 2021).

6.3.3 Pathogens

Enterobacteriaceae removal is significantly carried by vermicomposting and this can be verified by the experiments carried out on urban sewage substrate. *Salmonella* spp. concentration is reduced over the vermicomposting process (Dermendzhieva et al. 2021). *Clostridium Perfringens* is the spore-forming bacteria secreted in the gut of worms that pass in the compost through faeces. The gut-related activities of the earthworms are a major factor that results in the decrease of pathogens in the matured compost (Domínguez et al. 2021). The number of pathogens decreases via vermicomposting; it is largely affected by factors like intestinal earthworm activities, antibacterial properties contained in coelomic fluids, and also completion between different microbial groups for survival (Parseh et al. 2021).

6.3.4 NFB and PSM

Nitrogen-fixing bacteria (NFB) increase the nitrogen content and phosphate-solubilizing microorganisms (PSM) increase the phosphate content in the final compost (Mal et al. 2021). NFB reduces the C/N ratio in the substrate and thus increases the rate of decomposition of the organic matter. On the other hand, the microbial population is positively altered by the presence of PSM. *Azotobacter* chroococcum and *Pseudomonas fluorescens* are the commonly used NFB and PSM. With the progress of the vermicomposting process, the colony-forming units of NFB slowly add in number. The phosphorus content from the PSM initiates NFB growth, which eventually improves the quality of the compost by fixing atmospheric N_2 (Mal et al. 2021).

7. FUTURE PROSPECTS

Because of a lack of understanding of the resource management of organic waste, waste management technologies have not been implemented effectively and efficiently over the world. This necessitates developing a new system of production, consumption, reuse, recycling, and waste avoidance. Governmental organisations, nations, and individuals should be aware of the benefits of composting as an eco-friendly method for managing organic waste and the potential of compost in a circular economy.

Vermicomposting has been practiced for ages. But its technical aspects require a lot of studies even now. Microbial actions, which result in the production of humic acids during composting, require more addressing. Plant growth hormones plays a very important role in the composting process. The mechanism involved in their production needs further study. The study of earthworms and their microbial fluid content needs attention. The interaction of worms with microbes needs more detailed and careful study. Vermicomposting takes less time as compared to composting, but to be more efficient in dealing with the large quantity of waste and using vermicomposting on different substrates, there's an urgent need to analyse this process more, involving research and development. Aeration is a must for vermicomposting to be carried on easily, but as the depth of substrate increases anaerobic conditions start prevailing which hinders the process in many aspects. So to avoid this condition, new measures are to be formulated. Municipal sludge has varied composition based on the season, temperature, region, moisture content, etc., applying vermicomposting on different samples produces different results and hence has more scope for study. To assimilate the production of GHGs in composting and vermicomposting processes, more research is to be carried out.

Based on this analysis of composting procedures, more study should concentrate on the following areas: (1) the production and release of greenhouse gases as well as volatile fatty acids, which is another subject of intense investigation, are drawbacks of composting. Compound additives based on microorganisms have demonstrated improved nitrogen conservation and may eliminate the requirement for physical and chemical additions. More effort should be put towards minimising these drawbacks without compromising the quality of the compost; (2) only once the compost has become mature and stable can the nutrients produced by the composting process be used. Expanding study on compost maturation is important to increase compost applicability; (3) despite having less nutrients than commercial fertilisers, compost still contributes essential organic matter to soil. To provide nutrients to fields and crops in a targeted way, it is crucial to examine the features of the soil and fertiliser prior to their integration.

Further studies are to be carried out to investigate the time of application of C-rich amendments in the substrate. The composition and effects of additives for different substrates are to be studied in detail to enquire about their effects on the earthworms and also on the microbial activities. Vermicomposting has been practiced to a large extent to treat different substrates. As the entire study suggests, it proves to be a feasible technique in maximum cases; with specific amendments, yields can be improved except in certain cases. However, the final product qualities depend on many factors and require further study to come up with better results.

REFERENCES

Ales Hanc, Tereza Hrebeckova, Petr Pliva et al. 2020. Vermicomposting of sludge from a malt house 118: 232–240. https://doi.org/10.1016/j.wasman.2020.08.027.

Ales Hanc, Tereza Hrebeckova, Alena Grasserova et al. 2021. Conversion of spent coffee grounds into vermicompost. DOI: 10.1016/j.biortech.2021.125925. https://pubmed.ncbi.nlm.nih.gov/34614558/.

Alshehrei, F. and Ameen, F. 2021. Vermicomposting: A management tool to mitigate solid waste. Saudi Journal of Biological Sciences 28(6): 3284–3293. DOI: 10.1016/j.sjbs.2021.02.072. https://doi.org/10.1016/j.sjbs.2021.02.072.

Ananthanarayanan Yuvaraj, Ramasundaram Thangaraj, Balasubramani Ravindran et al. 2020. Centrality of cattle solid wastes in vermicomposting technology – A cleaner resource recovery and bio-waste recycling option for agricultural and environmental sustainability DOI: 10.1016/j.envpol.2020.115688. https://doi.org/10.1016/j.envpol.2020.115688.

Asteria Aili, Ndiipohamba Katakulaac, Bethold Handuraa et al. 2021. Optimized vermicomposting of a goat manure-vegetable food waste mixture for enhanced nutrient release 12: e00727. https://doi.org/10.1016/j.sciaf.2021.e00727.

Baoyi Lv, Di Zhanga, Yuxue Cui et al. 2018. Effects of C/N ratio and earthworms on greenhouse gas emissions during vermicomposting of sewage sludge. Bioresource Technology 268: 408–414. https://doi.org/10.1016/j.biortech.2018.08.004.

Bhawana Sohal, Sharanpreet Singh, Soubam Indra Kumar Singh et al. 2021. Comparing the nutrient changes, heavy metals, and genotoxicity assessment before and after vermicomposting of thermal fly ash using Eisenia fetida. Environmental Science and Pollution Research 28: 48154–48170. https://doi.org/10.1007/s11356-021-13726-8.

Chand Malav, L, Yadav, K.K., Gupta, N. et al. 2020. A review on municipal solid waste as a renewable source for waste-to-energy project in India: Current practices, challenges, and future opportunities. Journal of Cleaner Production (241): 118413. https://doi.org/10.1016/j.jclepro.2020.123227.

Diyana Dermendzhieva, Toncho Dinev, Gergana Kostadinova et al. 2021. Agro-ecological characterization of vermicomposted sewage sludge from municipal and poultry enterprise wastewater treatment plants. Sains Malaysiana 50(8): 2167–2178. http://doi.org/10.17576/jsm-2021-5008-03.

Erick, C., Fernandes, M. and Erika Styger. 2008. Sustainable Land Management: SOURCEBOOK DOI:10.1596/978-0-8213-7432-0. https://www.researchgate.net/publication/273886185.

Fatimah Alshehrei, Nouf M. Al-Enazi and Fuad Ameen. 2021. Vermicomposting amended with microalgal biomass and biochar produce phytopathogen-resistant seedbeds for vegetables. https://doi.org/10.1007/s13399-021-01770-w.

Fu-Sheng Sun, Guang-Hui Yua, Jing-Yuan Ning et al. 2020. Biological removal of cadmium from biogas residues during vermicomposting, and the effect of earthworm hydrolysates on Trichoderma guizhouense sporulation 312: 123635. https://doi.org/10.1016/j.biortech.2020.123635.

Hřebečkováa, T., Wiesnerováb, H. and Hanča, A. 2020. Changes in layers of laboratory vermicomposting using spent mushroom substrate of Agaricus subrufescens P 276: 111340. https://doi.org/10.1016/j.jenvman.2020.111340.

Iman Parseh, Keyvan Mousavi, Ahmad Badieenejad et al. 2021. Microbial and Composition Changes during Vermicomposting Process Resulting from Decomposable Domestic Waste, Cow Manure and Dewatered Sludge IP: 14.139.85.200. DOI: 10.4103/ijehe.ijehe_56_20. http://www.ijehe.org.

Jefferson Campos Silva, Andreza Jayane Nunes, Siqueira Hermogenes Bezerra et al. 2021. Vermicomposting corn waste under cultural and climatic conditions of the Brazilian Backwoods 15: 100730. https://doi.org/10.1016/j.biteb.2021.100730.

Jiangang Chea, Weifen Lina, Jie Yea et al. 2020. Insights into compositional changes of dissolved organic matter during a full-scale vermicomposting of cow dung by combined spectroscopic and electrochemical techniques 301: 122757. https://doi.org/10.1016/j.biortech.2020.122757.

Jie Qin, Xiaoyong Fu, Xuemin Chen et al. 2021. Changes in physicochemical properties and microfauna community during vermicomposting of municipal sludge under different moisture conditions. Environmental Science and Pollution Research 28: 31539–31548. https://doi.org/10.1007/s11356-021-12846-5.

Jorge Domínguez, Manuel Aira, Keith A. Crandall et al. 2021. Earthworms drastically change fungal and bacterial communities during vermicomposting of sewage sludge 11: 15556. https://doi.org/10.1038/s41598-021-95099-z.

Kavita Sharma, V.K. Garg. 2020. Conversion of a toxic weed into vermicompost by Eisenia fetida: Nutrient content and earthworm fecundity 11: 100530. https://doi.org/10.1016/j.biteb.2020.100530.

Lipika Pandita, Debadatta Sethib, Sushanta Kumar Pattanayak et al. 2020. Bioconversion of lignocellulosic organic wastes into nutrient rich vermicompost by Eudrilus eugeniae 12: 100580. https://doi.org/10.1016/j.biteb.2020.100580.

María Gómez-Brandón, Marta Lores, Hugo Martínez-Cordeiro et al. 2020. Effectiveness of vermicomposting for bioconversion of grape marc derived from red winemaking into a value-added product. Environmental Science and Pollution Research 27: 33438–33445. https://doi.org/10.1007/s11356-019-04820-z.

Maria Gómez-Brandón, Hugo Martínez-Cordeiro and Jorge Domínguez. 2021. Changes in the nutrient dynamics and microbiological properties of grape marc in a continuous-feeding vermicomposting system. Waste Management (135): 1–10. https://doi.org/10.1016/j.wasman.2021.08.004.

Michael Nyame, Acquah Helen, Michelle Korkor et al. 2021. Degradation and accumulation rates of fresh human excreta during vermicomposting by Eisenia fetida and Eudrilus eugeniae 293: 112817. https://doi.org/10.1016/j.jenvman.2021.112817.

Mona Ghorbani, Mohammad Reza Sabour and Masoud Bidabadi. 2021. Vermicomposting Smart Closed Reactor Design and Performance Assessment by Using Sewage Sludge https://doi.org/10.1007/s12649-021-01426-w.

Nuhaa Soobhany. 2019. Insight into the recovery of nutrients from organic solid waste through biochemical conversion processes for fertilizer production: A review. Journal of Cleaner Production (241): 118413. https://doi.org/10.1016/j.jclepro.2019.118413.

Romeela Mohee and Nuhaa Soobhany. 2014. Comparison of heavy metals content in compost against vermicompost of organic solid waste: Past and present. Resources, Conservation and Recycling 92: 206–213. https://doi.org/10.1016/j.resconrec.2014.07.004.

Renu Negi and Surindra Suthar. 2018. Degradation of paper mill wastewater sludge and cow dung by brown-rot fungi Oligoporus placenta and earthworm (Eisenia fetida) during vermicomposting. Journal of Cleaner Production 201: 842e852. https://doi.org/10.1016/j.jclepro.2018.08.068.

Rosana Krauss Niedzialkoski, Ritieli Marostica, Felippe Martins Damaceno et al. 2021. Combination of biological processes for agro-industrial poultry waste management: Effects on vermicomposting and anaerobic digestion DOI: 10.1016/j.jenvman.2021.113127. https://doi.org/10.1016/j.jenvman.2021.113127.

Sartaj Ahmad Bhata, Sharanpreet Singha, Jaswinder Singhb et al. 2018. Bioremediation and detoxification of industrial wastes by earthworms: Vermicompost as powerful crop nutrient in sustainable agriculture 252: 172–179. https://doi.org/10.1016/j.biortech.2018.01.003.

Shumpei Hashimoto, Mai Furuya, Xiaodong You et al. 2021. Chemical and microbiological evaluation of vermicompost made from school food waste in Japan. JARQ 55(3): 225–232. https://www.jircas.go.jp.

Song Zhi-wei, Sheng Tao, Deng Wen-jing et al. 2019. Investigation of rice straw and kitchen waste degradation through vermicomposting 243: 269–272. https://doi.org/10.1016/j.jenvman.2019.04.126.

Sujit Mal, Chattopadhyay, G.N. and Kalyan Chakrabarti. 2021. Microbiological integration for qualitative improvement of vermicompost. International Journal of Recycling of Organic Waste in Agriculture 10: 157–166. Doi: 10.30486/IJROWA.2021.1902019.1087.

Vinay Kumar Badhwar, Sukhwinderpal Singh and Balihar Singh. 2020. Biotransformation of paper mill sludge and tea waste with cow dung using vermicomposting. Bioresour Technology (318): 124097. https://doi.org/10.1016/j.biortech.2020.124097.

Vinay Kumar Badhwar, Sukhwinderpal Singh and Balihar Singh. 2020. Biotransformation of paper mill sludge and tea waste with cow dung using vermicomposting DOI: 10.1016/j.biortech.2020.124097. https://doi.org/10.1016/j.biortech.2020.12409.

Xia Hui and Huang Kui. 2021. Effects of TiO2 and ZnO nanoparticles on vermicomposting of dewatered sludge: Studies based on the humification and microbial profiles of vermicompost. Environmental Science and Pollution Research 28: 38718–38729. https://doi.org/10.1007/s11356-021-13226-9.

Xue Liu, Bing Geng, Changxiong Zhu et al. 2021. An improved vermicomposting system provides more efficient wastewater use of dairy farms using Eisenia fetida. Agronomy 11: 833. https://doi.org/10.3390/agronomy11050833.

Yune Cao, Yongqiang Tian, Qing Wu et al. 2021. Vermicomposting of livestock manure as affected by carbon-rich additives (straw, biochar and nanocarbon): A comprehensive evaluation of earthworm performance, microbial activities, metabolic functions and vermicompost quality 320: 124404. https://doi.org/10.1016/j.biortech.2020.124404.

12

APPLICATION OF ANAEROBIC DIGESTION APPROACH FOR THE TREATMENT OF AGRO-RESIDUE WASTE

Veluswamy Venkatramanan,[1] *Shivani Maddirala,*[2] *Pritam Bardhan,*[3] *Rupam Kataki,*[3] *Dheeraj Rathore,*[4] *Shiv Prasad,*[5] *Anoop Singh*[6] and *Surajbhan Sevda*[2,*]

1. INTRODUCTION

The anaerobic bioreactor (digester) is widely used to generate clean energy (biogas and biohydrogen) using wastewater and other organic waste as raw material. The agro-residue waste provides higher organic material concentration than the monomer substrate; hence, a high amount of clean energy can be produced using this as substrates in the anaerobic co-digestion process. The agro-residue waste is rich in organic matter and has all the essential nutrients like proteins, carbohydrates, amino acids, and minerals. In addition, the agro-residues material supports cellular growth and can be used for cheaper substrates in both anaerobic co-digestion and fermentation processes to generate valuable products (Mahato et al. 2020).

[1] School of Interdisciplinary and Transdisciplinary Studies, Indira Gandhi National Open University, New Delhi 110068, Delhi, India.
[2] Environmental Bioprocess Laboratory, National Institute of Technology Warangal, Warangal, Telangana, India-506004.
[3] Department of Energy, Tezpur University, Tezpur 784028, Assam.
[4] School of Environment and Sustainable Development, Central University of Gujarat, Gandhinagar-382030, Gujarat, India.
[5] Division of Environment Science, ICAR-Indian Agricultural Research Institute, New Delhi- 110012, India.
[6] Department of Scientific and Industrial Research (DSIR), Ministry of Science and Technology, Government of India, Technology Bhawan, New Mehrauli Road, New Delhi- 110016 India.
*Corresponding author: sevdasuraj@gmail.com, sevdasuraj@nitw.ac.in

Biogas is generated from organic matter (agricultural wastes, organic wastes, animal manure and energy crops) through the anaerobic process (Herrmann et al. 2016). The anaerobic digestion process involves "hydrolysis", "acidogenesis", "acetogenesis", and "methanogenesis" (Kadam and Panwar 2017). Anaerobic digestion technology has proven application in biogas generation, nutrient recycling, rural employment generation (Herrmann et al. 2016, Styles et al. 2022), waste valorization and bioeconomy (Venkatramanan et al. 2021a, Shah et al. 2021), climate change mitigation and carbon neutrality (Venkatramanan et al. 2020, Styles et al. 2022). Anaerobic digestion technology is advocated as a viable, sustainable approach to valorize agricultural crop residues and achieve Sustainable Development Goals (Venkatramanan et al. 2021b), as it plays a vital role in climate- energy-food-waste nexus (Styles et al. 2022). The rising interest and development of biogas technology have been driven by the availability of feedstocks and also the policy support provided by the national governments. 90% of global biogas production is contributed by Europe, China and the USA (IEA 2020). Sun et al. (2021) recommended that biomethane production can be improved in China through government subsidies, particularly in northern parts of China, which generate a significant amount of crop residues.

The feedstocks for biogas production include crop residues, animal manure, municipal solid waste and wastewater sludge. Agricultural crop residues are materials that are left over in the farm field after harvesting and at post-harvest processing sites (Venkatramanan et al. 2021). Crop residues include mainly the leftover from the harvest of rice, wheat, sorghum, maize, bagasse, sugar beet, soybean and coarse cereals. The quantum of agro-residue wastes generated is increasing globally due to increased agricultural production and intensifying agricultural practices. In addition, many studies have reported the sustainable use of agricultural residues for bioenergy generation (Prasad et al. 2019, 2021).

Over the past two decades, increasing research attention has been given to hydrogen production from biomass using biological processes such as anaerobic fermentation, dark and photo-fermentation. In this regard, agricultural waste such as crop residue and energy crops (mainly grasses) has become an attractive raw material for biohydrogen production because it is abundantly available as a carbohydrate-rich, low-cost resource on the Earth. Moreover, using agro residues for biohydrogen production leads to concomitant benign disposal of potential wastes from the environment. However, the recalcitrance of lignocellulosic biomass to biodegradability is a significant hurdle that accounts for the low hydrogen production yields (Chatellard et al. 2017). Nevertheless, different pretreatment methods such as ultrasonication, acid hydrolysis or alkaline treatment, steam explosion, moist heat, microwave-assisted treatment and biological pretreatment using microbial cells or enzymes are used on agro-residues to enhance the availability of lignocellulose for hydrogen production (Saravanan et al. 2021). Figure 1 depicts a schematic representation of the production of biohydrogen and other value-added products from agricultural residue.

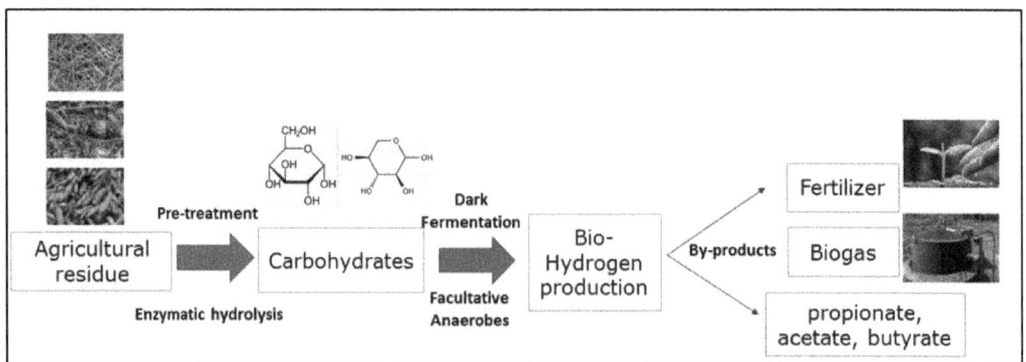

Figure 1. Schematic representation of bio-hydrogen production and value-added products from agricultural waste residue using dark fermentation.

2. BIOGAS COMPOSITION

The composition of biogas varies with feedstocks (organic waste) and production pathways (biodigesters, landfill gas recovery systems and wastewater treatment plants) (Yusuf and Almomani 2023) (Table 1). Biogas mainly contains methane, carbon dioxide and nitrogen. Certain chemical species like water vapour, sulphur compounds (hydrogen sulphide, mercaptans, sulphides), silicon (siloxanes, silanes), ammonia, hydrogen, halogenated compounds, and volatile organic compounds are present in the raw biogas in trace amounts (Ullah Khan et al. 2017, Calbry-Muzyka et al. 2022). The amount of trace compounds present in the raw biogas depends on feedstocks, co-substrates and digester conditions (operating temperature, type of digester, digestate/gas retention time) (Calbry-Muzyka et al. 2022). For instance, while siloxane concentration is low in agricultural biogas, its concentration is more in waste-based biogas (Calbry-Muzyka et al. 2022). Raw biogas contains methane (50–75%), carbon dioxide (25–50%), nitrogen (0–5%), water vapour (1–5%), hydrogen sulphide (0–5000 ppm) and siloxanes (0–50 mg/m^3) (Braun 2007).

The first step in biogas purification is to remove the compounds (hydrogen sulphide, silicon, ammonia, volatile organic compounds) from raw biogas. This step is called "biogas cleaning". The second step is to increase the heating value of the biogas. This step is called "biogas upgrading" (Angelidaki et al. 2018). Upgraded biogas with a composition of 95% methane is called biomethane (Omar et al. 2019). The process of biomethane production involves two essential steps: (a) Biogas production from organic waste materials through anaerobic digestion; (b) Upgrading and purification of biogas into biomethane and value-added products (Yusuf and Almomani 2023). Biomethane can

Table 1. Biogas composition and feedstocks.

	CH$_4$ (v%)	CO$_2$ (v%)	N$_2$ (v%)	O$_2$ (v%)	H$_2$S (ppm$_v$)	NH$_3$ (ppm$_v$)	Reference
Raw biogas	50–75	25–50	0–5	NA	0–5000	NA	(Braun 2007, Surendra et al. 2014)
Dairy cow manure only	54.9	39	5	0.9	400–1000	30–53	(Calbry-Muzyka et al. 2022)
Chicken, cattle manure, vegetable and green waste	53.0	43.3	3.2	0.4	4–10	< 0.25	(Calbry-Muzyka et al. 2022)
Cow, pig manure + wastes from the food, dairy, and beverage industry	57.7	40.8	1.2	0.3	18	70	(Calbry-Muzyka et al. 2022)
Organic waste	60–70	30–40	1	1–5	10–180	NA	(Ullah Khan et al. 2017)
MSW wastes, sewage sludge, etc.	44–67	30–44	0.1–6	0.1–3	2–3174	NA	(Calbry-Muzyka et al. 2022)
Landfill	47–57	37–41	< 1–17	< 1	36–115	NA	(Rasi et al. 2007)
Landfill	40–70	25–40	0–17	0–3	0–5143	NA	(Calbry-Muzyka et al. 2022)
Sewage digester	61–65	36–38	< 2	< 1	NA	NA	(Rasi et al. 2007)
Agricultural waste	49–69	29–44	0.6–13	0.2–3	7–6570	NA	(Calbry-Muzyka et al. 2022)
Farm biogas plant	55–58	37–38	< 1–2	< 1	32–169	NA	(Rasi et al. 2007)
Sugar beet silage	48.5	NA	NA	NA	NA	NA	(Herrmann et al. 2016)
Maize silage	55	NA	NA	NA	NA	NA	(Herrmann et al. 2016)
Potatoes silage	49.4	NA	NA	NA	NA	NA	(Herrmann et al. 2016)
Winter wheat silage	54.3	NA	NA	NA	NA	NA	(Herrmann et al. 2016)
Sunflower silage	56.6	NA	NA	NA	NA	NA	(Herrmann et al. 2016)

also be produced by gasifying solid organic biomass and methanation. The solid organic biomass at a high temperature of about 800°C and high pressure in an oxygen-limiting environment is converted into a gaseous mixture of methane, hydrogen and carbon monoxide. Therefore, the heating value and usability of biogas can be enhanced by upgrading and biogas purification (Yusuf and Almomani 2023). Purified biogas finds application in electricity generation, heat generation, vehicular transport and natural gas grid.

Ullah Khan et al. (2017) reviewed different biogas purification and upgrading technologies. The principles involved in biogas purification are absorption, adsorption, membrane separation and cryogenic separation (Angelidaki et al. 2018). The biogas purification and upgrading technologies include "pressure swing adsorption", "high-pressure water scrubbing", "organic solvent scrubbing", "amine scrubbing", "membrane separation", and "cryogenic separation" (Ullah Khan et al. 2017, Angelidaki et al. 2018). The main objective of these technologies includes carbon dioxide separation and removal of gaseous contaminants. The carbon dioxide present in the raw biogas acts as a dilutant, decreases the flame speed and reduces the calorific value/heating value (Verma et al. 2017). (Abdeen et al. 2016) reported that biogas with 60% methane and 40% carbon oxide has a heat value of 17.7 MJ/kg. On the other hand, pure methane (100%) has a heat value of 50.2 MJ/kg. Further, the carbon dioxide present in the biogas is converted into carbon monoxide during combustion. The carbon monoxide and hydrocarbon cause corrosion in the pipelines, limiting the value of the biogas. Studies have reported on the physico-chemical and biological methods for carbon dioxide separation (Angelidaki et al. 2018). Nevertheless, the biological separation of carbon dioxide is advantageous over the physical and chemical processes, as it converts carbon dioxide into value-added products like alcohol and fatty acids. Biogas is wetter as compared to natural gas. The moisture present in the raw biogas can cause the corrosion of pipelines. So, it is removed by the processes like condensation, adsorption and the use of chemical compounds like aluminium oxide, activated charcoal, etc. (Ullah Khan et al. 2017, Calbry-Muzyka et al. 2022).

3. BIO-METHANE POTENTIAL FROM AGRO-RESIDUE WASTES

The potential use and efficiency of feedstocks for anaerobic digestion are gauged by the biomethane potential (BMP) test. The standard methods for the BMP test are Automated Methane Potential Testing System (AMPTS) and the German DIN (Deutsches Institut für Normung) Standard method using eudiometers (Kleinheinz and Hernandez 2016). Biomethane potential test is applied to estimate the amount of methane produced from a substrate. Biomethane potential is stated as "the volume of dry methane gas under standard conditions (273.15 K and 101.33 KPa) per mass of volatile solids of substrate added". The unit is Nml_{CH4}/g_{VS} (Holliger et al. 2017). The biomethane potential of crop residues is the function of production of crop residues and "biochemical methane potential". The biochemical methane potential is "the measurement of the maximum methane volume per unit of total or volatile solids that a substrate can produce" (Thomas et al. 2019).

Several studies have highlighted the methane yield of diverse feedstocks, including cereal crops, energy crops and agricultural residues (Braun 2007). However, in the present context, methane yields of straw (242–324 m^3/t_{vs}), leaves (417–453 m^3/t_{vs}) and chaff (270–316 m^3/t_{vs}) reiterate the potential of agricultural residues for biomethane production (Braun 2007). The factors that influence the biomethane potential of agricultural residues include the feedstock type, biomass yield potential, seasonal availability of feedstocks, and issues related to biodegradability and optimization and sustainability of the process (Thomas et al. 2019, Singh and Kumar 2019).

The critical factors in assessing the biomethane potential are the choice of feedstock (Ammenberg and Feiz 2017) and biomass availability. The agricultural residues' availability should be looked at through the lens of crop type, biomass yield, seasonal availability of biomass, geographical location, and transport or logistics of transporting the agro-residues to the site of energy generation

(Venkatramanan et al. 2021c, Thomas et al. 2019). The gross agricultural residue generation varies with the region. For instance, India's annual gross crop residues are about 696.38 million tonnes (Venkatramanan et al. 2021c) while in China, the annual gross crop residue generated is about 970 million tonnes, and 495 million tonnes are available for biomethane production (Sun et al. 2021). Due to the variation in crop area and biomass productivity, the spatial variation in the availability of agricultural residues is observed.

Further, certain crop types may be preferred as feedstocks for biogas generation (Thomas et al. 2019). For example, certain crops like sorghum require low inputs and thrive in dryland environments. Thomas et al. (2019) observed variation in biomethane potential between sorghum genotypes and cultivation sites. The methane potential was observed to range between 200 ± 5 Nml_{CH4}/g_{TS} and 259 ± 12 Nml_{CH4}/g_{TS}. The spectrum of feedstocks has widened in the recent past, as the strategy is to obtain energy from waste materials and sustainably manage the resources. Spent Coffee Grounds (SCG) have emerged as a significant substrate for biogas generation due to two reasons: (a) SCG do not require pretreatment, and (b) SCG contains > 25% lipid content. Anaerobic digestion of liquid fraction of SCG and cattle manure produced 254 ml CH_4/g_{VS} (Luz et al. 2017).

Studies have highlighted the challenges to the anaerobic digestion of agriculture residues: recalcitrant lignocellulosic structure, high C/N ratio and silica content (Singh and Kumar 2019, Ngan et al. 2020). The issues relating to the low digestibility of feedstocks can be addressed through strategies such as pretreatment, co-digestion, use of cellulose/hemicellulose degrading microorganisms, bioaugmentation, and microbial inoculum (Singh and Kumar 2019).

3.1 Pretreatment

The rigidity of the lignocellulose complex, crystallinity of cellulose and spatial structure hinder the digestion by anaerobic microorganisms. These constraints can be addressed by appropriate pretreatment methods, which would change the chemical composition of wastes, increase the biodigestibility of feedstocks, efficiency of anaerobic digestion and increase the biogas and biomethane yields. The pretreatment methods include physical, chemical and biological methods. The physical treatment methods include "particle size reduction", high-temperature treatment methods and "hydrothermal treatment method".

Ngan et al. (2020) reported that pretreatment of rice straw through "physical methods (particle size reduction), chemical methods (acid and alkali treatment) and biological methods (use of fungi)" and co-digestion of "rice draw with animal manure" improves the biodegradation of lignocellulosic wastes. In addition, particle size reduction of agro-residues increases the specific surface area of biomass and digestibility (Singh and Kumar 2019).

Studies have reported the effect of high-temperature pretreatment on the anaerobic digestion of crop residues (Rajput et al. 2018). In a study to know the effect of high-temperature pretreatment on wheat straw digestion, it was found that high-temperature (180°C) treatment changed the chemical structure of lignocellulose, increased the cellulose content, yielded biogas of 615 Nml/gVS and enhanced the biogas production by 53% (Rajput et al. 2018). Fernandez-Cegri and co-workers reported that sunflower oilseed cake used in anaerobic digestion through hydrothermal pretreatment at 100°C improved the efficiency of biogas production (Fernández-Cegrí et al. 2012).

Chemical pretreatment methods involve the use of ionic liquids, solvents and chemicals. In a study by (Gao et al. 2013), the pretreatment of water hyacinth with ionic liquid (1-ethyl-3-methyl imidazolium acetate and 1-*N*-butyl-3-methyl imidazolium chloride) and co-solvent (dimethyl acetylamide and dimethyl sulfoxide) resulted in an increase of "biogas yield" and "methane concentration" by 16.3–97.6% and 13.2–28.3%, respectively. The biomethane production from wheat straw increased significantly on the adoption of calcium hydroxide pretreatment, "particle size reduction", and "enzyme addition". The methane potential increased from 48 NmL_{CH4}/g_{VS} to

202 NmL_{CH4}/g_{VS} when the wheat straw was treated with calcium hydroxide, and the particle size was also reduced (Reilly et al. 2015). Biological pretreatment methods involve the use of fungus, enzymes and lignocellulose-degrading organisms. Biological methods are advantageous over physical and chemical methods due to "substrate specificity" and "reaction specificity" (Singh and Kumar 2019).

3.2 Co-digestion

As compared to the mono-digestion process, co-digestion has added benefits in terms of biogas production and digester optimization. Co-digestion of carbon-rich and nitrogen-rich biomass optimizes the C/N ratio, meets the nutritional requirement of degrading microorganisms and establishes synergism between the constituents of the digester. Agricultural residues are co-digested with animal manure, sewage sludge, food waste, etc. For instance, agricultural residues like "dry fallen leaves" (DFL) and "fruit and vegetable wastes" (FVW) are co-digested with cow dung to produce biogas. Anaerobic digestion with a DFL to FVW ratio of 40:60 generated a maximum methane yield of 388 ± 131 ml_{CH4}/g_{VS}. In addition, the fast biodegradability of FVW and slow release of carbon from DFL stabilized the co-digestion process (d'Silva et al. 2022). Biogas produced from feedstocks such as Pongamia pinnata de-oiled cake and cow dung have a methane concentration of 73% by volume (Verma et al. 2017).

3.3 Use of efficient microbial inoculum

The anaerobic digestion process depends on the microbial consortium, which plays a pivotal role in the sequential reaction of feedstock digestion. In order to provide diversity among microbial communities, inoculum sources like animal manure, sludge, and digestate are used in the anaerobic digestion process. In addition, a bioaugmentation or co-inoculation strategy is recommended to enrich the microbial communities to perform a particular activity in the digestion process. For instance, co-inoculation with hydrolytic organisms is adopted to increase the efficiency of the anaerobic digestion (AD) of paddy straw (Singh and Kumar 2019). Mainly, AD process involves the decomposition of agro residues waste in biogas by four groups of microorganisms. The hydrolysis process is the initial stage, where hydrolytic enzymes produced by hydrolytic bacteria like *Bacteroides*, *Megasphaera*, *Bifidobacterium*, *Lactobacillus*, *Sporobacterium*, *Propionibacterium*, etc., to create a suitable environment and produce intermediate metabolites such as soluble and insoluble monomers. After that, acid-forming bacterial species such as *Clostridium*, *Peptococcus*, *Bifidobacterium*, *Desulphovibrio*, *Corynebacterium*, *Actinomyces*, and *Escherichia coli* convert monomers into fatty acids and H_2. Then organic acids such as propionic, butyric and small quantities of valeric acid are produced under acetogenesis. Under the Acetogenesis process, microbial strains such as *Acetobacterium, Acetoanaerobium, Acetogenium,* and *Butyribacterium produce organic acid converted into acetate, and finally, autotrophic methanogenic bacteria*, for example, *Methanosarcina* and *Methanosaeta* perform methanogenesis to convert acetate to biogas ((Prasad et al. 2019, 2021) as shown in Fig. 2.

However, the AD process is affected by various environmental factors such as the quality of organic substrates, temperature, pH, volatile fatty acids, C/N ratio, amount of various required nutrients and other operating conditions, including retention time, organic matter loading rate and mixing of the substrate. The sustainability of the anaerobic digestion process is a function of "supply-chain logistics" and sustained biomethane production. Agricultural residues are voluminous, and high cost is involved in the collection of residues, storage and transportation of residues. In order to reduce the cost of residue preparation, strategies like baling, briquetting and establishment of decentralized biogas plants can be adopted. First, the logistic cost can be significantly reduced by setting up the biogas plants at the village levels. Secondly, adopting advanced and innovative technologies

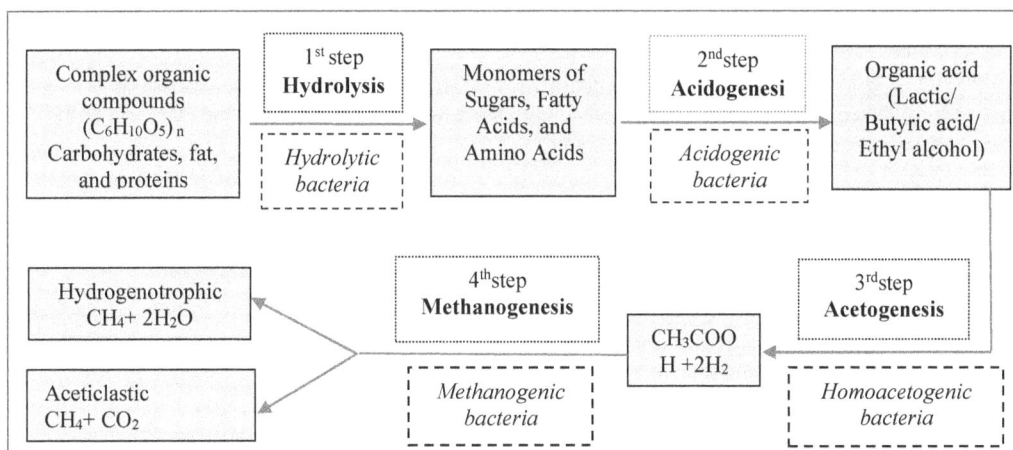

Figure 2. Various stages in the AD process for biogas production.

would upscale the anaerobic digestion process and increase biomethane production (Singh and Kumar 2019).

4. BIOHYDROGEN PRODUCTION FROM AGRO-RESIDUE WASTE

Biological methods provide a wide range of approaches to generate H_2, including photo-fermentation and dark-fermentation. Photosynthetic bacteria evolve molecular hydrogen catalyzed by nitrogenase under N-deficient conditions using light energy and reduced compounds like organic acids produced from agro-residues (Levin et al. 2004, Chatellard et al. 2017, Kumar et al. 2019). These bacteria themselves are not powerful enough to split water. However, under AD, these bacteria are able to use simple organic acids, like acetic acid, as electron donors. These electrons are transported to the nitrogenase by ferredoxin using energy in the form of ATP. When N is not present, this nitrogenase enzyme can reduce proton into H_2 gas again using extra energy in the form of ATP. The overall reaction of H_2 production can be given as follows:

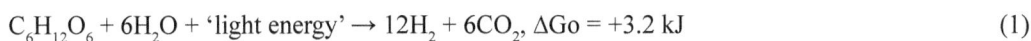

$$C_6H_{12}O_6 + 6H_2O + \text{'light energy'} \rightarrow 12H_2 + 6CO_2, \Delta Go = +3.2 \text{ kJ} \qquad (1)$$

The H_2 production rates of the order of 145–160 mmol/h per liter have been reported in the literature (Levin et al. 2004).

H_2 can be produced by anaerobic bacteria grown in the dark on the carbohydrate-rich substrate from agro-residues. Most of the microbial H_2 production is driven by the AD metabolism of pyruvate, formed during the catabolism of various substrates. The breakdown of pyruvate is catalyzed by one of two enzyme systems (Hallenbeck and Benemann 2002, Chatellard et al. 2017, Kumar et al. 2019):

(i). Pyruvate formatelyase

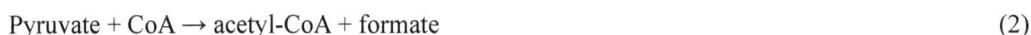

$$\text{Pyruvate} + \text{CoA} \rightarrow \text{acetyl-CoA} + \text{formate} \qquad (2)$$

(ii). Pyruvate ferredoxinoxido reductase

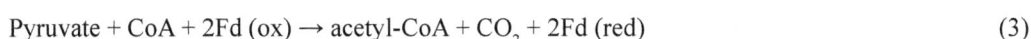

$$\text{Pyruvate} + \text{CoA} + 2\text{Fd (ox)} \rightarrow \text{acetyl-CoA} + CO_2 + 2\text{Fd (red)} \qquad (3)$$

Carbohydrates from crops/agro-residues are the preferred feedstock for H_2 production. The amount of H_2 produced from carbohydrates/glucose/pentose is affected by fermentation pathways and liquid end-products. Butyric acid ($C_4H_8O_2$) and acetic acid CH_3COOH constituted more than 80% of total end products. Theoretically, 4 mol of H_2 is formed from 1 mol of glucose in acetate-type fermentation. However, only 2 mol of H_2 is produced when $C_4H_8O_2$ is the essential fermentation product. There is a difference in strict anaerobic bacteria, a theoretical 4 moles of H_2 per mole of glucose is obtained, and in facultative anaerobes like *Escherichia coli* maximum of 2 moles of H_2 per mole of glucose is produced. In laboratory experiments, the H_2 production rate of 77 mmol/h per liter has been achieved (Kumar and Das 2002). AD of solid organic waste, including municipal and agricultural wastes and wastewater sludge, is a renewable source for fermentative H_2 production. However, continual H_2 production using this process has limitations, one of which is the low yields of H_2 currently realized from the fermentation of even the simplest sugars. A combination of dark and photo-fermentation in a two-stage hybrid system can improve the overall yield of H_2.

Microorganisms involved in dark fermentation are chemotrophic, which includes strict anaerobes (*Clostridia* sp.), facultative anaerobes (*Enterobacteraerogenes*) and aerobic bacteria (*Bacillus* sp.) (Kumar et al. 2019). In addition, photo-fermentation is carried out by purple non-sulphur photosynthetic bacteria (*Rhodospirillum* sp., *Rhodobacter* sp.). These microbes generate biohydrogen under photoheterotrophic conditions (in the presence of light, anaerobic conditions, and organic electron donors). Some recent critical reports on biohydrogen production from various agricultural wastes and crop residues using biological methods are listed in Table 2.

Table 2. Some recent reports (2018–2021) on fermentative H_2 production from various agricultural wastes and crop residues.

Agri-residue	Pretreatment/ hydrolysis method	Microbial culture/ inoculum	Bioconversion method	H_2 yield or volume	References
Fruit waste	Microwave (2.45 GHz, 15 min) Moist heat (121°C, 15 min)	*Clostridium* strain BOH3	Clostridial fermentation	359.97 ± 12.80 ml/g Total Solid	(Mahato et al. 2020)
Rice husk	Moist heat (121°C, 45 min) followed by enzymatic hydrolysis	Co-culture of *Clostridium termites* and *Clostridium intestinal* in the ratio of 5:1	Batch solid-state fermentation (SSF)	5.9 ml H_2/g substrate	(Tosuner et al. 2018)
Corn stalk, corn cob, rice straw, wheat straw	H_2SO_4 treatment (0.5 wt.%) followed by pretreatment (121°C, 60 min)	Anaerobic sludge (from wastewater treatment plant) enriched with H_2-producing bacteria	Dark fermentation (batch mode)	762.3 mL L^{-1}	(Ren et al. 2019)
Wheat straw	Ultrasound combined with dilute alkali cooking followed by enzymatic saccharification (cellulase)	*Enterobacter aerogenes* ATCC13048	Dark fermentation (Batch mode)	133.6 mL/g Total solid	(Zhu et al. 2022)
Potato peel waste	–	Mixed culture (abundance of *Clostridium* sp.)	Dark fermentation	71.0 mL/g-VS$_{added}$	(Cao et al. 2022)
Corncob	Deep eutectic solvent pretreatment followed by enzymatic hydrolysis	Purple non-sulphur bacteria	Photo-fermentation	151 mL/g	(Jing et al. 2022, Reungsang et al. 2018)
Sewage sludge and grass residue	Drying and milling (for grass residue)	Inoculum sampled from the anaerobic digester	Dark fermentation	45.6 mL/g-VS$_{added}$	(Yang and Wang 2019)

Mahato et al. (2020) investigated biohydrogen (H_2) generation using fruit waste using Clostridium strain BOH3 (Mahato et al. 2020). The biohydrogen can be produced using clostridium fermentation using substrates like wastewater, food waste and agro-residue waste (Mahato et al. 2020).

4. CONCLUSION

With the depletion of fossil fuels and increasing conventional energy prices, clean energy provides a suitable new energy option worldwide. The high amount of organic content in the agro-residue waste and their easy availability provides a suitable solution to use them as substrates compared to the monomer substrates in the anaerobic co-digestion and fermentation for the production of biogas and biohydrogen. In order to effectively utilize agro-residue waste in the AD and fermentation process, a suitable pretreatment process for agro-residues needs to be developed to maximize the biogas and biohydrogen yield. H_2 and CH_4 production is an exciting new area of technology development that offers the potential to produce various renewable energy resources to meet the nation's energy demands. More research is required to develop a suitable conversion process by applying an anaerobic digestion approach to treat agro-residue waste in the continuous mode with particular types of agro-residue waste in both anaerobic co-digestion and dark and photo-fermentation process for the production of biogas and biohydrogen in a cost-effective way.

REFERENCES

Abdeen, F.R.H., Mel, M., Jami, M.S. et al. 2016. A review of chemical absorption of carbon dioxide for biogas upgrading. Chin. J. Chem. Eng. 24: 693–702. https://doi.org/10.1016/J.CJCHE.2016.05.006.

Ammenberg, J. and Feiz, R. 2017. Assessment of feedstocks for biogas production, part II—Results for strategic decision making. Resour. Conserv. Recycl. 122: 388–404. https://doi.org/10.1016/J.RESCONREC.2017.01.020.

Angelidaki, I., Treu, L., Tsapekos, P. et al. 2018. Biogas upgrading and utilization: Current status and perspectives. Biotechnol. Adv. 36: 452–466. https://doi.org/10.1016/J.BIOTECHADV.2018.01.011.

Braun, R. 2007. Anaerobic digestion: A multi-faceted process for energy, environmental management and rural development. pp. 335–416. *In*: Ranalli, P. (ed.). Improvement of Crop Plants for Industrial End Uses. Springer Netherlands, Dordrecht. https://doi.org/10.1007/978-1-4020-5486-0_13.

Calbry-Muzyka, A., Madi, H., Rüsch-Pfund, F. et al. 2022. Biogas composition from agricultural sources and organic fraction of municipal solid waste. Renew. Energy 181: 1000–1007. https://doi.org/10.1016/J.RENENE.2021.09.100.

Cao, J., Xu, C., Zhou, R. et al. 2022. Potato peel waste for fermentative biohydrogen production using different pretreated culture. Bioresource Technology 362: 127866.

Chatellard, L., Marone, A., Carrère, H. et al. 2017. Trends and challenges in biohydrogen production from agricultural waste. Biohydrogen Production: Sustainability of Current Technology and Future Perspective, 69–95.

d'Silva, T.C., Isha, A., Verma, S. et al. 2022. Anaerobic co-digestion of dry fallen leaves, fruit/vegetable wastes and cow dung without an active inoculum—A biomethane potential study. Bioresour. Technol. Rep. 19: 101189. https://doi.org/10.1016/J.BITEB.2022.101189.

Fernández-Cegrí, V., Ángeles de la Rubia, M., Raposo, F. et al. 2012. Effect of hydrothermal pretreatment of sunflower oil cake on biomethane potential focusing on fibre composition. Bioresour. Technol. 123: 424–429. https://doi.org/10.1016/J.BIORTECH.2012.07.111.

Gao, J., Chen, L., Yan, Z. et al. 2013. Effect of ionic liquid pretreatment on the composition, structure and biogas production of water hyacinth (Eichhornia crassipes). Bioresour. Technol. 132: 361–364. https://doi.org/10.1016/J.BIORTECH.2012.10.136.

Hallenbeck, P.C. and Benemann, J.R. 2002. Biological hydrogen production: Fundamentals and limiting processes. Int. J. Hydrogen Energy 27: 1185–93.

Herrmann, C., Idler, C. and Heiermann, M. 2016. Biogas crops grown in energy crop rotations: Linking chemical composition and methane production characteristics. Bioresour. Technol. 206: 23–35. https://doi.org/10.1016/J.BIORTECH.2016.01.058.

Holliger, C., Fruteau de Laclos, H. and Hack, G. 2017. Methane production of full-scale anaerobic digestion plants calculated from substrate's biomethane potentials compares well with the one measured on-site. Frontiers in Energy Research.

IEA. 2020. Outlook for biogas and biomethane: Prospects for organic growth. World Energy Outlook Special Report 92. https://doi.org/10.1787/040c8cd2-en.

Jing, Y., Li, F., Li, Y. et al. 2022. Biohydrogen production by deep eutectic solvent delignification-driven enzymatic hydrolysis and photo-fermentation: Effect of liquid–solid ratio. Bioresource Technology 349: 126867.

Kadam, R. and Panwar, N.L. 2017. Recent advancement in biogas enrichment and its applications. Renewable and Sustainable Energy Reviews 73: 892–903. https://doi.org/10.1016/J.RSER.2017.01.167.

Kleinheinz, G. and Hernandez, J. 2016. Comparison of two laboratory methods for the determination of biomethane potential of organic feedstocks. J. Microbiol. Methods 130: 54–60. https://doi.org/10.1016/J.MIMET.2016.08.025.

Kumar, N. and Das, D. 2002. Continuous hydrogen production by immobilized Enterobacter cloacae IIT-BT 08 using lignocellulosic materials as solid matrices. Enzyme Microbiol. Technol. 29: 280–7.

Kumar, S., Sharma, S., Thakur, S. et al. 2019. Bioprospecting of microbes for biohydrogen production: Current status and future challenges. Bioprocessing for Biomolecules Production, 443–471.

Levin, D.B., Pitt, L. and Love, M. 2004. Biohydrogen production: Prospects and limitations to practical application. Int. J. Hydrogen Energy 29: 173–85.

Luz, F.C., Cordiner, S., Manni, A. et al. 2017. Anaerobic digestion of coffee grounds soluble fraction at laboratory scale: Evaluation of the biomethane potential. Appl. Energy 207: 166–175. https://doi.org/https://doi.org/10.1016/j.apenergy.2017.06.042.

Mahato, R.K., Kumar, D. and Rajagopalan, G. 2020. Biohydrogen production from fruit waste by Clostridium strain BOH3. Renewable Energy 153: 1368–1377.

Ngan, N.V.C., Chan, F.M.S., Nam, T.S. et al. 2020. Anaerobic digestion of rice straw for biogas production. pp. 65–92. *In*: Gummert, M., Hung, N. van, Chivenge, P. et al. (eds.). Sustainable Rice Straw Management. Springer International Publishing, Cham. https://doi.org/10.1007/978-3-030-32373-8_5

Omar, B., El-Gammal, M., Abou-Shanab, R. et al. 2019. Biogas upgrading and biochemical production from gas fermentation: Impact of microbial community and gas composition. Bioresour. Technol. 286: 121413. https://doi.org/10.1016/J.BIORTECH.2019.121413.

Prasad, S., Sheetal, K.R., Venkatramanan, V. et al. 2019. Sustainable energy: Challenges and perspectives. Sustainable Green Technologies for Environmental Management. https://doi.org/10.1007/978-981-13-2772-8_9.

Prasad, S., Venkatramanan, V. and Singh, A. 2021. Renewable energy for a low-carbon future: Policy perspectives. pp. 267–284. *In*: Venkatramanan, V., Shah, S. and Prasad, R. (eds.). Sustainable Bioeconomy: Pathways to Sustainable Development Goals. Springer Singapore, Singapore. https://doi.org/10.1007/978-981-15-7321-7_12.

Rajput, A.A. and Zeshan, Visvanathan, C. 2018. Effect of thermal pretreatment on chemical composition, physical structure and biogas production kinetics of wheat straw. J. Environ. Manage. 221: 45–52. https://doi.org/10.1016/J.JENVMAN.2018.05.011.

Rasi, S., Veijanen, A. and Rintala, J. 2007. Trace compounds of biogas from different biogas production plants. Energy 32: 1375–1380. https://doi.org/10.1016/J.ENERGY.2006.10.018.

Reilly, M., Dinsdale, R. and Guwy, A. 2015. Enhanced biomethane potential from wheat straw by low-temperature alkaline calcium hydroxide pretreatment. Bioresour. Technol. 189: 258–265. https://doi.org/10.1016/J.BIORTECH.2015.03.150.

Ren, H.Y., Kong, F., Zhao, L. et al. 2019. Enhanced co-production of biohydrogen and algal lipids from agricultural biomass residues in long-term operation. Bioresource Technology 289: 121774.

Reungsang, A., Zhong, N., Yang, Y. et al. 2018. Hydrogen from photo fermentation. In Bioreactors for microbial biomass and energy conversion (pp. 221–317). Springer, Singapore.

Saravanan, A., Kumar, P.S., Aron, N.S.M. et al. 2021. A review on bioconversion processes for hydrogen production from agro-industrial residues. International Journal of Hydrogen Energy 47: 37302–37320.

Shah, S., Venkatramanan, V. and Prasad, R. (eds.). 2021. Bio-valorization of waste: Trends and perspectives. Environmental and Microbial Biotechnology 347. https://doi.org/10.1007/978-981-15-9696-4.

Singh, R. and Kumar, S. 2019. A review on biomethane potential of paddy straw and diverse prospects to enhance its biodigestibility. J. Clean Prod. 217: 295–307. https://doi.org/10.1016/J.JCLEPRO.2019.01.207.

Styles, D., Yesufu, J., Bowman, M. et al. 2022. Climate mitigation efficacy of anaerobic digestion in a decarbonizing economy. J. Clean Prod. 338: 130441. https://doi.org/https://doi.org/10.1016/j.jclepro.2022.130441.

Sun, H., Wang, E., Li, X. et al. 2021. Potential biomethane production from crop residues in China: Contributions to carbon neutrality. Renewable and Sustainable Energy Reviews 148: 111360. https://doi.org/10.1016/J.RSER.2021.111360.

Surendra, K.C., Takara, D., Hashimoto, A.G. et al. 2014. Biogas as a sustainable energy source for developing countries: Opportunities and challenges. Renewable and Sustainable Energy Reviews 31: 846–859. https://doi.org/10.1016/J.RSER.2013.12.015.

Thomas, H.L., David Pot, Latrille, E. et al. 2019. Sorghum biomethane potential varies with the genotype and the cultivation site 10: 783–788. https://doi.org/10.1007/s12649-017-0099-3.

Tosuner, Z.V., Taylan, G.G. and Özmıhçı, S. 2019. Effects of rice husk particle size on biohydrogen production under solid state fermentation. International Journal of Hydrogen Energy 44(34): 18785–18791.

Ullah Khan, I., Hafiz Dzarfan Othman, M., Hashim, H. et al. 2017. Biogas as a renewable energy fuel—A review of biogas upgrading, utilization and storage. Energy Convers. Manag. 150: 277–294. https://doi.org/10.1016/J.ENCONMAN.2017.08.035.

Venkatramanan, V., Shah, S. and Prasad, R. 2021a. Sustainable Bioeconomy: Springer Singapore. https://doi.org/10.1007/978-981-15-7321-7.

Venkatramanan, V., Shah, S. and Prasad, R. 2020. Global climate change and environmental policy: Agriculture perspectives, Global Climate Change and Environmental Policy: Agriculture Perspectives. Springer Singapore. https://doi.org/10.1007/978-981-13-9570-3.

Venkatramanan, V., Shah, S. and Prasad, R. 2021a. Sustainable Bioeconomy: Springer Singapore. https://doi.org/10.1007/978-981-15-7321-7.

Venkatramanan, V., Shah, S., Rai, A.K. et al. 2021b. Nexus between crop residue burning, bioeconomy and sustainable development goals over North-Western India. Front. Energy Res. 8: 614212. https://doi.org/10.3389/fenrg.2020.614212.

Venkatramanan, V., Shah, S., Prasad, S. et al. 2021c. Assessment of bioenergy generation potential of agricultural crop residues in India. Circular Economy and Sustainability 1: 41, 1335–1348. https://doi.org/10.1007/S43615-021-00072-7.

Verma, S., Das, L.M. and Kaushik, S.C. 2017. Effects of varying composition of biogas on performance and emission characteristics of compression ignition engine using exergy analysis. Energy Convers. Manag. 138: 346–359. https://doi.org/10.1016/J.ENCONMAN.2017.01.066.

Yang, G. and Wang, J. 2019. Biohydrogen production by co-fermentation of sewage sludge and grass residue: Effect of various substrate concentrations. Fuel 237: 1203–1208.

Yusuf, N. and Almomani, F. 2023. Recent advances in biogas purifying technologies: Process design and economic considerations. Energy 265: 126163. https://doi.org/https://doi.org/10.1016/j.energy.2022.126163.

Zhu, J., Song, W., Chen, X. et al. 2022. Integrated process to produce biohydrogen from wheat straw by enzymatic saccharification and dark fermentation. International Journal of Hydrogen Energy.

13

APPLICATIONS OF MECHANICAL BIOLOGICAL PRE-TREATMENT METHODS OF MUNICIPAL SOLID WASTE

Navyashree Hanasoge Jagadeesha,[1] *Shivani Maddirala*[2]
and *Surajbhan Sevda*[2,*]

1. INTRODUCTION

The ever-increasing population, urbanization and industrialization has led to an increased waste generation. Proper handling of this enormous amount of solid waste generated by citizens is a burden for local governments. Municipal solid waste (MSW) is often waste from households, but it also includes waste of a similar sort from industries, institutions, and enterprises, as well as waste from maintaining public spaces. The MSW generated globally is approximately 2.01 billion tonnes/annum and is estimated to reach 3.40 billion tonnes/annum by 2050. Insufficient budget, infrastructure, political commitment, and public participation has led to poor management of MSW which has led to adverse effects on environment and public health. As the simplest, most affordable, and cost-effective method of waste disposal, landfills continue to be the preferred choice for managing MSW in many regions of the world. However, it doesn't stop environmental damage; instead, it raises the risk of leachate and landfill gas emissions, as well as the deterioration of soil and groundwater quality and the acceleration of global warming. When substantial amounts of putrescible waste are disposed of in landfills with inadequate technical and managerial controls, the problem is exacerbated. MBT plants, which handle unsorted solid waste intended for ultimate disposal, can be used as straightforward methods that instantaneously and significantly lessen environmental impacts in the vicinity of final disposal locations.

[1] Department of Civil Engineering, National Institute of Technology Warangal, Warangal-506004, India.
[2] Environmental Bioprocess Laboratory, Department of Biotechnology, National Institute of Technology Warangal, Warangal-506004, India.
* Corresponding author: sevdasuraj@gmail.com, sevdasuraj@nitw.ac.in

2. DESCRIPTION OF MECHANICAL BIOLOGICAL TREATMENT (MBT) PROCESS

Mechanical–biological pre-treatment (MBT) of municipal solid waste (MSW) plays a significant role in waste management and has gained considerable attention as it helps to achieve an environmentally sustainable landfilling. Based on the specific target to be achieved, in order to create a bio-stabilized product with fewer negative effects than raw waste being landfilled, MBT can be used in various combinations with mechanical sorting, bio-drying, and biological processes. These processes may be used as pre-treatments before incineration or pre-treatments before landfilling. MBT processes are capable of dividing the raw waste into different streams that include material recycling, energy recovery or disposal. They allow for the recovery of valuable resources and the expansion of the variety of energy recovery applications. MBT consists of several types mechanical pre-processing units which is usually followed by a biological treatment in which biodegradable matter is stabilized under controlled anaerobic and/or aerobic conditions. Mechanical separation systems in MBT use equipments such as bag openers, trommels, sieves, screw presses, shredders, magnetic sorters, mills and/or pulpers, etc., as shown in Fig. 1, to segregate the wastes into three fractions: **a fuel fraction** (known as RDF, refuse derived fuel), either used in specific dedicated plants or introduced into the general fuel market; **a wet fraction** (treated in aerobic or anaerobic condition and also a prospect for additional energy valorisation like biogas or compost) and **a mineral fraction** (for this, no reasonable option till now except disposal in a landfill).

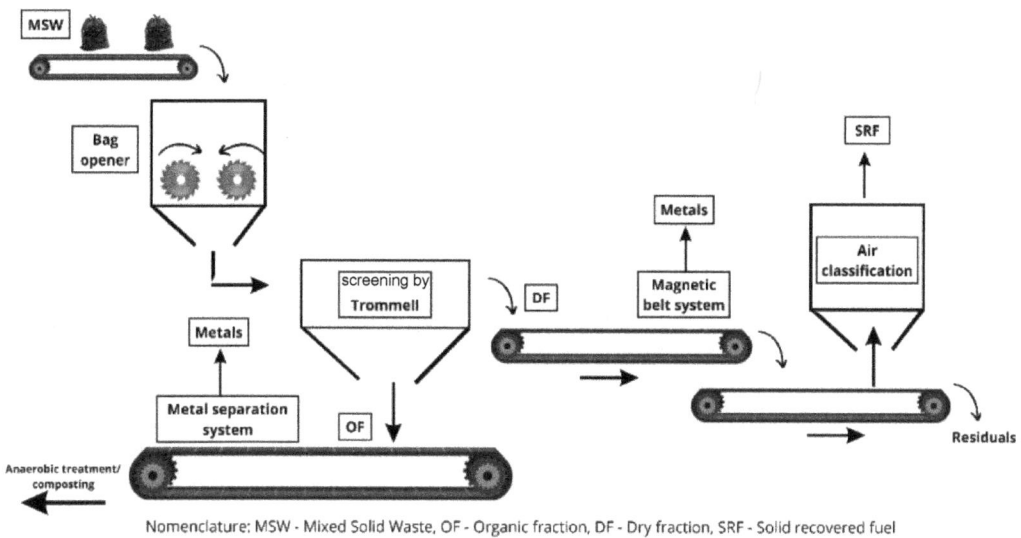

Nomenclature: MSW - Mixed Solid Waste, OF - Organic fraction, DF - Dry fraction, SRF - Solid recovered fuel

Figure 1. Diagram of the mechanical sorting section of the MBT plant (Di Maria et al. 2012).

Following mechanical separation, biological treatment (aerobic, anaerobic, or a combination of both) is conducted to produce a final stabilised material of the wet waste fraction with low methane potential (typically < 20 Nm^3 tonne^{-1} ww), allowing for landfilling with reduced environmental impacts, particularly with regard to gas and leachate generation. Depending on the period of the biological treatment, the product that has been biostabilized may display varying levels of biological stability. A first order kinetic model is a good fit to the organic matter (OM) degradation processes, and the ones lasting longer than 4 weeks do not significantly increase biological stability.

Depending on the duration of the process, biological pre-treatment of waste provides a significant reduction in potential biogas output. After the biological treatment, the biogas generation potential will be reduced. This is because of the reason that as organic matter degrades, there is a decline in the anaerobic biogasification potential (ABP) and dry matter content of MSW.

3. MBT WASTE MATERIAL

Mechanical biological treatment is designed to transform and separate the residual MSW into recyclables, combustibles (RDF), compost and/or biogas. MBT is employed to recover materials from waste and to stabilize the organic waste fraction in residual waste. By integrating MBT into current MSW systems, waste can be diverted in the concept of waste hierarchy. Mechanical pre-processing stage in MBT helps to recover the recyclable materials such as plastics, metals and glass materials. After recovery, waste stream is diverted into two streams: (a) one stream with oversized fraction which is further processed to be used as refuse-derived fuel (RDF) and (b) other stream with under-size fraction which is rich in putrescible organic material further will be sent for biological treatment using either aerobic or anaerobic process to stabilize the raw waste fraction.

3.1 Characterization of the raw waste

A collection truck is used for collection of a primary sample from street containers of mass about one tonne of waste. In order to make sure the sampling is representative of the total waste, a specific collection route was identified considering factors such as metropolitan morphology, residential typology and existence of industry production plants (Table 1). The days chosen for sampling were in the middle of the week in order to generalise the results as an example of typical circumstances. For the aim of keeping a steady nature of moisture content and temperature of waste during the analysis, a day in a spring season with a stable weather condition was taken into consideration. The primary sample was manually split into two sub-samples of 200 kg and 15 kg using the quartering method in order to obtain a representative sample. By manually sorting, the waste composition of a 200 kg sub-sample was determined. In order to determine the physical, chemical, and biological properties of waste, an additional sample of 15 kg was used in the laboratory analysis. The important physical, chemical and biological parameters that can be analysed are given in the table below.

Table 1. Characterization of raw waste.

Characteristics of raw waste	Parameters	Remarks
Physical properties of MSW	Density, Moisture content, Particle size and distribution, Field Capacity, Hydraulic Conductivity of Compacted Waste	The physical characterization of the solid waste is performed to appropriately select and operate the equipment. It also plays an important role in design of the disposal facilities
Chemical properties of MSW	Ultimate analysis, Proximate analysis, Energy content, Fusing point of ash	If waste is used as fuel or for biological conversion products, this plays a key role
Biological properties of MSW	OM of MSW can be classified as Sugars, Starches and organic acids, Proteins and amino acids, Hemicellulose, Cellulose and lignocellulose, Lignin, Fats, oils, and waxes	Information on the vital nutrients present in the waste materials is crucial since the organic matter from MSW is used as a feedstock for compost or biological conversion

3.2 Chemical and physical composition of the waste

The chemical, physical and biological characteristics of the waste fraction which contained similar fraction of materials is to be analysed in a laboratory. The samples will be evaluated for parameters like pH, moisture content, percentage of volatile solids, ash content, lower and upper calorific values, and carbon and nitrate content. The C/N ratio of the waste mixture is assessed to confirm the suitability of the waste material for the composting process. The composition study of waste carried out in a laboratory identifies whether the waste is suitable to be treated in biological system of MBT. Determining the physical composition of waste such as fraction of organics, paper, plastic, glass, metals, inert, etc., helps us to determine the best and suitable methodology and technology for treating MSW. Like physical characteristics, chemical characteristics are also useful in evaluating the final treatment options of MSW. Table 2 below provides an illustration of the physical properties of municipal solid waste in the top ten most popular nations. The chemical properties of MSW in various nations throughout the world are depicted in Table 3 below. Due to variables including economic level, climate, cultural values, and existing treatment options in their nation, the stated value is varies.

Table 2. Physical composition of MSW in the top 10 most populous countries (Source: Ria millati et al. 2019).

Country	Year	MSW composition (% mass)					
		Organic	Paper	Plastic	Glass	Metals	Textile and other
China	2002	59	8	10	3	1	19
India	2008	40	10	2	0.2	0	47.8
USA	2009	12.7	31	12	4.9	8.4	31
Brazil	2007	36.1	17.1	23.3	3.5	2.4	17.6
Indonesia	2007	74	10	8	2	2	4
Russia	1996	31.5	28	4	6	2.5	28
Japan	2003	42.6	22.3	11.4	1.6	9	13
Mexico	2000	52.4	14.1	4.4	5.9	2.9	20.3
Nigeria	2000	68	10	7	4	3	8
Germany	2005	30	24	13	10	1	22
Pakistan	2009	67	5	18	2	0	7

Table 3. Chemical composition of municipal solid waste in selected countries (Source: Ria millati et al. 2019).

Country	Elemental analysis (%)							Lower heating value (kJ/kg)
	C	H	O	N	S	Moisture (%)	Ash (%)	
Brazil	42	5.3	26.5	1.1	0.2	36	16	7376.5
China	29.4	3.9	28.7	1.6	0.5	55.4	18.9	4695.0
India	27.4	4.8	33.1	1.1	0.8	29.2	28.8	4270.0

4. MECHANICAL (GRINDING) PRE-TREATMENT AND PARTICLE SIZE DETERMINATION

To increase the surface area and pore size of the waste to be treated, mechanical grinding pre-treatment is used. Smaller particle sizes expose more surface area to enzyme attack, potentially improving carbon accessibility and hydrolysis of the handled material. The particle size reduction plays an important role in recovery of materials for reuse and for conversion to energy. Reduction in size provides larger surface area for the reaction in processes like composting and incineration. It makes recycling waste processing easier, promoting more frequent recycling. It enables more thorough component compaction in landfills, resulting in a more stable fill and enabling the disposal of more waste in a smaller space. Smaller particle sizes reported a faster biomass stabilization in a conventional anaerobic digestor. But very fine material of mean particle size of 2mm caused extreme foaming in wet anaerobic digestion, accumulation of volatile fatty acids and also lower biogas yields was reported due to the need to reduce organic loading rate. The most commonly used size reduction equipments are hammer mill, hydropulper, grinder, etc.

4.1 MSW management performance indicators

Performance indicators are used to evaluate and compare performance of various MSW management paths which helps to select the best available method that can be adopted. There are many indicators that are adopted and categorized for evaluating the performance of different components in an MSW management system. Some of the important performance indicators for MSW management include carbon footprint, material recovery rate, energy recovery rate, landfilling rate, specific fossil energy savings, waste management cost, specific fossil energy consumption, etc.

4.2 Mechanical biological treatment (MBT), plant description and system boundaries

MBT is composed of both mechanical units and biological treatment units. The mechanical units are used to mechanically separate MSW for the recovery of recyclables and the remaining biological wastes part are biologically (aerobic/anaerobic) digested to obtain stable end product. The mechanical-biological treatment (MBT) process is frequently employed as a pre-treatment before residual MSW is burned, recycled, or dumped in a landfill. MBT can be set up to serve a variety of needs and deal with different issues. Few benefits that can be obtained by adopting MBT are diverting biodegradable matter from landfills, sorting of recyclables mechanically, stabilizing the output to be used as compost or soil conditioner, producing biogas and producing combustibles (RDF) for energy recovery. The biowaste required for subsequent biological treatment are treated mechanically to obtain a clear bifurcation between wanted and unwanted organic components. Simplest way to pretreat biowaste is by using a shredder and a magnet. But this option is not feasible for waste with plastic and glass. For biowaste containing plastic and glass, a combination of screening and hand sorting of screened output is applied. But hand sorting is an unpleasant treatment and also does not guarantee efficiency of pretreatment. For more efficient treatment, screw press is used to separate wet parts from plastic and other unwanted materials.

For the purpose of recovering energy from MSW, a 100 tonne per day mechanical biological treatment plant was established in North Goa, India. The plant includes a material recovery facility (MRF), a bio-methanation and wastewater treatment plant, rotational and windrow composting, and a unit for converting biogas into power. The plant is able to produce 11.90% recyclables, 33% refused derived fuel (RDF), 5% compost of total waste received, 70 m³/day recyclable water and 0.435 MWh/day electricity.

5. MECHANICAL BIOLOGICAL TREATMENT PLANT DIRECT IMPACTS

Various Life Cycle Assessment (LCA) studies have been executed for different waste management options. LCA studies aim to evaluate various waste management options, as well as to improve and support decision-making in relation to various waste management systems. When compared with other treatment methods for MSW, especially incineration, MBT plant working in anaerobic condition to produce biogas had more benefits. But MBT facilities using aerobic treatment resulted in high impact on environmental point of view considering their high-energy intensive process, relative low yields and disposal of the system outflows into a landfill holding the main concern. Some scenarios suggest that all the material out from MBT generates most environmental benefits such as recyclables (metals) can be separated, biodegradable mass can be stabilized to compost that can be used as soil conditioner, and RDF can be used to recover energy. The direct impact that can possibly be reported depends on the units comprised in the MBT plant. Some of the basic impacts identified at MBT plant include CH_4 emissions produced during process operations, biogas combustion, transportation of the digestate for landfilling, etc. As MBT plants handle unsorted solid waste destined for final disposal, they can be used as swift and simple means to rapidly and significantly lessen the environmental effects close to final disposal locations. By using biological units to treat the waste, they can reduce the amount of waste entering disposal sites, prevent organic waste from decomposing in landfills, and extend the functional life of the disposal site. Leachate and landfill gas emissions have an adverse effect on the environment; however, MBT can mitigate this effect by reducing the waste's biodegradable organic content. The concentration of leachate COD can be reduced from 30,000 mg/L to 1000 mg/L, therefore offering significant reduction in the dimension of leachate treatment plant required. Furthermore, the operating life of the landfill can be enhanced and extended by more than 15 years by integrating MBT operations into MSW management systems.

The impact categories usually considered for evaluation include climate change, photochemical oxidant formation, terrestrial acidification, marine eutrophication, human toxicity, ecotoxicity, fossil resource depletion etc. When operators and planners are required to choose the best treatment schemes for an MBT plant, they can utilize the impact estimation of various MBT plant designs as a source.

5.1 Advances of Mechanical biological treatment

Mechanical biological treatment is a residual waste treatment process that comprises of both mechanical and biological treatment to handle the incoming waste. The waste stream is usually segregated into dry and wet portions. In the dry portion, the recyclables are separated and the portion having high calorific value is used as RDF to recover energy. The wet biodegradable portion is treated in a conventional composting or anaerobic units. Few advancements identified in MBT plants are stated below:

- Sorting of recyclables from mixed waste stream by incorporating sensors.
- By combining magnetic separation, eddy current, and optical sorting, residual waste is carefully sorted.
- Laser-induced breakdown spectroscopy (LIBS), X-ray sorting, optical sorting, spectral sorting, and Magnetic induction spectroscopy (MIS) are used to segregate non-ferrous metals, plastic, wood and glass.
- Pre-treatment of organic fraction of MSW (OFMSW) to improve biodegradability before bioconversion. Physical and mechanical (irradiation, high-pressure homogenization, electrohydrolysis), thermal (hydrothermal carbonization, hot compressed water, autoclave), chemical (acid, alkali, organic solvents) and types of biological methods (bacterial, fungal, enzymatic).
- Electricity generation from RDF fractions in advanced thermal treatment methods such as gasification, pyrolysis, and plasma-assisted gasification are seen as a progressive and optimistic technology.

> ➢ Co-digestion of OFMSW and wastewater sludge is one biological treatment approach that has demonstrated the concept of digestion in "combination" to be effective and promising.
> ➢ The mechanical pre-treatment of MSW for the separation of the organic fraction can be accomplished by high-pressure extrusion together with friction heating, mixing, crushing, and shearing activities.

5.2 Biogas production from Mechanical biological treatment waste

The organic fraction of MSW separated from mixed waste is either treated in a composting unit or anaerobic digestion. Anaerobic digestion (AD) exhibits an excellent potential for the treatment of organic fraction of solid waste including food waste, wastewater, sludges, etc. The conversion of the organic portion into methane-rich biogas and a highly stabilised digestate (bio-fertilizer) makes anaerobic digestion an effective treatment method for managing MSW with a high organic fraction. A community of microorganisms work together to break down and stabilise complex organic material, producing energy-dense biogas which represents a sustainable and renewable energy source that can be utilised to replace fossil fuels. The process is typically carried out in huge vessels where it can be managed to speed up reactions and collect the resulting biogas, which contains a high amount of methane and can be used for energy production. The AD process is divided into four main stages: hydrolysis, acidogenesis, acetogenesis, and methanogenesis. Methane (CH_4) and carbon dioxide (CO_2) are the major components of biogas produced by anaerobic digestion of organic material. When purified, the biogas generated by this method can be used to power vehicles and generate energy through a gas grid. It can also be used for cooking and heating. For sustainable waste management, anaerobic digestion is a suitable method which also offers an advantage of biofuel production. In order to improve the efficiency of bioconversion, several pretreatment techniques, including physicochemical, mechanical, and biological approaches, have been introduced. The current need to reduce the dependency on non-renewable energy sources highlights the importance of renewable energy produced (biogas: $\sim$ 70% CH_4) from anaerobic digestion.

6. CONCLUSIONS

The term "Mechanical Biological Treatment" (MBT) refers to the management of solid waste through the integration of both mechanical and biological treatment units. The first MBT facilities were built with the goal of minimizing the environmental effects of disposing of residual waste in landfills. As such, MBT as a part of an integrated waste management system complements rather than completely replacing other waste management technologies like recycling and composting. MBT plants usually consist of mechanical sorting to separate dry, wet and recyclable fraction. The wet part (organic portion) is composed or digested anaerobically to obtain stable end products. The dry part (RDF) is used for energy recovery. The main purpose of using MBT is diversion of biodegradable and non-biodegradable MSW from landfills and letting only the rejects into landfills. The biodegradation of organic portion anaerobically produces biogas which is a renewable resource and composting produces compost like stabilized output which can be used as soil conditioner. The dry fractions recovered are used to produce energy through advanced and conventional methods. This treatment has potential to manage various types of waste fractions in mixed stream of MSW, reflecting a sustainable and efficient waste management practice.

REFERENCES

Axel Zentner. 2018. From mechanical biological treatment to anaerobic digestion—Challenges in changing plant operations. Multidisciplinary Journal of Waste Resources & Residues 05: 46–56.

Beylot, A., Vaxelaire, S., Zdanevitch, I. et al. 2015. Life Cycle Assessment of mechanical biological pre-treatment of Municipal Solid Waste: A case study. Waste Management 39: 287–294.

Chantou, T., Feuillade, G., Mausset, D. et al. 2016. Application of stability indicators for the assessment of the degradation of residual household waste before landfilling. Waste Management & Research 34(12): 1283–1291.

Cimpan, C. and Wenzel, H. 2013. Energy implications of mechanical and mechanical–biological treatment compared to direct waste-to-energy. Waste Management 33(7): 1648–1658.

Dace Arina, Janis Kalnacs, Ruta Bendere et al. 2019. Mechanical pre-treatment for separation of bio-waste from municipal solid waste: Case study of district in Latvia.

De Gioannis, G. and Muntoni, A. 2007. Dynamic transformations of nitrogen during mechanical–biological pre-treatment of municipal solid waste. Waste Management 27(11): 1479–1485.

Di Maria, F., Sordi, A. and Micale, C. 2012. Optimization of solid state anaerobic digestion by inoculum recirculation: the case of an existing mechanical biological treatment plant. Applied Energy 97: 462–469.

Ewunie, G.A., Yigezu, Z.D.and Morken, J. 2021. Biochemical methane potential of Jatropha curcas fruit shell: Comparative effect of mechanical, steam explosion and alkaline pretreatments. Biomass Conversion and Biorefinery.

Fuss, M., Vergara-Araya, M., Barros, R.T.V. et al. 2020. Implementing mechanical biological treatment in an emerging waste management system predominated by waste pickers: A Brazilian case study. Resources, Conservation and Recycling 162: 105031.

Gourc, J.-P., Staub, M.J. and Conte, M. 2010. Decoupling MSW settlement into mechanical and biochemical processes – Modelling and validation on large-scale setups. Waste Management 30(8-9): 1556–1568.

Jank, A., Müller, W., Schneider, I. et al. 2015. Waste Separation Press (WSP): A mechanical pretreatment option for organic waste from source separation. Waste Management 39: 71–77.

Millati, R., Wikandari, R., Ariyanto, T., Putri, R.U., Taherzadeh, M.J., 2020. Pretreatment technologies for anaerobic digestion of lignocelluloses and toxic feedstocks. Bioresource Technology 304: 122998. https://doi.org/10.1016/j.biortech.2020.122998

Montejo, C., Tonini, D., Márquez, M. et al. 2013. Mechanical–biological treatment: Performance and potentials. An LCA of 8 MBT plants including waste characterization. Journal of Environmental Management 128: 661–673.

Panepinto, D., Blengini, G.A. and Genon, G. 2015. Economic and environmental comparison between two scenarios of waste management: MBT vs thermal treatment. Resources, Conservation and Recycling 97: 16–23.

Pantini, S., Verginelli, I., Lombardi, F. et al. 2015. Assessment of biogas production from MBT waste under different operating conditions. Waste Management 43: 37–49.

Ponsá, S., Gea, T. and Sánchez, A. 2010. The effect of storage and mechanical pretreatment on the biological stability of municipal solid wastes. Waste Management 30(3): 441–445.

Romero-Güiza, M.S., Peces, M., Astals, S. et al. 2014. Implementation of a prototypal optical sorter as core of the new pre-treatment configuration of a mechanical–biological treatment plant treating OFMSW through anaerobic digestion. Applied Energy 135: 63–70.

Scaglia, B., Salati, S., Di Gregorio, A. et al. 2013. Short mechanical biological treatment of municipal solid waste allows landfill impact reduction saving waste energy content. Bioresource Technology 143: 131–138.

Sergio Juarez-Hernandez. 2021. Energy, environmental, resource recovery, and economic dimensions of municipal solid waste management paths in Mexico city. Waste Management 136: 321–336.

Sivakumar Babu, G.L., Lakshmikanthan, P. and Santhosh, L.G. 2015. Shear strength characteristics of mechanically biologically treated municipal solid waste (MBT-MSW) from Bangalore. Waste Management 39: 63–70.

Trois, C. and Simelane, O.T. 2010. Implementing separate waste collection and mechanical biological waste treatment in South Africa: A comparison with Austria and England. Waste Management 30(8-9): 1457–1463.

Trulli, E., Ferronato, N., Torretta, V. et al. 2018. Sustainable mechanical biological treatment of solid waste in urbanized areas with low recycling rates. Waste Management 71: 556–564.

Tyagi, V.K., Kapoor, A., Arora, P. et al. 2021. Mechanical-biological treatment of municipal solid waste: Case study of 100 TPD Goa plant, India. Journal of Environmental Management 292: 112741.

14

INTEGRATION OF SITE-SPECIFIC DATA INTO LIFE CYCLE ASSESSMENT METHOD, WITH A FOCUS ON MUNICIPAL SOLID WASTE TREATMENT TECHNOLOGIES

Chejarla Venkatesh Reddy[1],* *Ashmita Kundu*[2] and
Ajay S. Kalamdhad[2]

1. INTRODUCTION

Asian countries have both developed and emerging economies. Due to rapid urbanization, the growth of the population and the consumption of goods has risen. As a result, the increasing municipal solid waste generation rate causes environmental, economic, and social issues. However, it can still address various demands (Li et al. 2019). The country's socioeconomic status defines the composition of municipal solid waste (MSW) and its generation rate. The value of waste generation is proportionate to national income levels, i.e., countries with higher GDP generate more plastic and packaging waste. Biodegradable waste is more persistent in countries with lower GDP (Shekdar 2009). In 2016, India ranked fifth globally with 0.12 million tonnes of solid waste produced per day, trailing only the United States, China, Brazil, and Japan (Bank 2005). Global municipal solid waste (MSW) generation is projected to reach more than 2200 million tonnes per year by 2025 (Tyagi et al. 2018). India's 1.36 billion population is expected to generate more than 56 million MSW tonnes each year at a 5% annual growth rate (CPCB 2016). The mixture of residential and commercial waste in India is typically conquered by organic waste (70–80%), remaining waste is inorganic

[1] Department of Civil Engineering, National Institute of Technology Goa, Farmagudi- 403401, Goa, India.
[2] Department of Civil Engineering, Indian Institute of Technology Guwahati, Guwahati - 781039, Assam, India.
* Corresponding author: venkatesh.venky1090@gmail.com

(Gupta et al. 2015, Ramachandra et al. 2018). As a result, direct waste disposal in open dumpsites or landfills without pretreatment of MSW has several geo-environmental consequences such as greenhouse gas (GHG) emissions and organic and inorganic compounds leaching from solid waste pollutes surface and groundwater (Gouveia and Prado 2010, Patil and Singh 2017). The total quantity of methane estimated in India was 4612.69 MT/d due to an open dumping system and stands as the third-largest source of GHGs (Singh et al. 2011). About 20% of waste is being processed, and the remaining 80% is being disposed of in an unscientific way, either inland or in water bodies, causing soil surface and surface water contamination (Gupta et al. 2018). Except for a few cities in India, the direct combustion of waste and dumping of garbage in low-lying areas without any treatment practices are common in the Indian scenario (Rana et al. 2019, Rathore et al. 2020). Significant problems are seen as being responsible for inadequate MSW management, i.e., collection, segregation, and recycling, as well as a shortage of treatment options (Babu et al. 2014, Botello-Álvarez et al. 2018). Uncontrolled dumping and open waste burning are standard waste disposal practices in India and other developing countries, contributing to environmental toxicity to human health and aquatic life (Yadav and Samadder 2017). Therefore, one of India's and other developing countries' most challenging problems is finding an MSW alternative with minimal environmental effects. Still, the problem can be solved sustainably to handle the tremendous amount of waste in India using life cycle assessment (LCA). In the Indian context, a case study from Delhi reported that recycling has the most negligible impact on the environment (Srivastava and Nema 2011). According to a case study from Mumbai, the open dump scenario was the least favoured choice because of the major environmental implications. Therefore, open dumps, landfills with and without gas recovery, bioreactor landfills, and bioreactor landfills were more influential than other approaches compared to four scenarios (Babu et al. 2014). Six different scenarios were chosen to determine the effects of MSW and the potential for global warming, acidification, eutrophication, and human toxicity as environmental impact categories. The findings indicated that while recycling, composting, anaerobic digestion, and landfilling are mixed, the environmental effect is negligible (Sharma and Chandel 2017). Waste to energy technology has a potential energy recovery optimal from landfills in developed countries, but unsanitary landfills are the most common practice in developing countries (Kumar and Samadder 2017). The cost of dumping waste in open areas, sanitary deposits with and without leachate treatment, and composting-landfilling were identified in a case study from Mumbai using a cost analysis of the life cycle. The study found that the one-tonne cost of waste disposal was INR 344, equivalent to USD 5.17. Mehta et al. (2018) reported that the waste gas flaring decreased global warming potential by 32 percent compared to open disposal, and treatment with leachate lowered its exposure to freshwater by 20 percent and human toxicity by 60 percent. However, composting-landfilling is the preferable option for waste management and decreases global warming potential by 79%. Khandelwal et al. (2019) stated that four possibilities were considered, i.e., S1 – composting + landfilling, S2 – material recovery facility (MRF) and composting combined with landfilling, S3 – MRF and anaerobic digestion combined with landfilling, and S4 – MRF, composting, combined with landfilling. The findings of the study indicated that S2 has a lesser adverse influence on the environment, namely global warming, acidification, eutrophication, and human toxicity. The current study was conducted in Guwahati city and capital of Assam. This is a small representative city in India that is concerned with various solid waste management issues. Besides that, previous research has shown that the environmental impacts of multiple technologies differ from city to city and region to region due to waste composition and environmental conditions. Guwahati city is currently experiencing problems resulting from the continuous increase in solid waste, and there is no scientific waste disposal process for transportation, collection, and disposal. As a result, not all cities can use the same technology. The analysis aims to use the LCA methodology to assess environmental consequences across various scenarios for waste management and to identify scenarios for reducing the environmental impact of Guwahati city. The scenarios include open dumping, composting, sanitary landfill, Refuse derived fuel (RDF), and manual sorting and impact categories such as global warming potential, acidification, eutrophication, freshwater toxicity, photochemical ozone creation potential, and fossil depletion potential were also analysed. The method of life cycle costing is used to assess the economic performance of various situations.

2. MATERIALS AND METHODS

2.1 Study area and composition of MSW

The Indian state of Assam is famous for its one-horned rhinoceros and functions as the entry point to north-eastern India as shown in Fig. 1. Guwahati is the state's primary commercial capital and had a population of 9.69 lakh people as per 2011 census. The city is located between 26.1445° N, 91.7362° E. The area covered all 31 wards administratively, and the area is around 216 km² (Chandramouli and General 2011). About 15–20 kg of waste was gathered from the overall amount of waste, which was thoroughly mixed and then quartered before samples of a size that could be treated in the laboratory were produced. The composition of MSW generated in Guwahati city is 37.42% food waste, 17.44% plastic, 16.41% paper, 4.94% textiles, 4.14% glass, 1.97% leather, 0.45% rubber, 0.37% metals, 5.25% lawn, 2.45% wood scraps and 9.16% miscellaneous. The major portion of waste generated in Guwahati is organic, plastic, and paper.

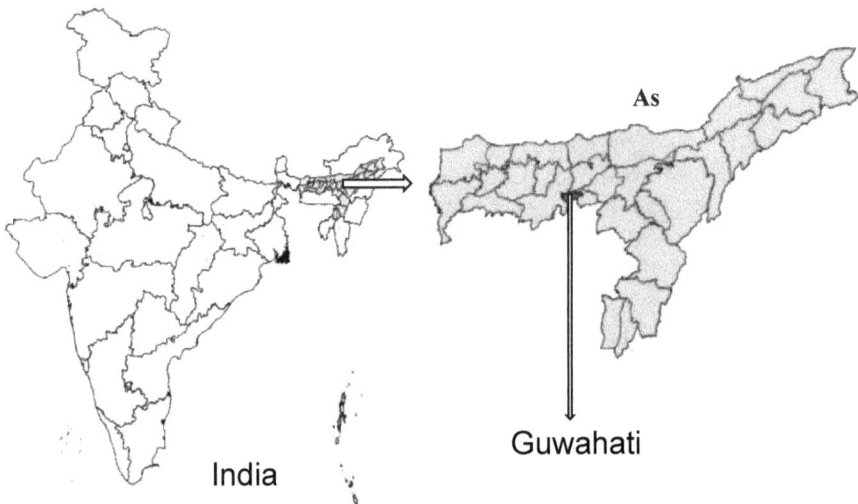

Figure 1. Study area and map of Assam.

2.1.1 Chemical characteristics of MSW

The chemical composition of MSW components aids in the evaluation of alternative processing and recovery options. The most crucial properties to understand are proximate analysis, ultimate analysis, and energy content, critical in assessing the combustion properties of waste or waste-derived fuel (refuse-derived fuel). The moisture content, ash content, hazardous matter content, and fixed carbon content were determined. Based on the results of bomb calorimeter experiments, mixed solid waste analysis was performed to determine the energy content of organic components in MSW, and metal analysis was also performed. The constituents of waste used in waste management technologies are listed in Table 1.

Table 1. Proximate analysis, calorific value analysis, and heavy metals of mixed solid waste.

Characteristics	Average (%)
Moisture content	64.83 ± 6.05
Ash content	45.27 ± 12.52
Volatile matter	51.31 ± 10.84
Fixed carbon	3.61 ± 3.01
Parameters	**Average values**
Calorific value (kcal/kg)	2531.70 ± 553.4
Energy content (MJ/kg)	10.60 ± 2.30
Energy content (BTU/lb)	4557.1 ± 996.0
Metals	**Content (mg/kg of waste)**
Zinc (Zn)	116.60 ± 50.8
Copper (Cu)	38.40 ± 14.5
Lead (Pd)	69.20 ± 25.2
Arsenic (Ar)	1555.20 ± 364.60
Cadmium (Cd)	5.70 ± 9.6
Iron (Fe)	10201.20 ± 3326.30

The ultimate analysis is important as mass balances for chemical, and thermal processes are calculated. The ultimate analysis of waste is performed to determine the carbon, hydrogen, oxygen, nitrogen, and sulphur content (C, H, O, N and S). Metal content (e.g., Cd, Cr, Hg, Ni, Zn, Mn, etc.) was calculated due to the possibility of adverse environmental effects. Figure 2 shows the C, H, O, N, S portions with percentage concentrations in the waste.

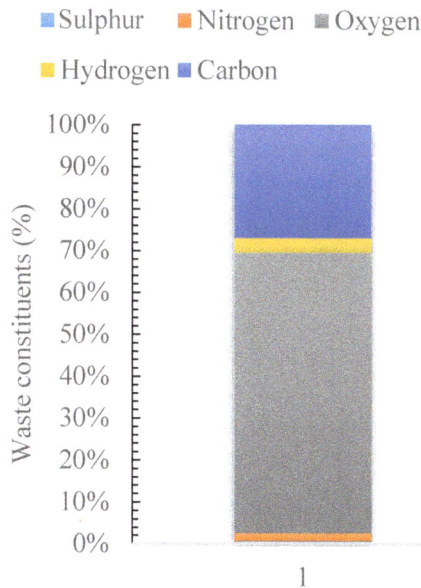

Figure 2. Ultimate analysis of mixed solid waste.

2.2 Life cycle assessment

LCA is used as an environmental protection method to measure the material from the cradle to the grave, as defined in ISO 14,040 and ISO 14044. The life cycle of a product begins with raw material acquisition, includes production and use of a product, and ends with waste disposal and decommissioning activities (Khandelwal et al. 2019). The current research work involves four phases: goal and scope definition, inventory analysis, impact evaluation, and interpretation.

2.2.1 Goal and scope definition

The study's primary goal is to measure and evaluate the environmental effects of four different disposal strategies and perform an economic assessment. The first scenario was chosen based on Guwahati's current MSW composition, which includes a lot of wet waste that is composting and open dumping. The second scenario was windrow composting and sanitary landfill with gas utilization. Based on the physicochemical properties of solid waste, the third scenario was chosen: refuse-derived fuel, windrow composting, and sanitary landfill with gas consumption. Selection of the fourth scenario was based on waste composition, physicochemical properties of solid waste: manual sorting, windrow composting, and sanitary landfill with gas utilization. The analysis was conducted using the Gabi program and the CML 2001 system.

To illustrate the impact of the present waste system on the environment, the economy used LCA and costing analysis for the Guwahati waste management system. Four commonly used MSW treatment technologies are compared to analyze the efficiency and suitability best disposal option. To evaluate the alternative scenarios, the management of one metric tonne of MSW in Guwahati, Assam, India, was selected as the functional unit.

  i) *Scenario 1 (S1):* This is the baseline scenario, which consists of landfilling into an open dump at the disposal site and windrow composting. Composting would be favoured due to the high moisture content of food waste and the remaining waste was transported to open dumping in this study.

 ii) *Scenario 2 (S2):* This scenario involves segregating the MSW into two different streams and including a composting process for the biodegradable organic waste while the non-biodegradable inorganic waste and biodegradable organic waste are disposed of in a sanitary landfill for gas utilization. However, gas is produced when a high amount of organic waste in a landfill

iii) *Scenario 3 (S3):* This scenario will include, in addition to composting and sanitary landfilling, a refused derived fuel (RDF). High calorific value products were used to make RDF pellets and transported to waste to energy plants.

iv) *Scenario 4 (S4):* In this scenario, the biodegradable fraction of MSW is sent for composting. At the same time, the non-biodegradable waste like plastic waste, wood wastes, and textile waste is segregated by manual sorting, and the resulting residue is landfilled along with other inerts.

2.3 System boundaries

The emission levels associated with different waste management options were calculated using the system boundary framework, a critical component of the LCA methodology. A subjective system boundary was used to establish the system boundary for this analysis. Figure 3 showed the system boundary accounts for all inputs of natural resources and outputs into the air, water, and land, as well as economic analysis (fixed cost and variable cost). It starts with the disposal strategies of MSW and the application of different MSW technologies to provide the best disposal method to Guwahati municipal solid waste management.

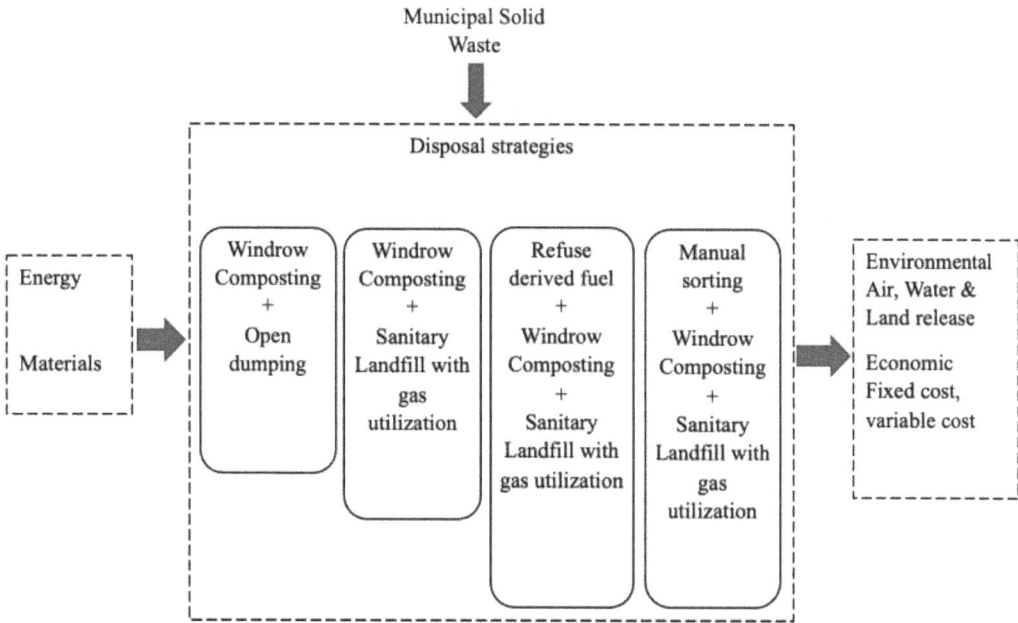

Figure 3. The system boundary of scenarios—*Scenario 1 (S1)* to *Scenario 4 (S4)*- of waste disposal strategies.

2.4 Life cycle inventory

2.4.1 Composting

2.4.1.1 Environmental impact (Air emissions)

Composting is the third most preferred choice in the hierarchy of advanced solid waste management (Hao et al. 2001). Aerobic compost occurs in the presence of oxygen which provides macro and micro-nutrients for the plants. MSW has a low heating value due to its high moisture content, which reduces its combustion ability. Windrow composting was included in all scenarios, since waste contains more amount of biodegradable waste. The composting method assumptions were made using data from the IPCC Guidelines for National Greenhouse Gas Inventories and an overview of the waste composition in the study area. The environmental effect of windrow composting (air emissions) was measured using the fractional mass of the generated waste. CO_2 and CH_4 concentrations are maximum in the centre of the windrows, where O_2 is scarce (Hao et al. 2001). The researcher stated that methanotrophic bacteria oxidize 46–98% of CH_4 formed in windrows before exiting the window composting (Jackel et al. 2005). Table 2 shows the default emission factors for CH4 and N2O emissions from waste biological treatment. Data on vehicular emissions collected from vehicular movement to compost plants as shown in Table 3.

Table 2. Default factors for CH_4 and N_2O emissions from biological treatment for the tier 1 method (GWMCPL 2008).

Composting	CH_4 emission factors (g CH_4/kg waste treated)	N_2O emission factors (g N_2O/kg waste treated)
Dry weight basis	10 (0.08 - 20)	0.6 (0.2–1.6)
Wet weight basis	4 (0.03–8)	0.3 (0.06–0.6)

Table 3. Vehicular emissions from the movement of vehicles to compost plant (GWMCPL 2008).

Parameters	Indicators	Emission factors/Amount	Converted amount (kg/tonne of waste)
Plant capacity		50 TPD	–
Material consumption	Diesel Fuel economy	50 –	41.6 kg –
	Land use area		7000 m²
Direct gas emissions	CH_4	4	6.424×10^{-7}
	N_2O	0.3	4.818×10^{-8}
	CO_2	515.2 g/km	2.06
	CO	3.6 g/km	0.014
	NOx	6.3 g/km	0.25
	CH_4	0.09 g/km	3.6×10^{-4}
	SO_2	1.42 g/km	5.68×10^{-3}
	HC	0.28 g/km	1.12×10^{-3}
	PM	0.87 g/km	3.48×10^{-3}
	NMVOC	–	7.2×10^{-3}
	N_2O	–	2.16×10^{-5}
	1,3-Butadiene	0.0074 mg/km	2.9×10^{-8}
	Benzene	0.0049 mg/km	1.9×10^{-8}
	Formaldehyde	0.0610 mg/km	2.4×10^{-7}
	Acetaldehyde	0.0	–
	Total Aldehyde	0.0837 mg/km	3.3×10^{-7}
	Total PAH	3.9707 mg/km	1.5×10^{-5}

2.4.2 Refuse derived fuel (RDF)

2.4.2.1 Environmental impact (Air emissions)

RDF is used as an alternate fuel in the cement industry to reduce the usage of fossil fuels and the overall effect on electricity production, GHG pollution, and environmental impacts (Lima et al. 2018). RDF manufacturing and usage demand less energy (4.7 GJ/ton clinker) than hard coal, resulting in CO_2 emissions' reductions ranging from 863–888 kg per tonne of clinker output, based on waste composition (Reza et al. 2013). Fuel emissions are measured using annual pollutant emissions as provided in Eq. 1.

$$E_{j,i} = Q_f \times (\text{pollutant concentration in fuel}/100) \times (MW_p/EW_f) \times \text{Opt hrs} \qquad \text{Eq. (1)}$$

where

$E_{j,i}$ = Annual emissions of pollutant i, kg/day

Q_f = Fuel use, kg/hr

Opt hrs = Operating hours, hrs/day

MW_p = Molecular weight of pollutant emitted, kg/kg-mole

EW_f = Elemental weight of pollutant in fuel, kg/kg-mole

i = Concentration of pollutant i in fuel expressed as weight percent, %

The emissions inventory of the RDF plant was calculated using Eq. 1. The plant capacity was assumed to be 200 TPD; the material consumption is diesel, and the area required is 14000 m². Potential air emissions were estimated and presented in Table 4.

Table 4. Emission inventory of RDF plant (GWMCPL 2008).

Parameters	Indicators	Emission factors/Amount	Converted amount (kg/tonne of waste)
Plant capacity		200 TPD	
Material consumption	Diesel	50	194.6 kg
	Land use area	–	14,000 m^2
Direct gas emissions	CH_4	56	0.18
	CO	98	0.32
	CO_2	154	0.51
	SO_2	1.8	6×10^{-3}
	N_2O	9.4	0.03
	CO_2	515.2 g/km	4.90
	CO	3.6 g/km	0.033
	NO_x	6.3 g/km	0.59
	CH_4	0.09 g/km	8.56×10^{-4}
	SO_2	1.42 g/km	1.35×10^{-2}
	HC	0.28 g/km	2.66×10^{-3}
	PM	0.87 g/km	8.28×10^{-3}
	NMVOC	–	1.71×10^{-2}
	N_2O	–	5.14×10^{-5}
	1,3-Butadiene	0.0074 mg/km	6.9×10^{-8}
	Benzene	0.0049 mg/km	4.52×10^{-8}
	Formaldehyde	0.0610 mg/km	5.71×10^{-7}
	Acetaldehyde	0.0	–
	Total Aldehyde	0.0837 mg/km	7.8×10^{-7}
	Total PAH	3.9707 mg/km	3.57×10^{-5}

2.4.3 Sanitary landfill/open dumping

2.4.3.1 Environmental impact (Air emissions)

Landfilling is the most traditional and often used form of MSW treatment. Landfilling operations worldwide vary from uncontrolled dumpsites to heavily engineered facilities that store leachate and landfill gas. The organic fraction of waste decomposes in this environment, producing so-called landfill biogas. This is composed mostly of methane (CH_4) (58%) and carbon dioxide (CO_2) (41%) but can also include traces of hydrogen sulphide (H_2S), hydrochloride (HCl), hydrogen fluoride (HF), and other chemical compounds (Cherubini et al. 2009). Leachate is a liquid that percolates from landfills after the biodegradation of solid waste. Landfill gas and leachate are the two major issues of landfills and their impacts on water and air. Therefore, the landfill gas is estimated using the flux chamber method. The circular stainless-steel chamber has a width of 40 cm and a height of 35 cm, with one side closed and two holes, as shown in Fig. 4. One hole at the centre is made for sampling, and the other is close to rims for temperature measurement. The 10 cm of the 35 cm high chamber is inserted into the soil or waste bed, and the upper section is placed on top. Sampling points were established based on the age (2–4 years) of the MSW on the surface layer and deposition height (5–15 feet) (Jha et al. 2008). The chamber was embedded in the landfills, and appropriate installation ensured that the surrounding atmosphere was retained. A displacement procedure was used to take the gas in the sampling chamber and transferred it to vials. The chamber was positioned throughout the day; samples were taken at daily intervals of one hour and measured landfill temperature at 5 cm

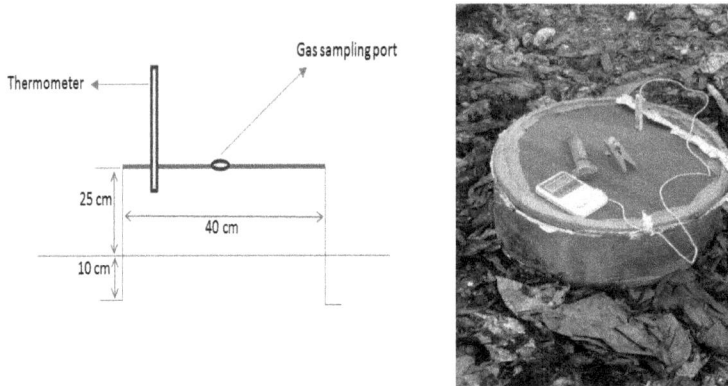

Figure 4. Field setup of stainless-steel chamber for gas collection.

depth and atmospheric temperature. Methane concentration was analyzed using the Porapak Q column of gas chromatograph. The emission flux rates were calculated using Eq. 2 (Gollapalli and Kota 2018).

$$EF = \frac{\Delta c}{\Delta t} \times h \qquad\qquad\qquad \text{Eq. (2)}$$

where
$\Delta C/\Delta t$ denotes the slope of the linear relationship between concentrations and sampling period, and h denotes the chamber's height above the soil level.

The gas samples were collected at different intervals, as shown in Table 5, and concentration vs. time was plotted to obtain the slope and presented in Fig. 5. The CO_2 emission was calculated using the above Eq. 2, and obtained value was 4.484 mg/ m^2 d.

Table 5. Gas samples from landfill.

Sample no.	Time	Volume	Temperature
Sample 1	0	0.0	25
Sample 2	20	3.259	21
Sample 3	40	3.073	22
Sample 4	60	4.059	23

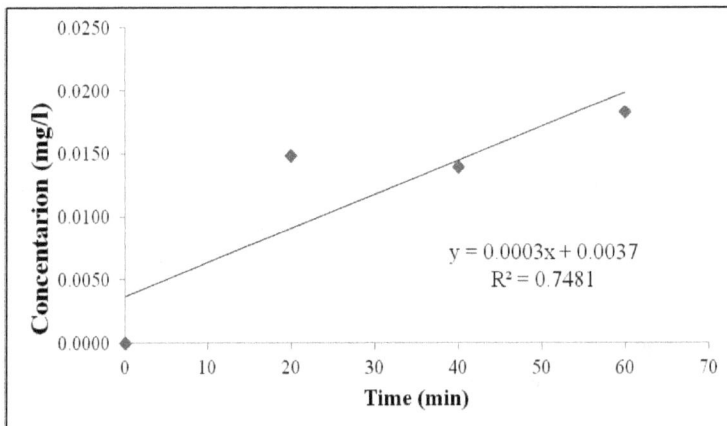

Figure 5. Linear diagram for CO_2 analysis.

(i) Determination of theoretical methane emissions using IPCC default method

The United Nations Intergovernmental Panel on Climate Change developed a multi-phase model in 2006 to measure global methane pollution for all nations. Waste generation rates, population, waste degradable organic carbon (DOC) composition, DOC_f fraction, waste decay rate (k), and CH_4 correction factor may all be used for default values or actual data. The IPCC model makes use of the first-order effects. The IPCC model relies on using first-order decay models, the DOC, and k-values for various kinds of waste, including paper, food, furniture, etc. (Gollapalli and Kota 2018). Emissions were estimated using Eqs. 3, 4, 5 and 6.

$$\text{Methane emission (Gg/yr)} = \left\{ \left(MSW_T \times MSW_f \times MCF \times DOC \times DOC_f \times F \times \left(\frac{16}{12} \right) \times (1 - OX) \right) \right\} \text{ Eq. (3)}$$

where:

MSW_T: total MSW generated (Gg/yr)

MSW_F: fraction of MSW disposed to solid waste disposal sites

MCF: methane correction factor (fraction)

DOC: degradable organic carbon (fraction) (kg C/kg SW)

DOC_f: fraction DOC dissimilated

F: fraction of CH_4 in landfill gas (IPCC default is 0.5)

16/12: conversion of C to CH_4

R: recovered CH_4 (Gg/yr)

OX: Oxidation factor (fraction – IPCC default is 0)

(ii) Methane emissions from landfills are theoretically calculated

$$\text{Methane emission (Gg/yr)} = \{(200 \text{ Gg/yr}) \times (0.7) \times (0.4) \times (0.157) \times (0.77) \times$$
$$(0.5) \times \left(\frac{16}{12} - 0 \right) \} \times (1 - 0) = 1.89 \text{ Gg/yr.} \qquad \text{Eq. (4)}$$

(iii) Volatile organic compound emission (as hexane)

$$Q_{VOC} = \left(1 + \left(\frac{C_{CO_2}(\%)}{C_{CH_4}(\%)} \right) \times Q_{CH_4} \times \left(\frac{C_{VOC}}{10^6} \right) = 1229.35 \frac{m^3}{yr} \right) \qquad \text{Eq. (5)}$$

where

Q_{VOC} = Emission rate of pollutant VOC,

Q_{CH_4} = methane generation rate, m³/yr (from Eq. 1)

C_{CH_4} (%) = the concentration of CH_4 as a percentage of the total landfill gas. If unknown, assume 55% CH_4

C_{CO_2} (%) = the concentration of CO_2 and other gas constituents as a percentage of the total landfill gas. If unknown, assume 45%

106 = conversion from ppmv

(iv) Mass calculation of VOC (as hexane)

$$M (kg/yr) = Q_{VOC} \left[\frac{MW \times 1\,atm}{8.205 \times 10^{-5}} \right] = 4332.9 \text{ kg} / \text{yr}$$

Eq. (6)

where

MW = Molecular weight (as hexane), g/mol

8.20510^{-5} = constant to convert emissions of VOC to kg/yr, m^3-atm/g mol-k.

- Air emissions (site-specific data) were estimated from landfills using the above equations, as shown in Table 6.
- The leachate generation from the landfill site was observed. Leachate and water samples are assessed to evaluate the contaminant flow to local water bodies. The parameters were tested for COD, BOD_5, and heavy metals like arsenic, lead, zinc, and cadmium and are illustrated in Table 7.

Table 6. Air emission inventory from landfill (GWMCPL 2008).

Parameters	Indicators	Emission factors/Amount	Converted amount (kg/tonne of waste)
Material consumption	Land use	86,000 m^2	–
Fuel consumption	Diesel	200	166.4 kg
Direct gas emission (air)	CO_2	4.484 mg/m^2 day	1.87×10^{-4}
	VOC	4332.9 kg/year (Hexane)	0.02
	CH_4	1.89 Gg/yr	9.41×10^{-6}
	CO_2	515.2 g/km	0.749
	CO	3.6 g/km	5.2×10^{-3}
	NO_x	6.3 g/km	9.1×10^{-3}
	CH_4	0.09 g/km	1.3×10^{-4}
	SO_2	1.42 g/km	2.06×10^{-3}
	HC	0.28 g/km	4.07×10^{-4}
	PM	0.87 g/km	1.26×10^{-3}
	NMVOC	–	2.52×10^{-3}
	N_2O	–	7.56×10^{-6}
	1,3-Butadiene	0.0074 mg/km	1.07×10^{-8}
	Benzene	0.0049 mg/km	7.12×10^{-9}
	Formaldehyde	0.0610 mg/km	8.8×10^{-8}
	Acetaldehyde	0.0	0
	Total Aldehyde	0.0837 mg/km	1.2×10^{-7}
	Total PAH	3.9707 mg/km	5.6×10^{-6}

Table 7. Leachate emission inventory from landfill.

Parameters	Indicators	Emission factors/Amount	Converted Amount (kg/tonne of waste)
Direct water emission	Colour	–	Black
	Sample volume	500 mL	–
	pH	6.5	–
	Alkalinity	72.5 mg/L	6.59×10^{-8}
	TDS	1394.5 mg/L	1.26×10^{-6}
	EC	652.25 mS/cm	–
	BOD_5	24,750 mg/L	2.25×10^{-5}
	COD	46,450 mg/L	4.22×10^{-5}
	BOD/COD	0.538	-
	TN	302.5 mg/ L	2.75×10^{-7}
	NH_4^+-N (converted as ammonia)	206.5 (251.081 as ammonia) mg/L	2.28×10^{-7} –
	Zn	443 µg/L	4.02×10^{-10}
	Cd	2.6925 µg/L	2.44×10^{-12}
	Pb	49.5 µg/L	4.5×10^{-11}
	Cu	115.75 µg/L	$1.01\ 5 \times 10^{-10}$
	Ni	0.29 µg/L	2.63×10^{-13}
	Cr	337 µg/L	3.06×10^{-10}

3. RESULTS AND DISCUSSION

3.1 Life cycle assessment of different proposed scenarios

3.1.1 Baseline impact analysis

The multiple scenarios' life cycle risk assessments (LCIA) are carried out using the CML 2001-April 2015 methodology for global warming, acidification, eutrophication, freshwater toxicity, and the ReCiPe 1.08 methodology impacts on fossil depletion. The LCIA outcome for the baseline scenario is based on different effect categories, as shown in Fig. 6. Climate change is caused due to CO_2. The baseline scenario 1 (S1) had the maximum release of CO_2 impact (7.30E+ 00 kg of CO_2 eq) compared with other categories. Pollutant contaminants such as particulate matter (PM), oxides of sulphur (SO_x), oxides of nitrogen (NO_x), and heavy metals cause human toxicity. S1 had a moderate impact on human toxicity (4.00E-01 kg of 1,4-dichlorobenzene (DCB) eq). Acidifying substances cause acidification like SO_x, NO_x. Scenario 1 had the least impact on acidification and freshwater aquatic toxicity (1.00E-02 kg SO2 Eq). NO_x and VOCs have the potential to generate photochemical ozone since ozone is found at lower levels of the atmosphere under sunlight. The baseline scenario shows maximum photochemical oxidation potential (5.00E-02 kg of NMVOC eq) since NO_x is the primary pollutant generated by diesel engines. Another study in Nagpur, India, explained that the baseline considered composting and remaining material to landfilling. The findings showed that the baseline situation had the least effect on acidification (1.92E-01 kg SO2 eq), eutrophication and photochemical reactions are the pre defined environmental impacts in the software (7.9E-02 kg PO4- eq). Photochemical had the maximum impact (3.04E-01 kg C_2H_4 eq) (Khandelwal et al. 2019)

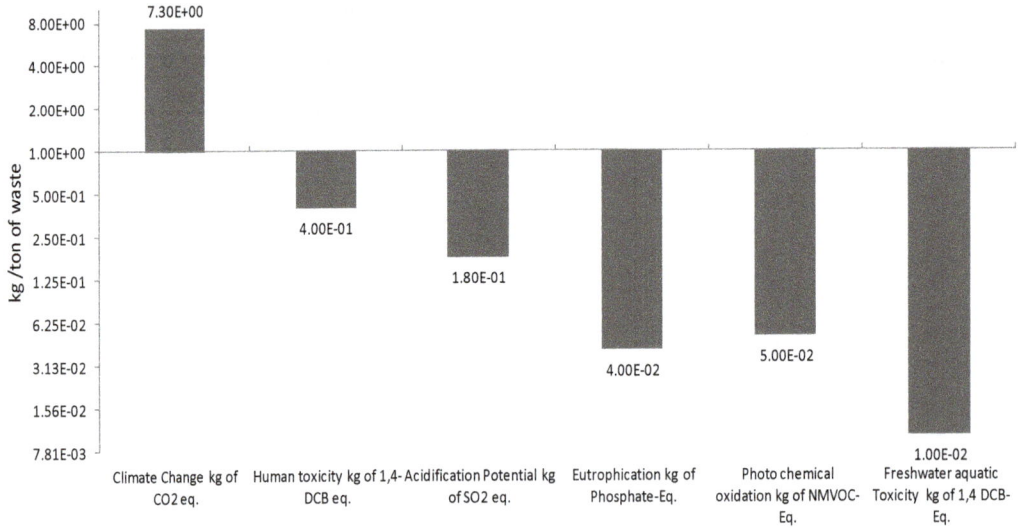

Figure 6. Impact assessment of baseline study using CML 2001-April 2013.

3.1.2 Environmental impacts of proposed scenarios for Guwahati MSW

The LCIA of the various scenarios was conducted using the CML 2001 – April 2015 methodology for global warming, acidification, eutrophication, freshwater toxicity, and the ReCiPe 1.08 methodology impacts on fossil depletion. The LCIA result for the different proposed scenarios according to the different impact categories is presented in Fig. 7. The MSW composition in Guwahati city is 37.42% food waste, 17.44% plastic, 16.41% paper, 4.94% textiles, 4.14% glass, 1.97% leather, 0.45% rubber, 0.37% metals, 5.25% lawn, 2.45% wood scraps and 9.16% miscellaneous. The majority of waste generated in Guwahati is organic, plastic, and paper found in significant amounts in the present study. Food waste had an overall moisture content of 64.83 percent, which was higher than that of

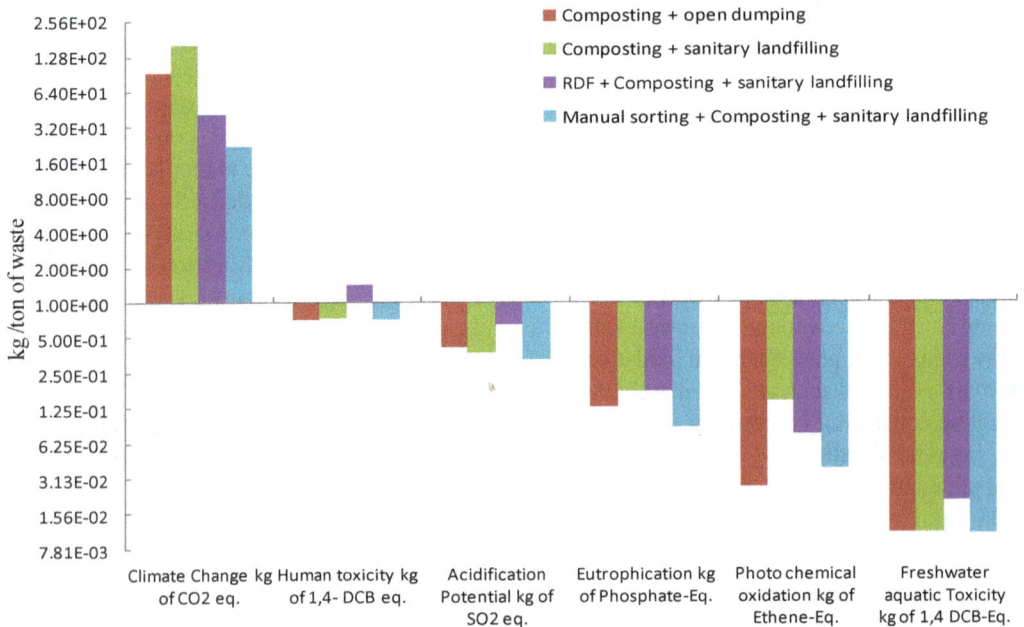

Figure 7. Environmental impact assessment of different scenarios using CML 2001-April 2013.

China (61%) (Zhen-Shan et al. 2009) and Turkey (57%) (Yay 2015), and Dhanbad (65%) (Yadav and Samadder 2018). Therefore, it concluded that the moisture content of waste depends on biodegradable material in the landfill. The proposed scenarios for Guwahati municipal solid waste are as follows:

Scenario 1 (S1): Windrow composting + open dumping (COMP_OD)

Scenario 2 (S2): Windrow composting + sanitary landfill with gas utilization (COMP_SLF)

Scenario 3 (S3): Refuse derived fuel + windrow composting + sanitary landfill with gas utilization (RDF_COMP_SLF)

Scenario 4 (S4): Manual sorting + windrow composting + sanitary landfill with gas utilization (MSP_COMP_SLF)

3.1.2.1 Global warming potential (GWP)

Global warming impacts will be mainly due to GHGs like CO_2, CH_4, and N_2O. The results for the global warming potential for the baseline scenario S2 indicate that compositing + sanitary landfilling system has a very high impact on climate change GWP 1.7E+02 kg CO_2 equivalent, which is mostly due to the large amounts of waste that are currently being landfilled without any prior treatment given, especially to the large amounts of organic waste that are disposed of. The emission of huge quantities of CH_4 is mostly from the degradation processes of organic waste in the landfill. Also, CH_4 is a highly potent greenhouse gas that accounts for up to 25 times the GWP than CO_2. The percentage contribution to GWP by the waste collection process is negligible compared to emissions from the landfill site. The results for the baseline scenario have been compared with the baseline emissions from the city of Nagpur, i.e., 1259.69 kg CO_2 eq. (Khandelwal et al. 2019) and city of Dhanbad, i.e., 1.24 E + 01 kg CO_2 eq., and are found to be much higher than Dhanbad and lower than the Nagpur city. This may be because the disposal site in Guwahati is quite old and subjected to heavy rainfall compared to the two cities, which affects the rate of decomposition in the landfills. Among the alternatives (S2, S3, and S4), while the emission control system in alternatives S2, S3, and S4 may cause any methane that is formed in the landfill to be utilized and also waste segregation can be done in other three scenarios, the emissions in both cases will be much lower (1.70 E+02 kg CO_2 eq, 4.31E+01 kg CO_2 eq and 23.29E+01 kg CO_2 eq., respectively) than the baseline due to the organic waste being diverted for biological treatment through composting. Although composting is a completely biological process, the net emissions of GHGs and fuel consumed are required by the different types of machinery in the compost plant. The GWP in the case of alternative S2 will be higher (1.70 E+02 kg CO_2 eq.) than the two previous scenarios due to the usage of fossil fuels such as diesel in various moving vehicles. The S3 (RDF_COMP_SLF) and S4 (MSP_COMP_SLF) had the least impact on global warming due to waste segregation, separation of organic waste, and utilization of landfill gases.

3.1.2.2 Acidification potential

The acidification potential arises mainly due to ammonia (NH_3), NO_x and SO_2 emissions. The acidification potential of the baseline scenario is about 3.48E-01 kg SO_2 eq. However, when comparing the different alternate scenarios with the baseline, there will be significant impacts from the three scenarios compared to the baseline. In comparison, scenarios S3 and S4 had the greatest impact on acidification potential. However, in scenario S2 (COMP_SLF) is about 3.93E-01 kg SO2 eq, mainly due to releases of acidifying compounds into the atmosphere during the composting process. S2 (3.93E-01 kg SO_2 eq), S3 (6.67 E-01 kg SO_2 eq), and S4 (4.29E-01 kg SO_2 eq) have a higher impact on acidification compared with baseline scenario S1 due to compost and movement of vehicles to open dumping and the lowest acidification among all the scenarios because, in this case, sulphur and nitrogen compounds get oxidized in the lesser amount resulting in the lower emissions of SOx and NOx. In another study in Dhanbad, S3 (RDF_COMP_SLF) had shows the same impact on acidification potential (2.72E-06 kg SO_2 eq.) compared with other alternative scenarios (Yadav and Samadder 2018).

3.1.2.3 Eutrophication potential

Eutrophication of aquatic systems due to nutrient enrichment by compounds such as NH_3- N and total phosphorus will occur mainly from leachate pollution from landfills. As seen from the results, the eutrophication potential of the baseline scenario1 is about 1.31E-01 kg PO_4^- eq.; S2 (1.87E-01 kg PO_4^- eq) and S3 (1.87E-01 kg PO_4^- eq) show the maximum potential for nutrient enrichment, due to the presence of impermeable synthetic bottom liners in sanitary landfills, S4 has the least impact on (8.44E-02 kg PO4- eq) eutrophication of the alternative scenarios. An explanation for the eutrophication potential in S2 may be attributed to the excessive amount of nitrogen and phosphorus nutrient present in the compost and NH_3 gas both from the composting and landfill. However, with proper collection facilities and energy recovery, as seen in S3, the impact can be compensated.

3.1.2.4 Fresh water toxicity

Toxicity-related impacts to water will be mainly in organic, inorganic, and heavy metal emissions measured in terms of kg 1,4-DCB equivalents. The baseline scenario's water toxicity impacts are much higher (1.05E-02 kg 1,4-DCB eq) than all the three alternate scenarios. This shows that heavy metals in the leachate samples analyzed significantly affect the quality of water bodies compared to inorganic or organic emissions, respectively. Among the alternate scenarios, S1, S2, and S3 had almost the same impact on water toxicity which may be mainly due to the emission of heavy metals from the landfill, but scenario S4 (1.00E-02 kg 1,4-DCB eq) had a better performance due to recycling and separating recyclables from the main waste stream.

3.1.2.5 Photochemical ozone creation potential

VOCs, CO, and CH4 are the main contributors to photochemical ozone creation potential. The S2 (1.37E-01 kg C_2H_4 eq had the greatest impact mainly due to the release of VOCs, methane, carbon monoxide, and other inorganic air pollutants. The major cause of VOCs is landfilling and disposal of waste. Comparing the results of POCP for the baseline scenario (2.55E-02 kg C_2H_4 eq) showed a low impact on POCP of the present study with the S2; however, S4 (3.77E-02 kg C_2H_4 eq) is found to be much better than the other scenarios. The S3 (7.74E-02 kg C_2H_4 eq) and S4 (5.47 E-02 kg C_2H_4 eq) were less impacted than S2 due to sanitary landfilling, manual sorting of materials, and use of high calorific value products. This reduction in effect in S4 is attributed to avoiding methane pollution from the system by composting the food and residual in a sanitary landfill.

4. CONCLUSIONS

The Guwahati waste management involves only the landfilling without any energy recovery leading to high environmental burdens like global warming potential, acidification potential, eutrophication potential, human toxicity, freshwater Aquatic toxicity, and photochemical oxidation. The study compared the environmental impact of four different proposed waste treatment scenarios. Composting and the sanitary landfill had a lower environmental impact than S1 because biodegradable waste is separated and the remaining material is disposed of, whereas S4 (manual sorting + windrow composting + sanitary landfill with gas utilization) is the best choice from the perspective of the environmental impacts because methane emissions, SOX, NOX, and heavy metal pollution can be avoided by separating waste materials and found to be less polluting. The environmental aspect analysis will give administration and policymakers more open options for the suitable system to reduce environmental impact over society and government. The current urban solid waste management system in Guwahati is ineffective, as the environmental impacts on the main impact categories were large compared to the proposed scenario.

REFERENCES

Babu, G.S., Lakshmikanthan, P. and Santhosh, L. 2014. Life cycle analysis of municipal solid waste (MSW) land disposal options in Bangalore City, ICSI 2014: Creating Infrastructure for a Sustainable World, pp. 795–806.

Bank, W. 2005. Waste management in China: Issues and recommendations. Urban Development Working Papers, East Asia Infrastructure Department, World Bank–Working Paper No. 9.

Botello-Álvarez, J.E., Rivas-García, P., Fausto-Castro, L. et al. 2018. Informal collection, recycling and export of valuable waste as transcendent factor in the municipal solid waste management: A Latin-American reality. Journal of Cleaner Production 182: 485–495.

Chandramouli, C. and General, R. 2011. Census of India 2011. Provisional Population Totals. New Delhi: Government of India, 409–413.

Cherubini, F., Bargigli, S. and Ulgiati, S. 2009. Life cycle assessment (LCA) of waste management strategies: Landfilling, sorting plant and incineration. Energy 34: 2116–2123.

Cleary, J. 2009. Life cycle assessments of municipal solid waste management systems: A comparative analysis of selected peer-reviewed literature. Environment International 35: 1256–1266.

CPCB. 2016. Annual report https://cpcb.nic.in/uploads/MSW/MSW_AnnualReport_2016-17.pdf (Accessed at 31 March, 2021).

Gollapalli, M. and Kota, S.H. 2018. Methane emissions from a landfill in north-east India: Performance of various landfill gas emission models. Environmental Pollution 234: 174–180.

Gouveia, N. and Prado, R.R.d. 2010. Health risks in areas close to urban solid waste landfill sites. Revista de Saude Publica 44: 859–866.

Gupta, M., Srivastava, M., Agrahari, S.K. et al. 2018. Waste to energy technologies in India: A review. J. Energy Environ. Sustain. 6: 29–35.

Gupta, N., Yadav, K.K. and Kumar, V. 2015. A review on current status of municipal solid waste management in India. Journal of Environmental Sciences 37: 206–217.

Hao, X., Chang, C., Larney, F.J. et al. 2001. Greenhouse gas emissions during cattle feedlot manure composting. Journal of Environmental Quality 30: 376–386.

Jäckel, U., Thummes, K. and Kämpfer, P. 2005. Thermophilic methane production and oxidation in compost. FEMS Microbiology Ecology 52: 175–184.

Khandelwal, H., Thalla, A.K., Kumar, S. et al. 2019. Life cycle assessment of municipal solid waste management options for India. Bioresource Technology 288: 121515.

Koroneos, C.J. and Nanaki, E.A. 2012. Integrated solid waste management and energy production—A life cycle assessment approach: The case study of the city of Thessaloniki. Journal of Cleaner Production 27: 141–150.

Kumar, A. and Samadder, S.R. 2017. A review on technological options of waste to energy for effective management of municipal solid waste. Waste Management 69: 407–422.

Li, Y., Jin, Y., Borrion, A. et al. 2019. Current status of food waste generation and management in China. Bioresource Technology 273: 654–665.

Lima, P.D.M., Colvero, D.A., Gomes, A.P. et al. 2018. Environmental assessment of existing and alternative options for management of municipal solid waste in Brazil. Waste Management 78: 857–870.

Mehta, Y.D., Shastri, Y. and Joseph, B. 2018. Economic analysis and life cycle impact assessment of municipal solid waste (MSW) disposal: A case study of Mumbai, India. Waste Management & Research 36: 1177–1189.

Othman, S.N., Noor, Z.Z., Abba, A.H. et al. 2013. Review on life cycle assessment of integrated solid waste management in some Asian countries. Journal of Cleaner Production 41: 251–262.

Patil, B.S. and Singh, D.N. 2017. Simulation of municipal solid waste degradation in aerobic and anaerobic bioreactor landfills. Waste Management & Research 35: 301–312.

Ramachandra, T., Bharath, H., Kulkarni, G. et al. 2018. Municipal solid waste: Generation, composition and GHG emissions in Bangalore, India. Renewable and Sustainable Energy Reviews 82: 1122–1136.

Rana, R., Ganguly, R. and Gupta, A.K. 2019. Life-cycle assessment of municipal solid-waste management strategies in Tricity region of India. Journal of Material Cycles and Waste Management 21: 606–623.

Rathore, P., Sarmah, S.P. and Singh, A. 2020. Location–allocation of bins in urban solid waste management: A case study of Bilaspur city, India. Environment, Development and Sustainability 22: 3309–3331.

Reza, B., Soltani, A., Ruparathna, R. et al. 2013. Environmental and economic aspects of production and utilization of RDF as alternative fuel in cement plants: A case study of Metro Vancouver Waste Management. Resources, Conservation and Recycling 81: 105–114.

Sharma, B.K. and Chandel, M.K. 2017. Life cycle assessment of potential municipal solid waste management strategies for Mumbai, India. Waste Management & Research 35: 79–91.

Shekdar, A.V. 2009. Sustainable solid waste management: An integrated approach for Asian countries. Waste Management 29: 1438–1448.

Silva, V., Contreras, F. and Bortoleto, A.P. 2021. Life-cycle assessment of municipal solid waste management options: A case study of refuse derived fuel production in the city of Brasilia, Brazil. Journal of Cleaner Production 279: 123696.

Singh, R., Tyagi, V., Allen, T. et al. 2011. An overview for exploring the possibilities of energy generation from municipal solid waste (MSW) in Indian scenario. Renewable and Sustainable Energy Reviews 15: 4797–4808.

Srivastava, A.K. and Nema, A.K. 2011. Life cycle assessment of integrated solid waste management system of Delhi, Towards Life Cycle Sustainability Management. Springer, pp. 267–276.

Tarantini, M., Loprieno, A.D., Cucchi, E. et al. 2009. Life Cycle Assessment of waste management systems in Italian industrial areas: Case study of 1st Macrolotto of Prato. Energy 34: 613–622.

Tyagi, V.K., Fdez-Güelfo, L., Zhou, Y. et al. 2018. Anaerobic co-digestion of organic fraction of municipal solid waste (OFMSW): Progress and challenges. Renewable and Sustainable Energy Reviews 93: 380–399.

Yadav, P. and Samadder, S. 2017. A global prospective of income distribution and its effect on life cycle assessment of municipal solid waste management: A review. Environmental Science and Pollution Research 24: 9123–9141.

Yadav, P. and Samadder, S.R. 2018. Environmental impact assessment of municipal solid waste management options using life cycle assessment: a case study. Environmental Science and Pollution Research 25: 838–854.

Yay, A.S.E. 2015. Application of life cycle assessment (LCA) for municipal solid waste management: A case study of Sakarya. Journal of Cleaner Production 94: 284–293.

Zaman, A.U. 2010. Comparative study of municipal solid waste treatment technologies using life cycle assessment method. International Journal of Environmental Science & Technology 7: 225–234.

Zhao, Y., Wang, H.-T., Lu, W.-J. et al. 2009. Life-cycle assessment of the municipal solid waste management system in Hangzhou, China (EASEWASTE). Waste Management & Research 27: 399–406.

Zhen-Shan, L., Lei, Y., Xiao-Yan, Q. et al. 2009. Municipal solid waste management in Beijing City. Waste Management 29: 2596–2599.

15

BIOTECHNOLOGICAL APPROACHES FOR ADDRESSING THE PLASTIC WASTE TREATMENT

Tejovardhan Pulipati, Sai Teja A., Shivani Maddirala, Prem Vuyyuru, Pradeep Kari, Sudipa Bhadra and *Surajbhan Sevda**

1. INTRODUCTION

The utilization of biological processes, organisms, or systems for the production of beneficial products or services is known as biotechnology. The world is currently experiencing a major issue with plastic waste, which presents a significant challenge. Conventional means of treating plastic waste such as landfilling, incineration, and mechanical recycling have their limitations, thus necessitating the search for alternative solutions. Biotechnology presents a promising solution to address the issue of plastic waste (Barrila et al. 2018, Hemdan et al. 2021, Lear et al. 2021, Nikolaivits et al. 2021). Biotechnological methods make use of microorganisms, enzymes, and other biological processes to decompose plastic waste into its component parts. These methods offer several benefits over traditional strategies, such as low energy consumption, high effectiveness, and the ability to generate useful products (Al Rayaan 2021, Bai et al. 2021, Jung et al. 2020, Wang et al. 2022a).

To degrade plastic waste, one method involves utilizing microorganisms like bacteria, fungi, and algae. These microorganisms use biodegradation to simplify plastic waste into more straightforward compounds. To optimize the process of biodegradation, environmental conditions like temperature, pH, and nutrients are adjusted to encourage the growth of the microorganisms. Another technique to break down the plastic waste is through the use of enzymes, including proteases, esterases, and lipases. Enzymes accelerate the breakdown of plastic into smaller molecules, which can be further degraded by microorganisms. These enzymes can be produced from microorganisms or synthesized using recombinant DNA technology (Kim et al. 2022, Liao and Chen 2021, Lopez et al. 2018, Muhammad et al. 2015).

Environmental Bioprocess Laboratory, Department of Biotechnology, National Institute of Technology Warangal, Warangal (T.S.) – India, 506001.
* Corresponding author: sevdasuraj@gmail.com, sevdasuraj@nitw.ac.in

Additionally, there is the possibility of using genetic engineering to create microorganisms with specific abilities to degrade plastic waste. By identifying the genes responsible for plastic degradation and transferring them to other microorganisms, scientists can create more efficient and specialized microorganisms for plastic waste treatment. In conclusion, biotechnology offers a range of approaches for addressing the plastic waste treatment. These approaches have the potential to reduce the environmental impact of plastic waste and create new opportunities for sustainable products and energy sources (Byrne et al. 2022, Liu et al. 2020, 2006). However, further research and development are needed to optimize these technologies and ensure their viability on a larger scale.

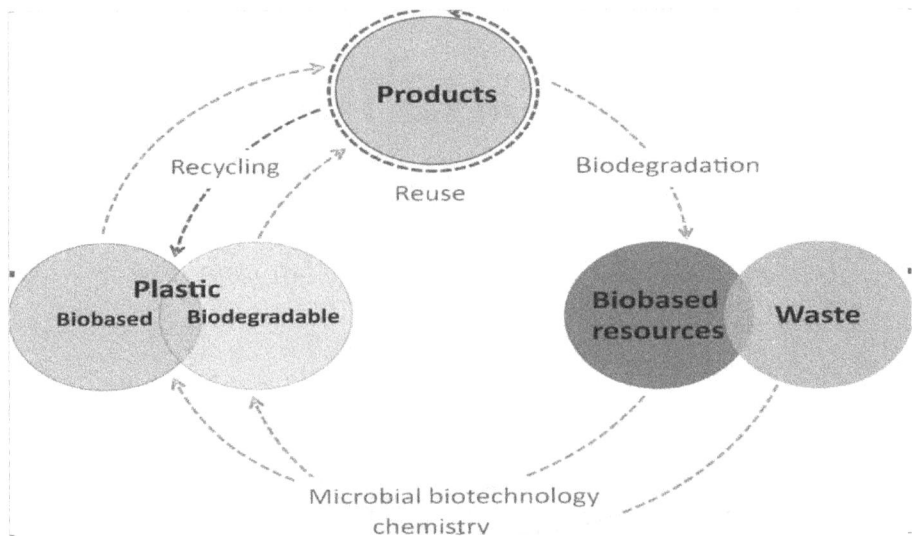

2. TYPES OF PLASTICS

Plastics are synthetic materials that are commonly used in various industries, including packaging, construction, and transportation. There are different types of plastics, each with its own unique properties and characteristics. In this, we will discuss the various types of plastics and their uses.

1. Polyethylene (PE): Polyethylene is the most common type of plastic, accounting for approximately 34% of all plastic produced. It is a versatile material that is used in various applications, including packaging, agriculture, and construction. Polyethylene is divided into two types: high-density polyethylene (HDPE) and low-density polyethylene (LDPE). HDPE is strong and durable and is used for products such as milk jugs and detergent bottles. LDPE is flexible and is used in products such as plastic bags and shrink wrap (Lear et al. 2021, Nikolaivits et al. 2021, Pinedo et al. 2022, Seeley et al. 2020).

2. Polypropylene (PP): Polypropylene is a versatile plastic that is used in a variety of applications, including packaging, automotive parts, and textiles. It is durable and resistant to weather and chemicals, making it ideal for outdoor applications. PVC is also used in products such as toys, shower curtains, and medical tubing (Al-asadi et al. 2020, Miloloža et al. 2022, Raphael and Yang 2013, Seeley et al. 2020, Zheng et al. 2006).

3. Polystyrene (PS): Polystyrene is a lightweight plastic that is used in a variety of applications, including packaging, insulation, and disposable cups and plates. It is easy to mold and can be

made into various shapes and sizes. Polystyrene is divided into two types- expanded polystyrene (EPS) and extruded polystyrene (XPS). EPS is used in products such as foam cups and packaging materials, while XPS is used in insulation and construction materials (Jaafar et al. 2022, Li et al. 2022, Muhammad et al. 2015, Raphael and Yang 2013, Salar-Garcia et al. 2020, Taghavi et al. 2021).

4. Polyethylene terephthalate (PET): Polyethylene terephthalate is a strong and lightweight plastic that is commonly used in packagings, such as water bottles, soda bottles, and food containers. It is also used in clothing, such as polyester fabric. PET is recyclable and is commonly used in the production of recycled plastic products (Kwon et al. 2022, Morgan-Sagastume et al. 2014, Singh Jadaun et al. 2022, Singh and Ruj 2016).

5. Acrylonitrile-butadiene-styrene (ABS): ABS is a strong and durable plastic that is commonly used in automotive parts, such as dashboard components and trim. It is also used in electronic housings and toys. ABS is resistant to impact and high temperatures, making it ideal for products that require strength and durability (Ieropoulos et al. 2010, Muhammad et al. 2015).

In conclusion, plastics are widely used in various industries and applications, and there are different types of plastics, each with its own unique properties and characteristics. Understanding the different types of plastics is important for the proper disposal and recycling of plastic waste. It is also important for manufacturers to choose the appropriate type of plastic for their products based on their specific requirements and properties.

3. BIODEGRADABLE PLASTIC IDENTIFICATION

Biodegradable plastics are a type of plastic that can be broken down by microorganisms into natural substances such as water, carbon dioxide, and biomass. These plastics are often marketed as a more sustainable alternative to traditional plastics, which can take hundreds of years to break down in the environment (Costa et al. 2019, Deng et al. 2021, Ibrahim et al. 2021). Here are some ways to identify biodegradable plastics:

Look for certification labels: Biodegradable plastics are often certified by third-party organizations, such as the Biodegradable Products Institute (BPI) or the European Bioplastics Association (EBA). Look for the labels on the packaging or product to verify that it is biodegradable.

1. Check the composition: Biodegradable plastics are typically made from natural materials, such as corn starch, sugarcane, or potato starch. Look for these materials on the label or packaging to identify a biodegradable plastic.

2. Look for instructions: Some biodegradable plastics require specific disposal methods, such as composting or recycling in a special facility. Look for instructions on the packaging or product to ensure proper disposal.

3. Conduct a test: If you are unsure if a plastic is biodegradable, you can conduct a simple test. Place the plastic in soil or compost and observe if it breaks down over time. Biodegradable plastics should break down within a few months, while traditional plastics will remain intact.

It is important to note that not all plastics labeled as "biodegradable" or "compostable" are created equal. Some may only break down under specific conditions or may release harmful substances as they degrade (Byrne et al. 2022, Liao and Chen 2021, Solano et al. 2022). Always do your research and verify the claims made by the manufacturer before choosing a biodegradable plastic product.

List of different microorganisms reported to degrade different types of plastics

Polymer blends	
Plastic	Microorganism
Starch/polyethylene	Aspergillus niger
	Penicillium funiculosum
	Phanerochaete
	Chrysosporium
Starch/polyester	Streptomyces
	Phanerochaete chrysosporium
Natural plastics	
Poly(3-hydroxybutyrate-co-3- mercaptopropionate))	Schlegelella thermodepolymerans
Poly(3-hydroxybutyrate)	Pseudomonas lemoignei
Poly(3-hydroxybutyrate-co-3-mercaptopropionate)	Pseudomonas indica K2
Poly(3-hydroxybutyrate)	Streptomyces sp. SNG9
Poly(3-hydroxybutyrate-co-3-hydroxyvalerate)	Ralstonia pikettii T1
Poly(3-hydroxybutyrate-co-3-hydroxypropionate)	Acidovorax sp. TP4
Poly(3-hydroxybutyrate-co-3-hydroxypropionate)	Alcaligenes faecalis
Poly(3-hydroxybutyrate)poly(3-hydroxypropionate)poly(4-hydroxybutyrate)	Pseudomonas stutzeri
poly(ethylene succinate)poly(ethylene adipate)	Comamonas acidovorans
	Alcaligenes faecalis
Poly(3-hydroxybutyrate)	Schlegelella thermodepolymerans
Poly(3-hydroxybutyrate)	Caenibacterium thermophilum
Poly(3-hydroxybutyrate-co-3-hydroxyvalerate)	Clostridium botulinum
	Clostridium acetobutylicum
Polycaprolactone	Clostridium botulinum
Polycaprolactone	Clostridium acetobutylicum
Polylactic acid	Fusarium solani
	Fusarium moniliforme
	Penicillium roquefort
	Amycolatopsis sp.
	Bacillus brevis
	Rhizopus delemer
Synthetic plastics	
Polyethylene	Brevibacillus borstelensis
	Rhodococcus rubber
	Penicillium simplicissimum YK
Polyurethane	Comamonas acidovorans TB-35
	Curvularia senegalensis
	Fusarium solani
	Aureobasidium pullulans
	Cladosporium sp.
	Pseudomonas chlororaphis
Polyvinyl chloride	Pseudomonas putida AJ
	Ochrobactrum TD
	Pseudomonas fluorescens B-22
	Aspergillus niger
	van Tieghem F-1119
Plasticized Polyvinyl chloride	Aureobasidium pullulans
BTA-copolyester	Thermomonspora fusca

3.1 Effects of plastics on soil microbial habitat

Plastics have become a ubiquitous material in modern society, with an estimated 359 million tons produced globally in 2018. Nevertheless, their widespread use has resulted in significant environmental concerns, including their negative impact on soil microbes. Studies have shown that plastics can negatively impact soil microbial communities, which play a critical role in soil health and nutrient cycling (Kundu et al. 2021, Kwak and An 2021, Liao and Chen 2021, Solano et al. 2022). Plastics can physically alter soil structure, reducing soil pore size and limiting air and water movement, which can inhibit microbial growth and activity. In addition, plastics can release chemicals and additives, such as phthalates and bisphenol A (BPA), which can be toxic to soil microbes. These chemicals can disrupt the balance of the soil microbial community, causing declines in the abundance and diversity of microorganisms.

Furthermore, plastics can serve as a substrate for microbial growth, creating biofilms that can impede soil nutrient availability and cycling. The accumulation of plastic waste in the soil can also lead to the build-up of anaerobic conditions, which can favor the growth of certain microbial populations, such as those responsible for methane production. Overall, the effects of plastics on soil microbial habitats can have cascading effects on soil health and ecosystem functioning (Kundu et al. 2021, Liao and Chen 2021). Further research is needed to better understand the long-term impacts of plastics on soil microbial communities and to develop strategies to mitigate these effects.

3.2 Biodegradation of plastics by soil microorganisms

Plastics have become a major environmental concern due to their persistence in the environment. However, some microorganisms have the ability to degrade certain types of plastics, a process known as biodegradation (Liao and Chen 2021, Singh Jadaun et al. 2022). Soil microorganisms, such as bacteria and fungi, have been shown to degrade a range of plastics, including polyethylene, polypropylene, and polystyrene (Liao and Chen 2021). These microorganisms produce enzymes that break down the long chains of plastic polymers into smaller molecules, which can be used as a source of carbon and energy. The biodegradation of plastics by soil microorganisms is influenced by a variety of factors, including temperature, moisture, oxygen availability, and the chemical structure of the plastic. For example, the biodegradation of polyethylene is facilitated by microorganisms that produce the enzyme polyethylene-degrading enzyme (PETase).

While biodegradation offers a potential solution for plastic waste management, the process is often slow and incomplete. In addition, the breakdown of plastics can result in the release of harmful chemicals, such as phthalates and BPA, into the environment. Furthermore, plastic waste often contains additives and contaminants, such as heavy metals and persistent organic pollutants, which can inhibit the biodegradation process. Overall, the biodegradation of plastics by soil microorganisms is a promising avenue for plastic waste management (Anuar Sharuddin et al. 2016, Lebreton et al. 2017, Liu et al. 2019, Van Roijen and Miller 2022). However, further research is needed to better understand the factors that influence the process and to develop strategies to optimize biodegradation and minimize potential negative impacts.

3.3 Microbial degradation strategy for soil plastic pollution

Soil plastic pollution is becoming a significant environmental concern worldwide. The accumulation of plastic waste in the soil leads to a wide range of environmental problems, including reduced soil fertility, reduced water infiltration, and increased runoff (Solano et al. 2022). Microbial degradation is an emerging strategy for addressing soil plastic pollution, which involves using microorganisms to degrade plastic waste in the soil. In this essay, we will discuss the microbial degradation strategy for soil plastic pollution (Dvořák et al. 2017, Liu et al. 2019, Samadhiya et al. 2022, Solano et al.

2022, Wang et al. 2022b). Microbial degradation involves using microorganisms such as bacteria, fungi, and actinomycetes to break down plastic waste in the soil. These microorganisms produce enzymes that break down the polymer chains of plastic waste into smaller, more manageable molecules. The smaller molecules are then used as a source of energy by microorganisms for their growth and metabolism. The end products of microbial degradation are carbon dioxide, water, and biomass, which are not harmful to the environment. Several factors affect the microbial degradation of plastic waste in soil, including temperature, moisture, pH, and nutrient availability. Optimum conditions such as a temperature range of 30 to 40°C, pH range of 5 to 8, moisture content of 40 to 60%, and the presence of nutrients such as nitrogen and phosphorus can enhance the microbial degradation process (Malinconico et al. 2008). One approach for enhancing microbial degradation is through the use of bioaugmentation and biostimulation. Bioaugmentation involves adding specific microbial cultures that are capable of degrading plastic waste into soil (Raghavulu et al. 2012, Ren et al. 2009). Biostimulation involves adding nutrients or organic compounds that enhance the growth and metabolism of indigenous microorganisms in the soil, which in turn can degrade plastic waste (Pradhan et al. 2017, Zhang and Lo 2015). These approaches have shown promising results in laboratory and field trials.

However, microbial degradation has limitations in addressing soil plastic pollution. The degradation process can take a long time, and the effectiveness of the process may depend on the type of plastic, its physical and chemical properties, and the conditions in the soil. Additionally, the byproducts of microbial degradation may not be completely harmless, and the accumulation of degraded plastic waste may have negative impacts on soil quality. In conclusion, microbial degradation is an emerging strategy for addressing soil plastic pollution (Nikolaivits et al. 2021, Steinbüchel 2005, Sun et al. 2022a). The approach involves using microorganisms to break down plastic waste in soil, and its effectiveness can be enhanced by optimizing the conditions and using bioaugmentation and biostimulation. However, further research and development are needed to optimize this technology and ensure its viability for large-scale applications.

3.4 Formation of Microplastics during Degradation

Microplastics are tiny plastic particles that are less than 5 millimeters in size. They are formed due to the degradation of larger plastic items such as bags, bottles, and packaging materials. The degradation of plastic can occur through several processes such as mechanical abrasion, photo-degradation, thermal degradation, and chemical degradation. During these processes, microplastics can be formed due to several reasons.

Firstly, mechanical abrasion can occur due to the physical breakdown of plastic items caused by exposure to environmental factors such as wind, water, and sunlight. The mechanical force can cause the plastic to break into smaller pieces, resulting in the formation of microplastics (Pinto et al. 2018, Sun et al. 2022b).

Secondly, photo-degradation can occur due to exposure to sunlight, which causes the plastic to break down into smaller pieces due to the energy from the ultraviolet light. The process can also result in the formation of microplastics (Liao and Chen 2021).

Thirdly, thermal degradation can occur when plastic is exposed to high temperatures, which can cause the plastic to break down into smaller pieces. The process can occur during the incineration of plastic waste, resulting in the formation of microplastics (Solano et al. 2022).

Lastly, chemical degradation can occur due to exposure to chemicals, such as acids and alkalis that can break down the plastic into smaller pieces. The process can occur during the degradation of plastic waste in landfill sites, resulting in the formation of microplastics (Gulizia et al. 2022, Moreno-Bayona et al. 2019, Sun et al. 2022b). The formation of microplastics during degradation has several negative impacts on the environment. They can contaminate soil and water sources, and they can be ingested by aquatic organisms, leading to adverse health effects. Microplastics can also persist in the environment for a long time, leading to long-term environmental impacts. In conclusion, the formation of microplastics during degradation occurs due to several processes such as mechanical abrasion, photo-degradation, thermal degradation, and chemical degradation. The formation of microplastics has several negative impacts on the environment and calls for a need to reduce plastic waste and develop effective strategies for managing plastic waste.

4. FUTURE DIRECTIVES

Biotechnology approaches for addressing plastic waste treatment have shown promising results in recent years, but there is still a need for further research and development to optimize these approaches and ensure their effectiveness for large-scale applications. The following are some future directives that could be pursued to advance biotechnological approaches for addressing plastic waste treatment:

1. Identifying and characterizing plastic-degrading microorganisms: There is a need to identify and characterize microbial strains that are efficient in degrading different types of plastic waste. This will require a comprehensive understanding of the diversity of microorganisms present in different environments and their plastic-degrading abilities.

2. Developing more efficient bioreactors: Bioreactors are a key component of biotechnological approaches for plastic waste treatment. There is a need to develop more efficient bioreactors that can operate at a larger scale and handle a variety of plastic waste streams.

3. Enhancing the efficiency of microbial degradation: The efficiency of microbial degradation can be enhanced by optimizing the conditions such as temperature, pH, and nutrient availability. There is a need for further research to identify the optimal conditions for different types of plastic waste and to develop approaches for maintaining these conditions in bioreactors.

4. Developing new biotechnological approaches: There is a need to explore new biotechnological approaches for plastic waste treatment, such as genetically engineering microorganisms to enhance their plastic-degrading abilities or using enzymes produced by microorganisms for plastic degradation.

5. Addressing the environmental impacts of microbial degradation: While microbial degradation is a promising approach for plastic waste treatment, there is a need to ensure that the byproducts of degradation do not have negative impacts on the environment. Further research is needed to understand the environmental impacts of microbial degradation and to develop approaches for managing these impacts.

6. Collaboration and knowledge sharing: There is a need for collaboration and knowledge sharing between researchers, industry, and policymakers to advance biotechnology approaches for plastic waste treatment. This will help to accelerate the development and adoption of these approaches and ensure their sustainability in the long term.

In conclusion, biotechnology approaches have the potential to address plastic waste treatment, but there is still a need for further research and development to optimize these approaches and ensure their effectiveness for large-scale applications. Future directives should focus on identifying and characterizing plastic-degrading microorganisms, developing more efficient bioreactors, enhancing the efficiency of microbial degradation, developing new biotechnological approaches, addressing the environmental impacts of microbial degradation, and promoting collaboration and knowledge sharing.

REFERENCES

Al-asadi, M., Miskolczi, N. and Eller, Z. 2020. Pyrolysis-gasification of wastes plastics for syngas production using metal modified zeolite catalysts under different ratio of nitrogen/oxygen. J. Clean. Prod. 271. https://doi.org/10.1016/j.jclepro.2020.122186.

Al Rayaan, M.B. 2021. Recent advancements of thermochemical conversion of plastic waste to biofuel—A review. Clean. Eng. Technol. 2. https://doi.org/10.1016/j.clet.2021.100062.

Anuar Sharuddin, S.D., Abnisa, F., Wan Daud, W.M.A. et al. 2016. A review on pyrolysis of plastic wastes. Energy Convers. Manag. 115: 308–326. https://doi.org/10.1016/j.enconman.2016.02.037.

Bai, Z., Wang, N. and Wang, M. 2021. Effects of microplastics on marine copepods. Ecotoxicol. Environ. Saf. 217: 112243. https://doi.org/10.1016/j.ecoenv.2021.112243.

Barrila, J., Crabbé, A., Yang, J. et al. 2018. Modeling host-pathogen interactions in the context of the microenvironment: Three-dimensional cell culture comes of age. Infect. Immun. 86. https://doi.org/10.1128/IAI.00282-18.

Byrne, E., Schaerer, L.G., Kulas, D.G. et al. 2022. Pyrolysis-aided microbial biodegradation of high-density polyethylene plastic by environmental inocula enrichment cultures. ACS Sustain. Chem. Eng. 10: 2022–2033. https://doi.org/10.1021/acssuschemeng.1c05318.

Costa, S.S., Miranda, A.L., de Morais, M.G. et al. 2019. Microalgae as source of polyhydroxyalkanoates (PHAs)—A review. Int. J. Biol. Macromol. https://doi.org/10.1016/j.ijbiomac.2019.03.099.

Deng, H., Fu, Q., Li, D. et al. 2021. Microplastic-associated biofilm in an intensive mariculture pond: Temporal dynamics of microbial communities, extracellular polymeric substances and impacts on microplastics properties. J. Clean. Prod. 319: 128774. https://doi.org/10.1016/j.jclepro.2021.128774.

Dvořák, P., Nikel, P.I., Damborský, J. et al. 2017. Bioremediation 3.0: Engineering pollutant-removing bacteria in the times of systemic biology. Biotechnol. Adv. 35: 845–866. https://doi.org/10.1016/j.biotechadv.2017.08.001.

Gulizia, A.M., Brodie, E., Daumuller, R. et al. 2022. Evaluating the effect of chemical digestion treatments on polystyrene microplastics: Recommended updates to chemical digestion protocols. Macromol. Chem. Phys. 2100485. https://doi.org/10.1002/macp.202100485.

Hemdan, B., Garlapati, V.K., Sharma, S. et al. 2021. Bioelectrochemical systems-based metal recovery: Resource, conservation and recycling of metallic industrial effluents. Environ. Res. 112346. https://doi.org/10.1016/j.envres.2021.112346.

Ibrahim, N.I., Shahar, F.S., Sultan, M.T.H. et al. 2021. Overview of bioplastic introduction and its applications in product packaging. Coatings 11: 1423. https://doi.org/10.3390/coatings11111423.

Ieropoulos, I., Greenman, J. and Melhuish, C. 2010. Improved energy output levels from small-scale microbial fuel cells. Bioelectrochemistry 78: 44–50. https://doi.org/10.1016/j.bioelechem.2009.05.009.

Jaafar, Y., Abdelouahed, L., Hage, R. El et al. 2022. Pyrolysis of common plastics and their mixtures to produce valuable petroleum-like products. Polym. Degrad. Stab. 195: 109770. https://doi.org/10.1016/j.polymdegradstab.2021.109770.

Jung, S., Choi, D., Park, Y.K. et al. 2020. Functional use of CO_2 for environmentally benign production of hydrogen through catalytic pyrolysis of polymeric waste. Chem. Eng. J. 399. https://doi.org/10.1016/j.cej.2020.125889.

Kim, J., Kim, K.Y., Ko, J.K. et al. 2022. Characterization of a novel acetogen Clostridium sp. JS66 for production of acids and alcohols: Focusing on hexanoic acid production from syngas. Biotechnol. Bioprocess Eng. 27: 89–98. https://doi.org/10.1007/s12257-021-0122-1.

Kundu, A., Shetti, N.P., Basu, S. et al. 2021. Identification and removal of micro- and nano-plastics: Efficient and cost-effective methods. Chem. Eng. J. 421: 129816. https://doi.org/10.1016/j.cej.2021.129816.

Kwak, J.Il and An, Y.J. 2021. Post COVID-19 pandemic: Biofragmentation and soil ecotoxicological effects of microplastics derived from face masks. J. Hazard. Mater. 416. https://doi.org/10.1016/j.jhazmat.2021.126169.

Kwon, D., Jung, S., Moon, D.H. et al. 2022. Strategic management of harmful chemicals produced from pyrolysis of plastic cup waste using CO_2 as a reaction medium. Chem. Eng. J. 437: 135524. https://doi.org/10.1016/j.cej.2022.135524.

Lear, G., Kingsbury, J.M., Franchini, S. et al. 2021. Plastics and the microbiome: Impacts and solutions. Environ. Microbiomes 16: 1–19. https://doi.org/10.1186/s40793-020-00371-w.

Lebreton, L.C.M., Van Der Zwet, J., Damsteeg, J.W. et al. 2017. River plastic emissions to the world's oceans. Nat. Commun. 8: 1–10. https://doi.org/10.1038/ncomms15611.

Li, Y., Li, J., Ding, J. et al. 2022. Degradation of nano-sized polystyrene plastics by ozonation or chlorination in drinking water disinfection processes. Chem. Eng. J. 427: 1–8. https://doi.org/10.1016/j.cej.2021.131690.

Liao, J. and Chen, Q. 2021. Biodegradable plastics in the air and soil environment: Low degradation rate and high microplastics formation. J. Hazard. Mater. 418: 126329. https://doi.org/10.1016/j.jhazmat.2021.126329.

Liu, C.-L., Dong, H.-G., Xue, K. et al. 2020. Biosynthesis of poly-γ-glutamic acid in Escherichia coli by heterologous expression of pgsBCAE operon from Bacillus. J. Appl. Microbiol. 128: 1390–1399. https://doi.org/10.1111/jam.14552.

Liu, F., Li, J. and Zhang, X.L. 2019. Bioplastic production from wastewater sludge and application. IOP Conf. Ser. Earth Environ. Sci. 344. https://doi.org/10.1088/1755-1315/344/1/012071.

Liu, X., Zhu, Y. and Yang, S.-T. 2006. Construction and characterization of ack deleted mutant of Clostridium tyrobutyricum for enhanced butyric acid and hydrogen production. Biotechnol. Prog. 22: 1265–75. https://doi.org/10.1021/bp060082g.

Lopez, G., Artetxe, M., Amutio, M. et al. 2018. Recent advances in the gasification of waste plastics. A critical overview. Renew. Sustain. Energy Rev. 82: 576–596. https://doi.org/10.1016/j.rser.2017.09.032.

Malinconico, M., Immirzi, B., Santagata, G. et al. 2008. An overview on innovative biodegradable. Progress in Polymer Degradation and Stability Research.

Miloloža, M., Cvetnić, M., Kučić Grgić, D. et al. 2022. Biotreatment strategies for the removal of microplastics from freshwater systems. A review. Environ. Chem. Lett. 20: 1377–1402. https://doi.org/10.1007/s10311-021-01370-0.

Moreno-Bayona, D.A., Gómez-Méndez, L.D., Blanco-Vargas, A. et al. 2019. Simultaneous bioconversion of lignocellulosic residues and oxodegradable polyethylene by Pleurotus ostreatus for biochar production, enriched with phosphate solubilizing bacteria for agricultural use. PLoS One 14: 1–25. https://doi.org/10.1371/journal.pone.0217100.

Morgan-Sagastume, F., Valentino, F., Hjort, M. et al. 2014. Polyhydroxyalkanoate (PHA) production from sludge and municipal wastewater treatment. Water Sci. Technol. 69: 177–184. https://doi.org/10.2166/wst.2013.643.

Muhammad, C., Onwudili, J.A. and Williams, P.T. 2015. Catalytic pyrolysis of waste plastic from electrical and electronic equipment. J. Anal. Appl. Pyrolysis 113: 332–339. https://doi.org/10.1016/j.jaap.2015.02.016.

Nikolaivits, E., Pantelic, B., Azeem, M. et al. 2021. Progressing plastics circularity: A review of mechano-biocatalytic approaches for waste plastic (re)valorization. Front. Bioeng. Biotechnol. 9: 1–31. https://doi.org/10.3389/fbioe.2021.696040.

Pinedo, J., García Prieto, C.V., D'Alessandro, A.A. et al. 2022. Multi-objective superstructure optimization of a microalgae biorefinery considering economic and environmental aspects. Bioresour. Technol. 333: 122462. https://doi.org/10.1016/j.compchemeng.2022.107894.

Pinto, F., Paradela, F., Carvalheiro, F. et al. 2018. Effect of experimental conditions on co-pyrolysis of pre-treated eucalyptus blended with plastic wastes. Chem. Eng. Trans. 70: 793–798. https://doi.org/10.3303/CET1870133.

Pradhan, D., Sukla, L.B., Sawyer, M. et al. 2017. Recent bioreduction of hexavalent chromium in wastewater treatment: A review. J. Ind. Eng. Chem. https://doi.org/10.1016/j.jiec.2017.06.040.

Raghavulu, S.V., Babu, P.S., Goud, R.K. et al. 2012. Bioaugmentation of an electrochemically active strain to enhance the electron discharge of mixed culture: Process evaluation through electro-kinetic analysis. RSC Adv. 2: 677–688. https://doi.org/10.1039/C1RA00540E.

Raphael, I. and Yang, A. 2013. Plastics production from biomass: Assessing feedstock requirement. Biomass Convers. Biorefinery 3: 319–326. https://doi.org/10.1007/s13399-013-0094-2.

Ren, N., Wang, A., Cao, G. et al. 2009. Bioconversion of lignocellulosic biomass to hydrogen: Potential and challenges. Biotechnol. Adv. 27: 1051–60. https://doi.org/10.1016/j.biotechadv.2009.05.007.

Salar-Garcia, M.J., Montilla, F., Quijada, C. et al. 2020. Improving the power performance of urine-fed microbial fuel cells using PEDOT-PSS modified anodes. Appl. Energy 278: 115528. https://doi.org/10.1016/j.apenergy.2020.115528.

Samadhiya, K., Ghosh, A., Nogueira, R. et al. 2022. Newly isolated native microalgal strains producing polyhydroxybutyrate and energy storage precursors simultaneously: Targeting microalgal biorefinery. Algal Res. 62: 102625. https://doi.org/10.1016/j.algal.2021.102625.

Seeley, M.E., Song, B., Passie, R. et al. 2020. Microplastics affect sedimentary microbial communities and nitrogen cycling. Nat. Commun. 11: 1–10. https://doi.org/10.1038/s41467-020-16235-3.

Singh Jadaun, J., Bansal, S., Sonthalia, A. et al. 2022. Biodegradation of plastics for sustainable environment. Bioresour. Technol. 347: 126697. https://doi.org/10.1016/j.biortech.2022.126697.

Singh, R.K. and Ruj, B. 2016. Time and temperature depended fuel gas generation from pyrolysis of real world municipal plastic waste. Fuel 174: 164–171. https://doi.org/10.1016/j.fuel.2016.01.049.

Solano, G., Rojas-Gätjens, D., Rojas-Jimenez, K. et al. 2022. Biodegradation of plastics at home composting conditions. Environ. Challenges 7: 100500. https://doi.org/10.1016/j.envc.2022.100500.

Steinbüchel, A. 2005. Non-biodegradable biopolymers from renewable resources: Perspectives and impacts. Curr. Opin. Biotechnol. 16: 607–613. https://doi.org/10.1016/j.copbio.2005.10.011.

Sun, Yong, Hu, J., Yusuf, A. et al. 2022a. A critical review on microbial degradation of petroleum-based plastics: Quantitatively effects of chemical addition in cultivation media on biodegradation efficiency. Biodegradation 33: 1–16. https://doi.org/10.1007/s10532-021-09969-4.

Sun, Ying, Wang, X., Xia, S. et al. 2022b. Cu(II) adsorption on poly(lactic acid) microplastics: Significance of microbial colonization and degradation. Chem. Eng. J. 429: 132306. https://doi.org/10.1016/j.cej.2021.132306. https://doi.org/10.1016/j.cej.2021.132306.

Taghavi, N., Udugama, I.A., Zhuang, W.Q. et al. 2021. Challenges in biodegradation of non-degradable thermoplastic waste: From environmental impact to operational readiness. Biotechnol. Adv. 49: 107731. https://doi.org/10.1016/j.biotechadv.2021.107731.

Van Roijen, E.C. and Miller, S.A. 2022. A review of bioplastics at end-of-life: Linking experimental biodegradation studies and life cycle impact assessments. Resour. Conserv. Recycl. 181: 106236. https://doi.org/10.1016/j.resconrec.2022.106236.

Wang, C., Han, H., Wu, Y. et al. 2022a. Nanocatalyzed upcycling of the plastic wastes for a circular economy. Coord. Chem. Rev. 458: 214422. https://doi.org/10.1016/j.ccr.2022.214422.

Wang, P., Huang, Z., Chen, S. et al. 2022b. Sustainable removal of nano/microplastics in water by solar energy. Chem. Eng. J. 428: 131196. https://doi.org/10.1016/j.cej.2021.131196.

Zhang, Z. and Lo, I.M.C. 2015. Biostimulation of petroleum-hydrocarbon-contaminated marine sediment with co-substrate: Involved metabolic process and microbial community. https://doi.org/10.1007/s00253-015-6420-9.

Zheng, P., Feng, X., Qian, F. et al. 2006. Water system integration of a chemical plant. Energy Convers. Manag. 47: 2470–2478. https://doi.org/10.1016/j.enconman.2005.11.001.

Index